NANOMEDICINE

A Systems Engineering Approach

NANOMEDICINE

A Systems Engineering Approach

edited by

Mingjun Zhang
The University of Tennessee, USA

Ning Xi
Michigan State University, USA

PAN STANFORD PUBLISHING

Published by

Pan Stanford Publishing Pte. Ltd.
5 Toh Tuck Link
Singapore 596224

Distributed by

World Scientific Publishing Co. Pte. Ltd.
5 Toh Tuck Link, Singapore 596224
USA office: 27 Warren Street, Suite 401-402, Hackensack, NJ 07601
UK office: 57 Shelton Street, Covent Garden, London WC2H 9HE

British Library Cataloguing-in-Publication Data
A catalogue record for this book is available from the British Library.

NANOMEDICINE
A Systems Engineering Approach

ISBN-13 978-981-4241-36-6
ISBN-10 981-4241-36-9

Typeset by Research Publishing Services
E-mail: enquiries@rpsonline.com.sg

Printed in Singapore by B & JO Enterprise

Preface

The purpose of this edited volume is to prompt quantitative sciences to the fast emerging field of nanomedicine.

Nanomedicine currently faces two challenges. The first is to quantitatively describe drug effects at the micro-/nano-scale, particularly, by taking into consideration dynamic interactions of drugs with biological systems at the molecular level. The second is to precisely control the drug effects, especially the coupled effects of drugs with the complex biological system.

Systems engineering approach can help address the above challenges. Dynamics modeling, control and optimization are important components of system engineering. They have been applied successfully in many engineering disciplinary. Unfortunately, applications of systems engineering approach to nanomedicine has just started.

This book puts together case studies of systems engineering to the field of nanomedicine. It contains eight chapters ranging from introduction to nanomedicine, fundamentals of mathematical modeling, to applications in signal pathway, tumor therapy, multi-scale biological system modeling and controlled drug delivery.

It is challenging to edit such a highly interdisciplinary book. We are lucky to have a dedicated team of outstanding contributors, and a visionary editorial team. We would like to take this opportunity to thank all the contributors for their excellent contributions, Mr. Stanford Chong and Mr. Rhaimie Wahap for their outstanding support. They are the direct contributors to the success of this book.

Down to the road, we firmly believe personalized and quantitative aspects of medicine are two fundamental challenges to nanomedicine. Systems engineering approach will play an important role in the endeavor.

We welcome any comments and suggestions from the readers.

Mingjun Zhang and Ning Xi

Contents

Nanomedicine: *A Systems Engineering Approach* edited by Mingjun Zhang & Ning Xi

www.panstanford.com
978-981-4241-36-6

Chapter One

Nanomedicine: Dynamic Integration of Nanotechnology with Biomedical Science

Ki-Bum Lee*,
Aniruddh Solanki,
John D. Kim and Jongjin Jung

1.1 INTRODUCTION

The recent emergence of nanotechnology is setting high expectations in biological science and medicine, and many scientists now predict that nanotechnology will solve many key questions of biological systems that transpire at the nanoscale. Nanomedicine, broadly defined as the approach of science and engineering at the nanometer scale toward biomedical applications, has been drawing considerable attention in the area of nanotechnology. Given that the sizes of functional elements in biology are at the nanometer scale range, it is not surprising for nanomaterials to interact with biological systems at the molecular level. In addition, nanomaterials have novel electronic, optical, magnetic, and structural properties that cannot be obtained from either individual molecules or bulk materials. These unique features can be precisely tuned in order for scientists to explore biological phenomena in many ways. For instance, extensive studies have been done with chip-based or solution-based bio-assays, drug delivery, molecular imaging, disease diagnosis, and

* Corresponding author.

Nanomedicine: *A Systems Engineering Approach* edited by Mingjun Zhang & Ning Xi

www.panstanford.com
978-981-4241-36-6

alkanethiols. The surfactants not only interact with the atoms of nanostructures by either chemisorption or physisorption at the surface of nanostructures, but also stabilize nanostructures and prevent interparticle aggregation. Using the exchange reactions, surfactant molecules attached on the nanoparticles can be replaced by biomolecules, making direct biomolecule-nanoparticle covalent bonds. For example, gold nanoparticles can be modified with proteins consisting of cysteine residues or with thiol functionalized DNA molecules.

There are different types of coupling methods, where the electrostatic forces between proteins and citrate stabilized nanoparticles are used for the coupling. For the nanoparticles that are relatively unstable, the core-shell strategies can be applied to stabilize the nanoparticles.[25,26] For example, silver core-shell nanoparticles coated by thin layer of gold can be successfully functionalized with thiol-functionalized DNA.[25] Many semiconductor nanoparticles can also be linked with proteins or DNA by adding a hydrophobic silica shell.[16,27,28] Silica surfaces can be tailored with biomolecules, utilizing well known cross-linking methodologies, such as silanization chemisty and self-assembly monolayer (SAM) chemistry.[29–32]

1.2.3 Fabrication of Nanoarrays and Biochips

The search for novel ways to explore and understand biomolecular interactions has been sought in many ways, since interactions between biomolecules are fundamentally intriguing. For example, how proteins such as fibrinogen, fibronectin, and retronectin influence the adhesion of cells and control their morphology and physiology has been a central question of cell biology.[33,34] Several approaches have been examined over the past few years to comprehend these phenomena, and one of these approaches is the assembly of interfacial proteins constructed on micro- or nanoscale.[35] The biomolecular patterned surfaces are not only useful for probing biochemical interactions within whole cells but also crucial for biosensing. However, the realization of these applications is challenging, since the control of the interactions between proteins and surfaces with respect to a binding direction and biomolecular density is technically and biologically not easy to achieve.[36] Therefore, patterning techniques capable of high resolution — molecular to submicron scale — and compatible with biomolecules will be

Chapter One

Nanomedicine: Dynamic Integration of Nanotechnology with Biomedical Science

Ki-Bum Lee*,
Aniruddh Solanki,
John D. Kim and Jongjin Jung

1.1 INTRODUCTION

The recent emergence of nanotechnology is setting high expectations in biological science and medicine, and many scientists now predict that nanotechnology will solve many key questions of biological systems that transpire at the nanoscale. Nanomedicine, broadly defined as the approach of science and engineering at the nanometer scale toward biomedical applications, has been drawing considerable attention in the area of nanotechnology. Given that the sizes of functional elements in biology are at the nanometer scale range, it is not surprising for nanomaterials to interact with biological systems at the molecular level. In addition, nanomaterials have novel electronic, optical, magnetic, and structural properties that cannot be obtained from either individual molecules or bulk materials. These unique features can be precisely tuned in order for scientists to explore biological phenomena in many ways. For instance, extensive studies have been done with chip-based or solution-based bio-assays, drug delivery, molecular imaging, disease diagnosis, and

* Corresponding author.

Nanomedicine: *A Systems Engineering Approach* edited by Mingjun Zhang & Ning Xi

www.panstanford.com
978-981-4241-36-6

pharmaceutical screening.[1–4] In order to realize these applications, it is crucial to develop methods that investigate and control the binding properties of individual biomolecules at the fundamental nanometer level. This will require enormous time, effort, and interdisciplinary expertise of physical sciences associated with both biology and engineering. The overall goal of nanomedicine is to develop safer and more effective therapeutics as well as novel diagnostic tools. To date, nanotechnology has revolutionized biomedical science step by step not only by improving efficiency and accuracy of current diagnostic techniques, but also by extending scopes for the better understanding of diseases at the molecular level.[5–8] In this chapter, nanomaterials and their applications in biomedical research will be discussed.

1.2 DESIGNING NANOMATERIALS FOR BIOLOGY AND MEDICINE

One of the important technological aspects in nanomedicine lies in the ability to tune materials in a way that their spatial and temporal scales are compatible with biomolecules. That said, materials and devices fabricated at the nanometer scale can investigate and control the interactions between biomolecules and their counterparts at almost the single molecule level. This, in turn, indicates that nanomaterials and nanodevices can be fabricated to show high sensitivity, selectivity, and control properties, which usually cannot be achieved in bulk materials. The wide range of the scale of biointeractions is described in Fig. 1.

Given that one of the major goals of biology is to address the spatial-temporal interactions of biomolecules at the cellular or integrated systems level, the integration of nanotechnology in biomedicine would bring a breakthrough in current biomedical research efforts. In order to apply nanotechnology to biology and medicine, several conditions must be considered: (i) nanomaterials should be designed to interact with proteins and cells without perturbing their biological activities, (ii) nanomaterials should maintain their physical properties after the surface conjugation chemistry, and (iii) nanomaterials should be biocompatible and non-toxic.

In general, there are two approaches to build nanostructures or nanomaterials: "top-down" and "bottom-up" methods. Typically, the bottom-up approach utilizes self-assembly of one or

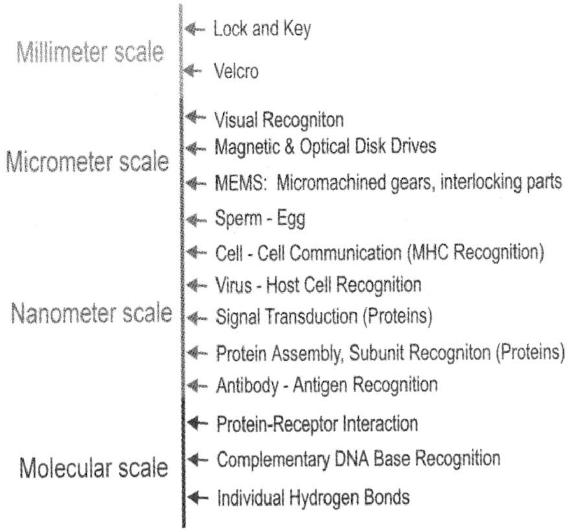

Figure 1. Scale of biomolecular interactions.

more defined molecular building blocks to create higher-ordered functional entities. For the bottom-up approach, the physical and chemical criterion, such as pH, concentration, temperature, and intrinsic properties of building blocks, must be fulfilled. On the other hand, the top-down approach usually involves processes such as lithography, etching, and lift-off techniques to fabricate micro- and nanoscopic structured materials from bulk materials. In many cases, nanomedicine strategies have been derived from what was originally a conventional biomedical application, with a certain degree of modification to address some scientific questions or technical limitations. As far as the applications of nanostructures are concerned, we will examine two examples, nanoparticles and nanoarrays/biochips, which are heavily used in biomedical applications.

1.2.1 *Inorganic Nanoparticles*

Deoxyribonucleic acid (DNA), ribonucleic acid (RNA), peptides, and proteins are nanometer scale components that are the best examples of nanomaterials found in nature.[9,10] For example, DNA has a double-stranded helical structure with a diameter of 2 nm, RNA has a single strand structure with a diameter of 1 nm, and most of protein sizes are less than 15 nm. Likewise, the sizes of functional elements in biology are at the nanoscale level, which inevitably generate significant interests at the intersection between nanotechnology and biological science. Even though much progress in the life sciences has been achieved over the last few decades, biological and physiological phenomena still remain beyond our understanding, because the interactions between elementary biomolecules and other higher components, such as viruses, bacteria, and cells, are complex and delicate. Moreover, the interactions of two biocomponents start from the single molecule level, where the recognition sites lie in a nanoscale domain. Thus, studies of these biological components require not only an ability to handle the biological properties, but also to develop highly advanced tools or techniques to analyze the biological systems.[11–14]

Bioconjugated nanomaterials have recently been used as cellular labeling agents to study the biological phenomena at the nanometer level. With significant advancements in synthetic and modification methodologies, nanomaterials can be tuned to desired sizes, shapes, compositions, and properties.[15] Inorganic nanoparticles are one of the most promising examples, since they can be synthesized easily in large quantities from various materials using relatively simple methods. Also, the dimensions of the nanoparticles can be tuned from one to a few hundred nanometers with monodispersed size distribution. Moreover, they can be made up of different metals, metal oxides, and semiconducting materials, whose compositions and sizes are listed in Table 1. Given many distinct properties, nanoparticles can be readily tailored with biomolecules via combined methodologies from bioorganic, bioinorganic, and surface chemistry.

Despite many significant advances in synthetic and surface modification methods, the fundamental development of bio-conjugation methods must first be achieved in order for the nanoparticles to be fully utilized. The bioconjugation strategies involve procedures for coupling biomolecules to nanomaterials, enabling the nanoparticles not only to be applied for clinical applications but also to ask and answer fundamental questions in cell biology. For the past few years, many methods have been developed for bio-labeled nanocomposites in various applications in cell biology: cell labeling, cell tracking,[19–22] and *in vivo* imaging.[23,24]

Table 1. Selection of available nanoparticle compositions, sizes, and shapes.

Particle Composition	Particle Size (nm)
Metals	
Au	2–250
Ag	1–80
Pt	1–20
Cu	1–50
Semiconductors	
CdX (X = S, Se, Te)	1–20
ZnX (X = S, Se, Te)	1–20
TiO_2	2–18
PbS	3–50
ZnO	1–30
GaAs, InP	1–15
Ge	6–30
Magnetic	
Fe_3O_4	6–40
Various polymer compositions	20 nm to 500 μm

1.2.2 Coupling of Nanoparticles with Biomolecules

Interdisciplinary knowledge from molecular biology, bioorganic chemistry, bioinorganic chemistry, and surface chemistry must be employed to functionalize nanostructures with biomolecules. Although nanostructures can be synthesized from various materials using several methods, the coupling and functionalization of nanostructures with biomolecules should be carried out in controlled manners such as a specific salt concentration or pH.[9] Three common methods of functionalizing nanoparticles with biomolecules are: (i) direct interaction between nanoparticles and biomolecules via electrostatic interactions or physical adsorptions, (ii) typical conjugation chemistry using organic linker molecules, and (iii) streptavidin-biotin affinity between functionalized nanoparticles.

Typically, solution phase synthesis of nanostructures is carried out in the presence of surfactants such as citrate, phosphates, and alkanethiols. The surfactants not only interact with the atoms of nanostructures by either chemisorption or physisorption at the surface of nanostructures, but also stabilize nanostructures and prevent interparticle aggregation. Using the exchange reactions, surfactant molecules attached on the nanoparticles can be replaced by biomolecules, making direct biomolecule-nanoparticle covalent bonds. For example, gold nanoparticles can be modified with proteins consisting of cysteine residues or with thiol functionalized DNA molecules.

There are different types of coupling methods, where the electrostatic forces between proteins and citrate stabilized nanoparticles are used for the coupling. For the nanoparticles that are relatively unstable, the core-shell strategies can be applied to stabilize the nanoparticles.[25,26] For example, silver core-shell nanoparticles coated by thin layer of gold can be successfully functionalized with thiol-functionalized DNA.[25] Many semiconductor nanoparticles can also be linked with proteins or DNA by adding a hydrophobic silica shell.[16,27,28] Silica surfaces can be tailored with biomolecules, utilizing well known cross-linking methodologies, such as silanization chemisty and self-assembly monolayer (SAM) chemistry.[29–32]

1.2.3 Fabrication of Nanoarrays and Biochips

The search for novel ways to explore and understand biomolecular interactions has been sought in many ways, since interactions between biomolecules are fundamentally intriguing. For example, how proteins such as fibrinogen, fibronectin, and retronectin influence the adhesion of cells and control their morphology and physiology has been a central question of cell biology.[33,34] Several approaches have been examined over the past few years to comprehend these phenomena, and one of these approaches is the assembly of interfacial proteins constructed on micro- or nanoscale.[35] The biomolecular patterned surfaces are not only useful for probing biochemical interactions within whole cells but also crucial for biosensing. However, the realization of these applications is challenging, since the control of the interactions between proteins and surfaces with respect to a binding direction and biomolecular density is technically and biologically not easy to achieve.[36] Therefore, patterning techniques capable of high resolution — molecular to submicron scale — and compatible with biomolecules will be required. Typically, current chip-based biodetection strategies pattern molecules with an analyte-capturing ability on the chip surfaces in the micro-/nano scale. This application of chip-based biosensing will allow scientists to detect various analytes at concentrations as low as picomolar in massively parallel ways. The aforementioned features are invaluable for the advancement in genomics and proteomics via generating DNA and protein arrays.[15] Yet, the key challenge lies in the fabrication of miniaturized surface structures in the form of nanoarrays that would allow for multi-magnitude orders of complex detection in the same chip and for improved detection sensitivity. There are many techniques available for patterning surfaces in terms of resolution and compatibility with soft materials (Table 2). Among different techniques, one primary distinction is whether the method uses a resist-based process or deposition of materials onto a surface directly. Although the indirect, resist-based patterning methods, such as photolithography, are multi-step processes that require specialized resists and etching protocols, they are currently by far the most widely used methods in industrial applications.

By contrast, direct-printing methods are typically useful for patterning soft materials such as small organic or biological molecules

Table 2. Features of selected lithography techniques.

Technique	Resolution Limit	Mode	Comments
Photolithography	Currently ~100 nm	Parallel only	Resist based, indirect
Electron beam lithography	5–10 nm	Serial only	Resist based, indirect
Indirect SPM methods	AFM 5–10 nm STM atomic to nm	Serial only	Resist based, indirect
Microcontact printing	~ 100 nm	Parallel only	Direct write or indirect
Ink-jet printing	6 μm	Serial	Direct write
Dip-pen nanolithography	5–10 nm	Parallel or serial	Direct write

with ease. Microcontact printing, developed by Whitesides and coworkers,[37–44] is a good example of a direct-printing method using elastomer stamps which can be "inked" with molecules and then used to transfer the inking molecules in a form of desired patterns to various substrates. This technique has been used to generate large area patterns of soft materials on surfaces with pattern resolutions approaching 100 nm. However, the technique is limited in its capability to generate multiple, chemically diverse, high-resolution patterns in alignment on a surface. Therefore, in terms of resolution to sub-100 nm, there is a high demand to develop methods of high-resolution lithographic techniques. Since the invention of scanning probe microscopes (SPMs), many scientists have realized that it might be possible to manipulate matter, atom-by-atom or molecule-by-molecule.[45,46] The early attempts to develop patterning methodologies from SPMs were able to demonstrate the high-resolution capabilities of these instruments.

The majority of SPM surface patterning methods have focused on either impressive but inherently slow scanning tunneling microscope (STM) based methods for moving individual atoms around on a surface in ultra high vacuum (UHV), or on indirect methods using atomic force microscope (AFM) and STM for stepwise etching of organic monolayers on a surface and backfilling with the molecule of interest.[37,47–50] However, resist-based approaches are inherently restricted to serial processes, and can only be used for a few molecule-substrate combinations. In 1999, Mirkin and his coworkers developed dip-pen nanolithography (DPN). DPN uses an AFM tip to transfer "ink" molecules onto a substrate through a water meniscus (Fig. 2).[3,51,52] DPN is simple, and its resolution is comparable to electron-beam lithography.

1.3 APPLICATION OF NANOMATERIALS IN BIOMEDICAL RESEARCH

Biomedical scientists have seen a great potential in the nanotechnology application. As a result, they have tried to incorporate the intrinsic properties of nanomaterials with conventional techniques in an attempt to improve detection methods and treatments for greater results.[53,54] Recently, many studies have reported that innovative nanotechnology has improved biomolecular detection sensitivity, diagnostic accuracy, and treatment efficiency.[55–59]

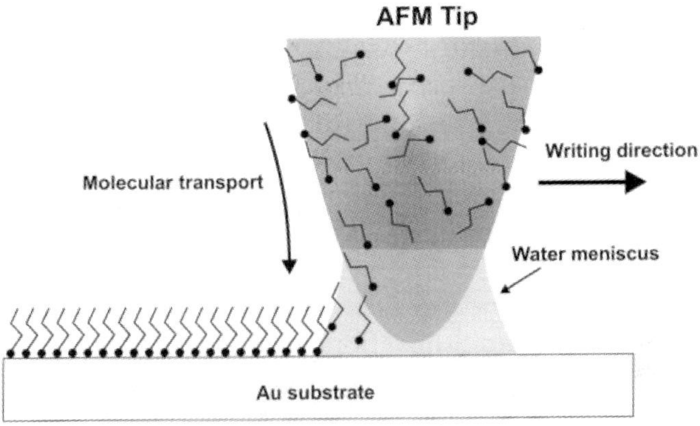

Figure 2. Schematic representation of Dip-Pen Nanolithography. "Ink" molecules coated on the AFM tip are transferred to the Au substrate through a water meniscus. Reprinted from Ref. 52 with permission from AAAS.

1.3.1 Molecular Imaging for Diagnosis and Detection

Optical imaging methods, such as fluorescence microscopy, differential interference contrast microscopy (DICM), and UV-Vis spectroscopy, are one of the most widely used methods for studying biological systems. These methods are simple and highly sensitive, yet tend to have high background noise that is mainly caused by cellular autofluorescence from labeled molecules.[60] Also the lack of quantitative data, requirement for the long observation time, and the loss of signals due to photobleaching in biological systems have initiated a need for new imaging agents. In order to complement conventional optical imaging methods, the approach using nanoprobes is one of the major research efforts in nanomedicine mainly due to their ability to recognize and characterize pre-symptomatic diseases.[61] Typically, the nanoprobes comprise of hybrid organic materials such as nanoliposomes and polymeric nanosystems, or inorganic nanoparticles such as quantum dots and magnetic nanomaterials. Depending on the composition and properties of nanomaterials, they can be further utilized *in vitro*, *ex vivo*, and*in vivo* imaging applications. Typical examples include vast arrays of functions such as cell or DNA labeling, molecular imaging, and angiogenesis as an imaging agent, particularly in tumor tissues.[15]

Quantum dots (QDs) are fluorescent semiconducting nanocrystals that can be used to overcome limitations associated with the more commonly used organic fluorophores. QDs offer many advantages such as high quantum yields, high molar extinction coefficients, wide range of absorption spectra from UV to near IR, narrow emission spectra, resistance to photobleaching and chemical degradation, and long fluorescence lifetimes (>10 ns). Their unique photophysical properties allow time-gated detection for separating their signal from that of the background noise resulting from cell autofluorescence.[62–65] Quantum dots are typically 2–8 nm in diameter and their shapes and sizes can be customized by modifying the variables — temperature, duration, and ligand molecules — used in their synthesis. Thus by changing their size and composition, it is possible to precisely tune the absorption and emission spectra to increase or decrease the band gap energy. Due to their narrow emission spectra, they can be used effectively in multiplexing experiments where multiple biological units can be labeled simultaneously. One such study was demonstrated by Jain and coworkers.[66] The QDs were applied *in vivo* to spectrally distinguish multiple species within the tumor tissue.[66] More specifically, they demonstrated that QDs can be customized to concurrently image and differentiate the tumor vessels. The group also examined the accessibility to tumor cells depending on the size of QDs. Sizes and compositions of QDs have been extensively studied by many groups.[59,67–70] Wright and coworkers[71] studied QDs with CdSe core with ZnS shell (CdSe/ZnS) and found that they have potent brightness which is advantageous for optical imaging. The group further studied the core/shell QDs conjugated with an antibody against the respiratory syncytial virus (RSV), a virus which is responsible for causing infections in the lower respiratory tract. It was shown that the use of QDs reduced the detection time from over four days to one hour. In fact, the results were very valuable from a therapeutic point of view as the available antiviral agent against the RSV is effective only when administered in the initial stages of the infection. Moreover, QDs enable scientists to study live cells and to track down the mechanism of biological processes in a real-time manner due to their resistance to photobleaching over long periods of time. The ability to track cells *in vivo* without having to sacrifice animals signifies a great improvement over the current techniques. Figures 3 and 4 show two examples of QD imaging: one is the high sensitivity and multicolor capability of QD imaging in live animals and the other is the detection of cancer marker Her2 with QD-streptavidin.[21]

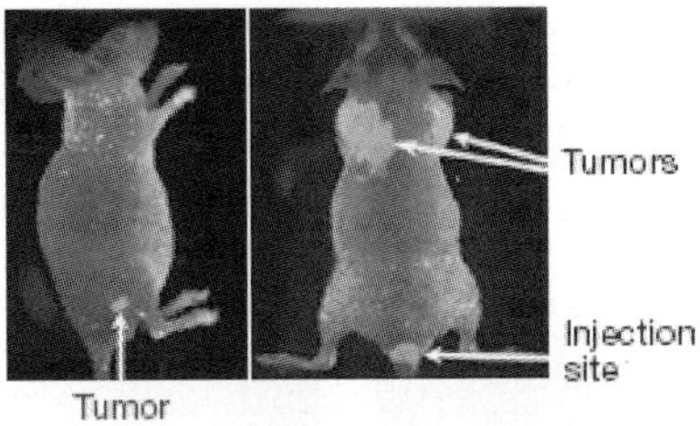

(a)

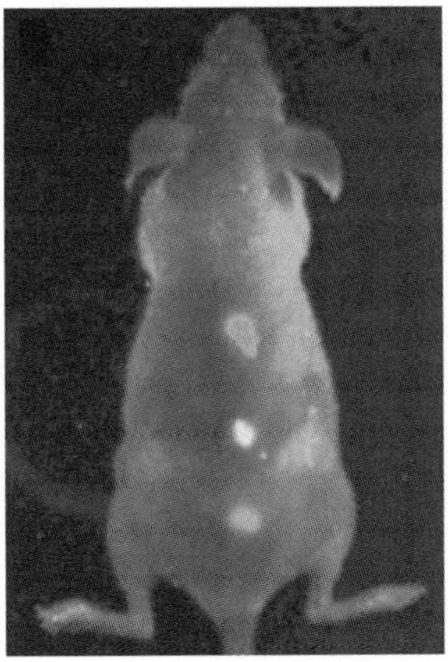

(b)

Figure 3. Imaging in live animals using quantum dots (QDs). (a) Molecular targeting and *in vivo* imaging using antibody-(QD) conjugate. (b) *In vivo* imaging of multicolored QD-encoded microbeads. Reprinted from Ref. 62 with permission from Nature Publishing Group. (See page 337 for color illustration.)

Cytotoxicity is a primary issue in QD applications,[72] because the release of Cd^{2+} and Se^{2-} ions from QDs could interfere with cell viability or function.[73,74] While the toxicity may not be critical at low concentrations optimized for labeling, it could be detrimental for the embryo development at higher concentrations. Yet, the problem can be solved by coating the QDs and making them biologically inert or by maintaining a safe concentration range. Several studies have reported that QDs can effectively accomplish their task without adversely affecting cellular processes.[75,76]

Other noble metal nanoparticles, such as Au, Ag, and Cu nanoparticles, are receiving as much attention as QDs, because they exhibit other unique and modifiable optical properties.[77] Typically,

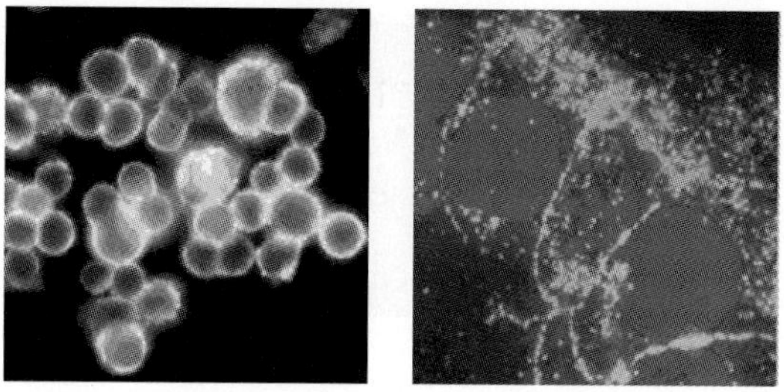

(a) (b)

Figure 4. Detection of cancer marker Her2 *in vitro* with QD-streptavidin. (a) Her2 detected on surface of free cells using QD 560-streptavidin (yellow). (b) Her2 detected on a section of mouse mammary tumor tissue using QD 630-streptavidin (red). Reprinted from Ref. 21 with permission from Nature Publishing Group. (See page 337 for color illustration.)

when spherical nanoparticles are exposed to electromagnetic field at a certain frequency, the free electrons on a metal surface, known as plasmons, undergo coherent oscillation, resulting in a strong enhancement of absorption and scattering of electromagnetic radiation. In case of noble metal nanoparticles, the surface plasmon resonance (SPR) of a metal surface especially yields intense colors and unique optical properties.[78] These noble metals have a greater potential for applications than other materials, because the resonance for Au, Ag, and Cu lie at visible frequencies and they have a high stability.[77] Furthermore, Au nanoparticles are insusceptible to photobleaching and can be easily synthesized in a wide range of sizes (4–80 nm) for tunable size-dependent optical properties. They are biocompatible and devoid of cytotoxicity, which are big advantages over QDs where cytotoxicity can be a limiting factor. The use of biocompatible, nontoxic capping material is critical for medical applications of Au nanoparticles.[79] Another reason that makes gold nanoparticles more attractive for optical imaging in biology is the well-defined chemistry between biomolecules with a gold surface. By modifying the surface of Au nanoparticles with an amine or thiol moiety,[80] for instance, the nanoparticles can mount antibodies and specifically target tumor cells and biomolecules, such as folic acid[81,82] and transferrin,[83,84] for imaging and drug/gene delivery. This was well demonstrated in a study by Richards-Kortum and coworkers,[85] where the group used 12 nm Au nanoparticles conjugated with anti-EGFR (epithelial growth factor receptor) monoclonal antibodies to image cervical epithelial cancer cell which exhibited an over expression of EGFR as compared to healthy cells. In their study, the conjugation was due to the electrostatic adsorption of the antibody molecules onto the citrate-capped and negatively charged Au nanoparticles.

The use of nanowires and nanotubes in the electrical detection method of analytes at extremely low concentration is one of the hot topics in nanomedicine. Their usage has two major advantages — high sensitivity and fast responses without tedious labeling steps. However, these nanostructures are not as readily functionalized as aforementioned quantum dots or nanoparticles. The unique advantages of these nanomaterials come from their one-dimensional morphological structures, and many researchers are trying to utilize them as a highly sensitive and selective signal transduction medium. For example, Lieber and coworkers[86] synthesized silicon nanowires with peptide nucleic acid (PNA) functionalization, and demonstrated how the synthetic material could detect DNA without labeling. A subsequent study modified silicon nanowires with biotin to detect picomolar concentrations of streptavidin and demonstrated high sensitivity to change in conductivity of the nanowires upon the biotin-streptavidin binding. Similar studies have also been carried out by Dai and coworkers,[87] where they focused on the application of carbon nanotubes as a material for the sensitive detection. The scheme (Fig. 5)[86] illustrates a basic structure of electrical detection of biomolecules with nanowire sensor.

Magnetic nanoparticles are emerging as novel contrast reagents for magnetic resonance imaging (MRI),[88–95] revolutionizing current diagnostic tools. Since their unique properties allow precise control of size and composition, magnetic nanocrystals offer great potential

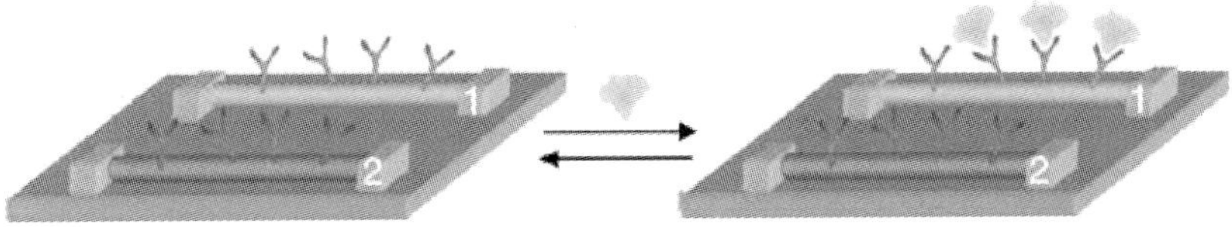

Figure 5. Schematic showing two nanowire devices, 1 and 2, within an array, where nanowires were modified with different (1, green; 2, red) antibody receptors. Reprinted from Ref. 86 with permission from Nature Publishing Group. (See page 338 for color illustration.)

for highly specific MRI for biological systems.[96–99] The nanocrystals tend to behave as a single magnetic domain in which all nuclear spins couple to create a single large magnetic domain. At certain temperatures and crystal sizes, these moments wander randomly (superparamagnetic), or become locked in one direction, making the material ferromagnetic.[88,94] Magnetic nanocrystals of differing compositions and sizes can be synthesized to generate ultra-sensitive molecular images. Cheon and coworkers[98] developed iron oxide magnetic nanoparticles which were doped with +2 cations such as Fe, Co, Ni, or Mn. It was observed that Mn-doped iron oxide nanoparticles were highly sensitive for detecting cancer cells. The nanoparticles even made it possible to image small tumors *in vivo*. In the same study, it was noted that 12 nm Mn-doped nanoparticles were bioconjugated with herceptin to have specificity towards the cancer cells. This approach is expected to improve the early diagnosis of diseases, which is critical for increasing survival rates.

1.3.2 Treatment of Diseases

The motivation to develop nanomedicine stemmed from the limitations of current treatments, such as surgery, radiation, or chemotherapy, which often damage healthy cells as well as tumor cells. To address the problem, novel therapeutics which could passively or actively target cancerous cells were developed. To date many nanomedicines are being administered to treat patients. For instance, nanomedicines such as liposomes (DaunoXome),[100–102] polymer coated liposomes (Doxil),[100,103,104] polymer-protein conjugates (Oncaspar),[105,106] antibodies (Herceptin),[107–109] and nanoparticles (Abraxane)[110] are already bringing clinical benefits to patients around the world. Most of the nanotherapeutics take two main paths to achieve their goals — passive targeting and active targeting.[111]

Rapid vascularization takes place within tumor tissues in order to supply nutrients for fast tumor growth. This inevitably causes the development of a defective, leaky architecture along with damaged lymphatic drainage. This leads to the enhanced permeation and retention (EPR) effect, which helps the injected nanoparticles to preferentially permeate and accumulate in the tumor tissue.[111,112] In order for the passive mechanism to be efficient, the size and surface properties of the nanoparticles should be well controlled. The nanoparticle size should be less than 100 nm, and the surface should be hydrophilic to prevent the uptake by the macrophages of the reticuloendothelial system (RES), which would significantly improve the circulation half-life of the nanoparticles. This is achieved by coating their surfaces with hydrophilic polymers such as polyethylene glycol (PEG), poloxamers, poloxamines, and polysaccharides[113] since a hydrophilic nanoparticle surface avoids the adsorption of plasma proteins.

Another passive targeting method utilizes the unique environment around the tumor, such as cancer-specific enzymes, high or low pH of the tumor tissue, to release the drug/biomolecules within the tumor tissue.[114] Otherwise inactive nanoparticle-drug conjugate molecules can specifically be activated by the tumor-specific environment when they reach the tumor site. The release takes place within the tumor tissue, considerably increasing the drug efficiency. Yet another method to passively target the tumor is by using direct local delivery of anticancer agents into the tumors. This is an effective technique, but can be highly invasive and the access to certain tumors such as lung cancers may be impossible.[114]

Active targeting is usually achieved by conjugation of targeting moieties such as proteins, peptides, antibodies, and carbohydrates with nanoparticles. These ligands act as guided missiles which deliver the nanoparticles to the specific cancer tissue and cells. For example, when Doxil (liposomal doxorubicin) is attached to an antibody against a growth factor overexpressed in breast cancer (ErbB2), it shows faster and greater tumor regression compared to unmodified Doxil.[115] Similarly, several targeting ligands are available for nanocarriers to deliver drugs at specified locations.[116]

1.3.3 Nanocarriers for Drug Delivery

Nanocarriers, 2 to 100 nm in diameter, are designed to deliver multiple drugs and/or imaging agents.[117] They have high surface to volume ratios which allow high density functionalization of their surfaces with targeting moieties. Promising nanocarriers summarized in Fig. 6 include liposomes, polymeric micelles, dendrimers, carbon nanotubes, and inorganic nanoparticles.[111]

Polymers such as poly lactic acid (PLA) and poly lactic coglycolic acid are the most extensively studied and commonly used materials for nanoparticle-based delivery systems.[57,118–122] Additionally, natural polymers such as chitosan and collagen have

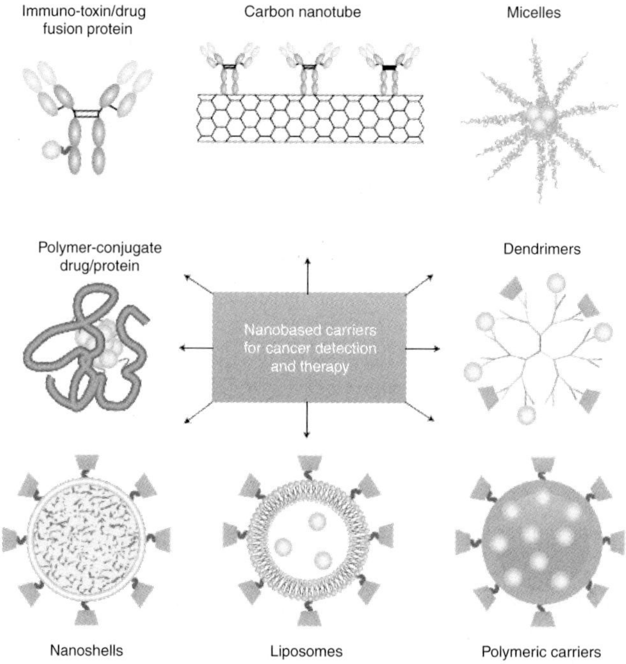

Figure 6. Examples of various drug delivery agents which typically include three components: a nanocarrier, a targeting moiety conjugated to the nanocarrier, and a therapeutic agent. Reprinted from Ref. 111 with permission from the Nature Publishing Group.

already been vigorously studied for encapsulation of drugs with their intrinsic biocompatibility.[123–126] The use of polymers has significant advantages because the drugs or biomolecules can be encapsulated easily without chemical modification processes, while maintaining a high cost efficiency and yield. Release of the encapsulated cargo of the polymeric nanoparticles takes place via diffusion through the polymer matrix, or swelling of the nanoparticles[127–132] due to changes in a local environment such as pH or specific enzyme. Efficiency can be further improved by attaching targeting moieties, and also by passivating the surfaces for longer circulation half-lives. For example, using PEG, Huang and coworkers formulated carbohydrate conjugated chitosan nanoparticles which effectively targeted the liver cancer cells (HepG2). In the study, the carbohydrate used for targeting was galactose, which specifically bound to the asialoglycoprotein (ASGP) receptors overexpressed in the HepG2 cells.[133]

Lipid-based carriers are attractive due to several properties such as biocompatibility, biodegradability, and ability to protect drugs from harsh environments. These properties came from the unique amphiphilic structure of liposomes, where the spherical amphiphiles consist of closed structures that contain one or more concentric lipid bilayers with an inner aqueous phase. Due to the amphiphilic nature, lipid-based carriers can be easily modified to entrap both hydrophobic and hydrophilic drugs.[134–136] However, Liposomes are rapidly cleared from circulation by the Kupffer cells in the liver. Thus, coating them with PEG could increase their circulation half-life.[137,138] Overall, liposomes have shown reduced toxicity and preferential accumulation in the tumor tissue by EPR effect.[127] They could also be actively targeted by attaching antibodies to their surface.

Dendrimers are well defined molecules which are built at a nanoscale with monodispersed systems with sizes ranging from 3 to 10 nm. They have unique molecular architectures and properties, both of which result in easy conjugation chemistry with targeting molecules, making them attractive for the development of nanomedicines.[127,139–142] Several studies have confirmed that conjugated dendrimers can be found concentrated in the tumor and liver tissue in contrast to non-targeted polymer folates. Increased therapeutic activity and marked reduction in toxicity was also observed.[143]

The synthesis of colloidal gold was first reported in 1857.[144] However, it was about 100 years later when scientists found out that the gold particles could bind proteins without altering their activity, paving a way for their application in biological systems. Several studies confirm that gold nanoparticles are easily taken up by cells.[83,145–147] The biocompatibility of gold nanoparticles can be further enhanced by PEG coating that prevents uptake by RES. For example, PEG-coated gold nanoparticle formulation, developed by CytImmune Sciences,[78] which carries the tumor necrosis factor (TNF), is currently under human trials. These gold nanoparticles showed preferential accumulation in MC-38 colon carcinoma tumors and no uptake by liver or spleen. It was further reported that this system is less toxic and more effective in reducing tumor burden than native TNF.[148] Gold nanoparticles are also very useful for delivering nucleic acids. As seen previously, their surface allows for easy functionalization with thiols which makes attachment of oligonucleotides on the surface relatively easy. In a recent study by Mirkin and coworkers, it was reported that the cellular uptake of gold nanoparticles increased with the increase in density of the oligonucleotides on the particle surface.[149]

Magnetic nanoparticles are extremely promising as novel drug delivery agents.[150] They show several advantages which are not seen in most nanoparticle-based systems. They can resolve problems related to polydispersity and irregular branching, both of which are common in polymeric nanocarriers. Magnetic nanoparticles, for example, can act as contrasting agents for MRI application.[151] Also, drug loaded magnetic nanoparticles can be guided or held in place with applied external magnetic field using their superb magnetic susceptibility. The magnetic nanoparticles can also be heated to induce hyperthermia of the tissue upon exposure to the external magnetic field. The magnetic nanoparticles can be further functionalized with targeting ligands to improve their uptake efficiency, minimize their toxicity, and prevent them from aggregating.[152,153] This is typically done by coating them with hydrophilic to neutral compounds such a PEG, dextran, or HSA, which not only stabilizes the particles but also increases their half-life in blood.

With photothermal cancer therapy, which uses optical heating for tumor ablation, doctors can deal with tumors in a non-invasive manner.[154–159] It could be a method of choice for tumors which are inaccessible for surgery or radiotherapy, both of which kill healthy cells along with tumor cells. For this specific application, gold nanoparticles are most widely used, because the gold nanoparticles have ability to effectively convert strongly absorbed NIR light to localized heat, and hence are useful in selective photothermal therapy. Using active and passive targeting, these nanoparticles can be localized into the tumor tissue, increasing the effectiveness of the heat produced and at the same time minimizing the non-specificity of treatment. In a recent study, Li and coworkers developed gold nanocages (< 45 nm) which were specifically designed to strongly absorb in the NIR region for photothermal therapy.[160] These nanocages were conjugated to anti-HER2 monoclonal antibodies against the epidermal growth factor receptor (EGFR), overexpressed on the surface of breast cancer cells.[160,161] They concluded that the death of cancer cells increased linearly with increase in irradiation power density, thus making the immunogold nanocages effective photothermal therapeutics agents.

1.4 NANOSYSTEMATIC APPROACH FOR CELL BIOLOGY

Cells are single living units of organisms which first receive the input perturbation signals from disease and injury, and then return the output signals to their microenvironments. Conventional experimental studies on particular cellular responses are typically conducted on a large cell population, which inevitably produces data measured from inhomogeneous distribution of cellular responses. Unless cellular behaviors and processes are isolated from inhomogeneous signals at the level of single cell, it would be extremely difficult to elucidate the intricate cellular systems and analyze the complex dynamic signaling transductions. In order to better study and control the responses of cells towards outer stimuli, scientists need to characterize the full range of cell behaviors, such as self-renewal, differentiation, migration, and apoptosis, from the single cell level or even the single molecule level. In particular, understanding how a genotypic aspect affects a cell phenotype is a complicated process, which can barely be revealed by conventional biomedical approaches. In order to understand these processes, the two distinct approaches – bottom-up and top-down – can be applied in a combinatorial way.

1.4.1 Microfluidics & Micropatterns

Microfluidic devices offer a robust analytical approach, allowing rapid analysis of cell assays in a parallel way to investigate complex cell behaviors (Fig. 7).[162–165] Microfluidic devices have advantages over the macroscopic setting, such as reduced sample/reagent volume, high surface-to-volume ratios, an improved control of the physical/chemical microenvironments, and high throughput/automatic capabilities.[166,167] These characteristics would be beneficial for understanding new aspects of complex cell dynamics (e.g. stem cell differentiation and cancer cell apoptosis) and tissue engineering. Although there have been a few examples of microfluidic systems used to culture and assay stem cells,[166–169] the stem cell assays in microfluidic gradients have not been fully explored. Generation of microfluidic gradients have been demonstrated by Whiteside and coworkers,[170] and further studies have been done to study cell behaviors in gradients (Fig. 8).

The use of micro- or nanometer scale patterned surfaces is also one of the useful approaches to study individual cells at

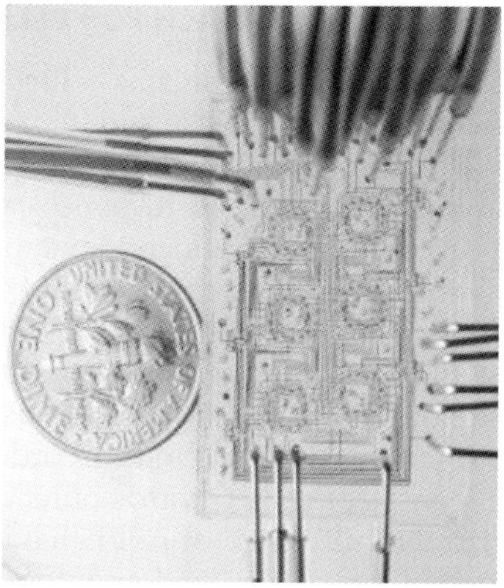

Figure 7. An optical micrograph of a microfluidic device. The coin is 18 mm in diameter. Reprinted from Ref. 171 with permission from AAAS.

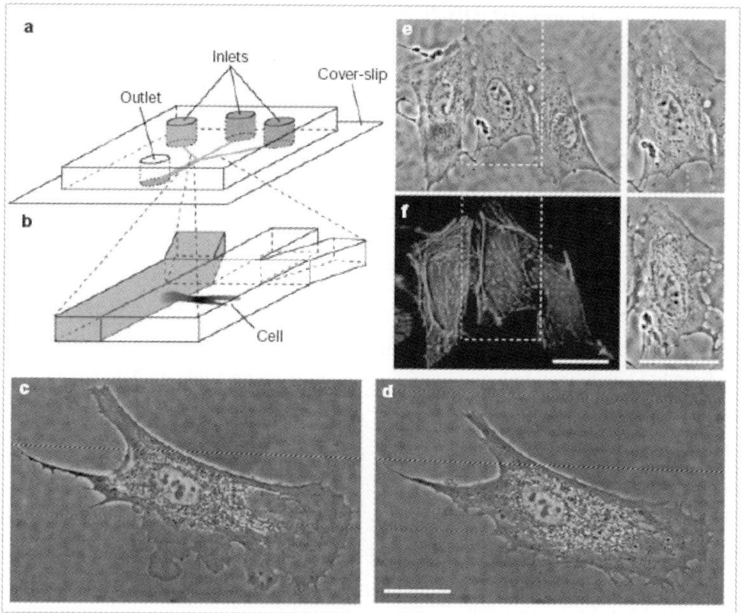

Figure 8. Microfluidics to study a single bovine endothelial cell using multiple laminar streams which deliver membrane permeable molecules to selected subcellular domains. Reprinted from Ref. 170 with permission from Nature Publishing Group.

the single cell level. Typically, the patterns are generated by microcontact printing, where a soft-lithographed material transfers molecules through conformal contact on substrates with planar or non-planar topographies to form SAMs. The key advantage of microcontact printing is that a variety of features can be generated simultaneously on a surface in a single step, thus making patterning easier and faster as compared to scanning probe techniques which require longer time and effort to generate patterns.

1.4.2 Nanopatterning for Stem Cell Research

Although stem cells hold great potential for the treatment of a number of devastating injuries and damages caused by degenerative diseases,[172,173] a better control of microenvironment that closely interacts with stem cells should be achieved before therapeutic applications can be fully realized. It is because stem cell fate is controlled by two prime factors; the intrinsic regulators, such as growth factors and signaling molecules, and the extracellular microenvironments, such as extracellular matrix (ECM). To date, there are few conventional methods available to study regulatory and extracellular microenvironmental cues that control stem cell fate at the single cellular level as well as in a combinatorial way. Stem cells normally reside within specific extracellular microenvironments,[174,175] known as "stem cell niches", comprising a complex mixture of soluble and insoluble ECM signals. The signals regulate stem cell behavior, such as self-renewal, migration, and differentiation.[176–178] For example, cell adhesion and ECM play important roles in early stem cell development. Cell adhesion process, which keeps the inner cell mass intact, is mediated by cadherins and integrins, which are further regulated post-translationally via protein kinase C and other signaling molecules (Fig. 9). The process determines cellular allocation and spatial organization of the inner cell mass (ICM) in the blastocyst.[179] Likewise, *in vivo* stem cells come in contact with various soluble and insoluble ECM components that affect their differentiation. *In vitro* studies also have shown how ECM components and growth factors regulate the differentiation of stem cells.[180] Several combinatorial high-throughput screening approaches on the function of soluble signal molecules on stem cell differentiations have been reported.[181]

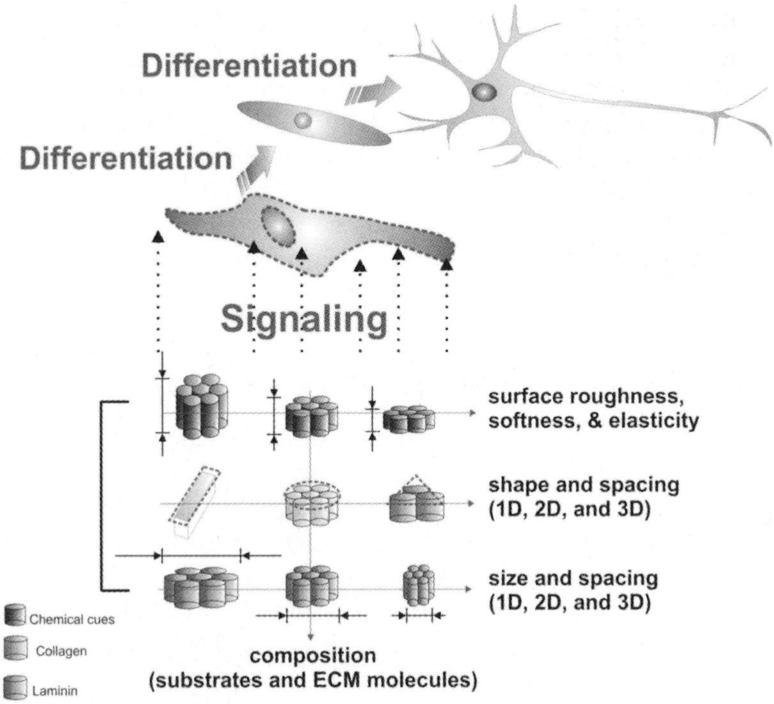

Figure 9. Application of micro-/nanoscale surface engineering in stem cells. Micro- and nanostructures that interact with stem cells at the molecular level can be utilized to control stem cell fate. (See page 338 for color illustration.)

However, similar approaches for screening the function of insoluble cues are limited due to the technical difficulty of identifying, isolating, and modulating individual stem cells from their surroundings. For instance, an extracellular matrix array format fabricated by conventional macro/micro patterning techniques has been used to probe cellular differentiation and migration.[182,183] Very recently, a combinatorial ECM microarray was used to investigate the role of the ECM components in the differentiation of mouse embryonic stem cells (mESCs) towards an early hepatic fate.[182] This technology required 1000 times less protein than a conventional method. There still is plenty of room for improvement in this approach in terms of pattern density, recognition sensitivity, and small sample requirement. In addition, relatively little is known about the subcellular interaction mechanisms between stem cells and ECM molecules and how such events eventually influence stem cell differentiation and migration. Moreover, it is still unknown how insoluble or soluble extrinsic signaling molecules identify their molecular target proteins and receptors at the single molecule level. The application of nanotechnology in stem cell biology will help to address those challenges.

1.5 CONCLUSION

Nanomedicine is where traditional biomedical science meets nanotechnology. The synergetic effect offers new possibilities in diagnosis and treatment of many malicious human diseases. Moreover, the advancement in nanomedicine allows scientists in cell biology and physiology to investigate targeted bio-interactions at the fundamental molecular level. However, for the potential of nanomedicine to be fully realized, multimodal technologies for cell recognition and molecular imaging, and targeted delivery should be developed. This challenge requires an interdisciplinary approach from a high level of expertise in each field of science. In the near future nanomedicine may bring revolutionary breakthroughs in not only the entire field of medicine, but also in fundamental biology.

ACKNOWLEDGEMENT

The authors gratefully acknowledge helpful comments on the manuscript from David Wang, Birju Shah, and Staci C. Brown.

REFERENCES

1. P. Alivisatos, "The use of nanocrystals in biological detection," *Nature Biotechnology*, **22**, 47–52 (2004).
2. W. C. W. Chan and S. M. Nie, "Quantum dot bioconjugates for ultrasensitive nonisotopic detection," *Science*, **281**, 2016–2018 (1998).
3. D. S. Ginger, H. Zhang, and C. A. Mirkin, "The evolution of dip-pen nanolithography," *Angewandte Chemie-International Edition*, **43**, 30–45 (2004).
4. S. G. Penn, L. He, and M. J. Natan, "Nanoparticles for bioanalysis," *Current Opinion in Chemical Biology*, **7**, 609–615 (2003).
5. A. P. Alivisatos, P. F. Barbara, A. W. Castleman, J. Chang, D. A. Dixon, M. L. Klein, G. L. McLendon, J. S. Miller, M. A. Ratner, P. J. Rossky,

S. I. Stupp, and M. E. Thompson, "From molecules to materials: Current trends and future directions," *Advanced Materials*, **10**, 1297–1336 (1998).

6. R. Elghanian, J. J. Storhoff, R. C. Mucic, R. L. Letsinger, and C. A. Mirkin, "Selective colorimetric detection of polynucleotides based on the distance-dependent optical properties of gold nanoparticles," *Science*, **277**, 1078–1081 (1997).
7. C. A. Mirkin, R. L. Letsinger, R. C. Mucic, and J. J. Storhoff, "A DNA-based method for rationally assembling nanoparticles into macroscopic materials," *Nature*, **382**, 607–609 (1996).
8. J. J. Storhoff and C. A. Mirkin, "Programmed materials synthesis with DNA," *Chemical Reviews*, **99**, 1849–1862 (1999).
9. C. M. Niemeyer, "Nanoparticles, proteins, and nucleic acids: Biotechnology meets materials science," *Angewandte Chemie-International Edition*, **40**, 4128–4158 (2001).
10. C. M. Niemeyer, "Nanotechnology: Tools for the biomolecular engineer," *Science*, **297**, 62–63 (2002).
11. Y. Weizmann, F. Patolsky, O. Lioubashevski, and I. Willner, "Magneto-mechanical detection of nucleic acids and telomerase activity in cancer cells," *Journal of the American Chemical Society*, **126**, 1073–1080 (2004).
12. I. Willner, "Photoswitchable biomaterials: En route to optobioelectronic systems," *Accounts of Chemical Research*, **30**, 347–356 (1997).
13. I. Willner, "Protein hinges for bioelectronics — Changes in protein conformation have been exploited to create chemoresponsive bioelectronic sensors," *Nature Biotechnology*, **19**, 1023–1024 (2001).
14. I. Willner, "Biomaterials for sensors, fuel cells, and circuitry," *Science*, **298**, 2407–2408 (2002).
15. N. L. Rosi and C. A. Mirkin, "Nanostructures in biodiagnostics," *Chemical Reviews*, **105**, 1547–1562 (2005).
16. M. Bruchez, M. Moronne, P. Gin, S. Weiss, and A. P. Alivisatos, "Semiconductor nanocrystals as fluorescent biological labels," *Science*, **281**, 2013–2016 (1998).
17. S. J. Clarke, C. A. Hollmann, Z. J. Zhang, D. Suffern, S. E. Bradforth, N. M. Dimitrijevic, W. G. Minarik, and J. L. Nadeau, "Photophysics of dopamine-modified quantumdots and effects on biological systems," *Nature Materials*, **5**, 409–417 (2006).
18. E. T. Ahrens, R. Flores, H. Y. Xu, and P. A. Morel, "In vivo imaging platform for tracking immunotherapeutic cells," *Nature Biotechnology*, **23**, 983–987 (2005).

19. T. A. Byassee, W. C. W. Chan, and S. M. Nie, "Probing single molecules in single living cells," *Analytical Chemistry*, **72**, 5606–5611 (2000).
20. J. K. Jaiswal, H. Mattoussi, J. M. Mauro, and S. M. Simon, "Long-term multiple color imaging of live cells using quantum dot bioconjugates," *Nature Biotechnology*, **21**, 47–51 (2003).
21. X. Y. Wu, H. J. Liu, J. Q. Liu, K. N. Haley, J. A. Treadway, J. P. Larson, N. F. Ge, F. Peale, and M. P. Bruchez, "Immunofluorescent labeling of cancer marker Her2 and other cellular targets with semiconductor quantum dots," *Nature Biotechnology*, **21**, 41–46 (2003).
22. J. W. M. Bulte, T. Douglas, B. Witwer, S. C. Zhang, E. Strable, B. K. Lewis, H. Zywicke, B. Miller, P. van Gelderen, B. M. Moskowitz, I. D. Duncan, and J. A. Frank, "Magnetodendrimers allow endosomal magnetic labeling and *in vivo* tracking of stem cells," *Nature Biotechnology*, **19**, 1141–1147 (2001).
23. W. C. W. Chan, D. J. Maxwell, X. H. Gao, R. E. Bailey, M. Y. Han, and S. M. Nie, "Luminescent quantum dots for multiplexed biological detection and imaging," *Current Opinion in Biotechnology*, **13**, 40–46 (2002).
24. M. Lewin, N. Carlesso, C. H. Tung, X. W. Tang, D. Cory, D. T. Scadden, and R. Weissleder, "Tat peptide-derivatized magnetic nanoparticles allow *in vivo* tracking and recovery of progenitor cells," *Nature Biotechnology*, **18**, 410–414 (2000).
25. M. Bailly, L. Yan, G. M. Whitesides, J. S. Condeelis, and J. E. Segall, "Regulation of protrusion shape and adhesion to the substratum during chemotactic responses of mammalian carcinoma cells," *Experimental Cell Research*, **241**, 285–299, (1998).
26. Y. W. Cao, R. Jin, and C. A. Mirkin, "DNA-modified core-shell Ag/Au nanoparticles," *Journal of the American Chemical Society*, **123**, 7961–7962, (2001).
27. X. H. Gao, W. C. W. Chan, and S. M. Nie, "Quantum-dot nanocrystals for ultrasensitive biological labeling and multicolor optical encoding," *Journal of Biomedical Optics*, **7**, 532–537 (2002).
28. J. T. Hu, T. W. Odom, and C. M. Lieber, "Chemistry and physics in one dimension: Synthesis and properties of nanowires and nanotubes," *Accounts of Chemical Research*, **32**, 435–445 (1999).
29. G. MacBeath, "Protein microarrays and proteomics," *Nature Genetics*, **32**, 526–532 (2002).
30. G. MacBeath, "Proteomics comes to the surface — Microarrays of purified proteins, representing most of the yeast genome, prove

useful for studying protein function on a genome-wide scale," *Nature Biotechnology*, **19**, 828–829 (2001).
31. G. MacBeath and S. L. Schreiber, "Printing proteins as microarrays for high-throughput function determination," *Science*, **289**, 1760–1763 (2000).
32. G. MacBeath, A. N. Koehler, and S. L. Schreiber, "Printing small molecules as microarrays and detecting protein-ligand interactions en masse," *Journal of the American Chemical Society*, **121**, 7967–7968 (1999).
33. C. S. Chen, M. Mrksich, S. Huang, G. M. Whitesides, and D. E. Ingber, "Geometric control of cell life and death," *Science*, **276**, 1425–1428 (1997).
34. M. Mrksich, "Tailored substrates for studies of attached cell culture," *Cellular and Molecular Life Sciences*, **54**, 653–662 (1998).
35. N. Sniadecki, R. A. Desai, S. A. Ruiz, and C. S. Chen, "Nanotechnology for cell-substrate interactions," *Annals of Biomedical Engineering*, **34**, 59–74 (2006).
36. Y. S. Lee and M. Mrksich, "Protein chips: from concept to practice," *Trends in Biotechnology*, **20**, S14–S18 (2002).
37. B. D. Gates, Q. B. Xu, J. C. Love, D. B. Wolfe, and G. M. Whitesides, "Unconventional nanofabrication," *Annual Review of Materials Research*, **34**, 339–372 (2004).
38. B. D. Gates, Q. B. Xu, M. Stewart, D. Ryan, C. G. Willson, and G. M. Whitesides, "New approaches to nanofabrication: Molding, printing, and other techniques," *Chemical Reviews*, **105**, 1171–1196 (2005).
39. M. Mrksich and G. M. Whitesides, "Using self-assembled monolayers to understand the interactions of man-made surfaces with proteins and cells," *Annual Review of Biophysics and Biomolecular Structure*, **25**, 55–78 (1996).
40. D. Qin, Y. N. Xia, J. A. Rogers, R. J. Jackman, X. M. Zhao, and G. M. Whitesides, "Microfabrication, microstructures and microsystems," *Microsystem Technology in Chemistry and Life Science*, **194**, 1–20 (1998).
41. D. B. Weibel, P. Garstecki, and G. M. Whitesides, "Combining microscience and neurobiology," *Current Opinion in Neurobiology*, **15**, 560–567 (2005).
42. G. M. Whitesides, E. Ostuni, S. Takayama, X. Y. Jiang, and D. E. Ingber, "Soft lithography in biology and biochemistry," *Annual Review of Biomedical Engineering*, **3**, 335–373 (2001).

43. Y. N. Xia, J. A. Rogers, K. E. Paul, and G. M. Whitesides, "Unconventional methods for fabricating and patterning nanostructures," *Chemical Reviews*, **99**, 1823–1848 (1999).
44. Y. N. Xia and G. M. Whitesides, "Soft lithography," *Angewandte Chemie-International Edition*, **37**, 551–575 (1998).
45. G. Y. Liu, S. Xu, and Y. L. Qian, "Nanofabrication of self-assembled monolayers using scanning probe lithography," *Accounts of Chemical Research*, **33**, 457–466 (2000).
46. D. Wouters and U. S. Schubert, "Nanolithography and nanochemistry: Probe-related patterning techniques and chemical modification for nanometer-sized devices," *Angewandte Chemie-International Edition*, **43**, 2480–2495 (2004).
47. K. Salaita, Y. H. Wang, and C. A. Mirkin, "Applications of dip-pen nanolithography," *Nature Nanotechnology*, **2**, 145–155 (2007).
48. J. Loos, "The art of SPM: Scanning probe microscopy in materials science," *Advanced Materials*, **17**, 1821–1833 (2005).
49. Q. Tang, S. Q. Shi, and L. M. Zhou, "Nanofabrication with atomic force microscopy," *Journal of Nanoscience and Nanotechnology*, **4**, 948–963 (2004).
50. D. Wouters and U. S. Schubert, "Nanolithography and nanochemistry: Probe-related patterning techniques and chemical modification for nanometer-sized devices," *Angewandte Chemie-International Edition*, **43**, 2480–2495 (2004).
51. S. H. Hong, J. Zhu, and C. A. Mirkin, "Multiple ink nanolithography: Toward a multiple-pen nano-plotter," *Science*, **286**, 523–525 (1999).
52. R. D. Piner, J. Zhu, F. Xu, S. H. Hong, and C. A. Mirkin, "Dip-pen nanolithography," *Science*, **283**, 661–663 (1999).
53. E. Katz and I. Willner, "Integrated nanoparticle-biomolecule hybrid systems: Synthesis, properties, and applications," *Angewandte Chemie-International Edition*, **43**, 6042–6108, (2004).
54. H. W. Liao, C. L. Nehl, and J. H. Hafner, "Biomedical applications of plasmon resonant metal nanoparticles," *Nanomedicine*, **1**, 201–208 (2006).
55. F. Hyafil, J. C. Cornily, J. E. Feig, R. Gordon, E. Vucic, V. Amirbekian, E. A. Fisher, V. Fuster, L. J. Feldman, and Z. A. Fayad, "Noninvasive detection of macrophages using a nanoparticulate contrast agent for computed tomography," *Nature Medicine*, **13**, 636–641 (2007).
56. J. M. Karp and R. Langer, "Development and therapeutic applications of advanced biomaterials," *Current Opinion in Biotechnology*, **18**, 454–459 (2007).

57. R. Langer and D. A. Tirrell, "Designing materials for biology and medicine," *Nature*, **428**, 487–492 (2004).
58. D. A. LaVan, T. McGuire, and R. Langer, "Small-scale systems for *in vivo* drug delivery," *Nature Biotechnology*, **21**, 1184–1191 (2003).
59. S. J. Son, X. Bai, and S. Lee, "Inorganic hollow nanoparticles and nanotubes in nanomedicine. Part 2: Imaging, diagnostic, and therapeutic applications," *Drug Discovery Today*, **12**, 657–663 (2007).
60. W. W. Wu and A. D. Li, "Optically switchable nanoparticles for biological imaging," *Nanomedicine*, **2**, 523–531 (2007).
61. Q. L. de Chermont, C. Chaneac, J. Seguin, F. Pelle, S. Maitrejean, J. P. Jolivet, D. Gourier, M. Bessodes, and D. Scherman, "Nanoprobes with near-infrared persistent luminescence for *in vivo* imaging," *Proceedings of the National Academy of Sciences of the United States of America*, **104**, 9266–9271 (2007).
62. X. H. Gao, L. L. Yang, J. A. Petros, F. F. Marshal, J. W. Simons, and S. M. Nie, "In vivo molecular and cellular imaging with quantum dots," *Current Opinion in Biotechnology*, **16**, 63–72, (2005).
63. X. Michalet, F. F. Pinaud, L. A. Bentolila, J. M. Tsay, S. Doose, J. J. Li, G. Sundaresan, A. M. Wu, S. S. Gambhir, and S. Weiss, "Quantum dots for live cells, *in vivo* imaging, and diagnostics," *Science*, **307**, 538–544 (2005).
64. S. M. Nie, Y. Xing, G. J. Kim, and J. W. Simons, "Nanotechnology applications in cancer," *Annual Review of Biomedical Engineering*, **9**, 257–288 (2007).
65. M. N. Rhyner, A. M. Smith, X. H. Gao, H. Mao, L. L. Yang, and S. M. Nie, "Quantum dots and multifunctional nanoparticles: New contrast agents for tumor imaging," *Nanomedicine*, **1**, 209–217 (2006).
66. M. Stroh, J. P. Zimmer, D. G. Duda, T. S. Levchenko, K. S. Cohen, E. B. Brown, D. T. Scadden, V. P. Torchilin, M. G. Bawendi, D. Fukumura, and R. K. Jain, "Quantum dots spectrally distinguish multiple species within the tumor milieu *in vivo*," *Nature Medicine*, **11**, 678–682 (2005).
67. R. E. Bailey and S. M. Nie, "Alloyed semiconductor quantum dots: Tuning the optical properties without changing the particle size," *Journal of the American Chemical Society*, **125**, 7100–7106, (2003).
68. S. J. Rosenthal, J. McBride, S. J. Pennycook, and L. C. Feldman, "Synthesis, surface studies, composition and structural characterization of CdSe, core/shell and biologically active nanocrystals," *Surface Science Reports*, **62**, 111–157, (2007).
69. D. S. Wang, J. B. He, N. Rosenzweig, and Z. Rosenzweig, "Superparamagnetic Fe_2O_3 Beads-CdSe/ZnS quantum dots core-shell

nanocomposite particles for cell separation," *Nano Letters*, **4**, 409–413 (2004).

70. Y. Yin and A. P. Alivisatos, "Colloidal nanocrystal synthesis and the organic-inorganic interface," *Nature*, **437**, 664–670, (2005).
71. E. L. Bentzen, F. House, T. J. Utley, J. E. Crowe, and D. W. Wright, "Progression of respiratory syncytial virus infection monitored by fluorescent quantum dot probes," *Nano Letters*, **5**, 591–595 (2005).
72. T. Jamieson, R. Bakhshi, D. Petrova, R. Pocock, M. Imani, and A. M. Seifalian, "Biological applications of quantum dots," *Biomaterials*, **28**, 4717–4732 (2007).
73. B. I. Ipe, M. Lehnig, and C. M. Niemeyer, "On the generation of free radical species from quantum dots," *Small*, **1**, 706–709 (2005).
74. C. Kirchner, T. Liedl, S. Kudera, T. Pellegrino, A. M. Javier, H. E. Gaub, S. Stolzle, N. Fertig, and W. J. Parak, "Cytotoxicity of colloidal CdSe and CdSe/ZnS nanoparticles," *Nano Letters*, **5**, 331–338 (2005).
75. A. M. Derfus, W. C. W. Chan, and S. N. Bhatia, "Probing the cytotoxicity of semiconductor quantum dots," *Nano Letters*, **4**, 11–18 (2004).
76. A. Hoshino, K. Fujioka, T. Oku, M. Suga, Y. F. Sasaki, T. Ohta, M. Yasuhara, K. Suzuki, and K. Yamamoto, "Physicochemical properties and cellular toxicity of nanocrystal quantum dots depend on their surface modification," *Nano Letters*, **4**, 2163–2169, 2004.
77. S. Link and M. A. El-Sayed, "Optical properties and ultrafast dynamics of metallic nanocrystals" *Annual Review of Physical Chemistry*, **54**, 331–366 (2003).
78. P. K. Jain, I. H. El-Sayed, and M. A. El-Sayed, "Au nanoparticles target cancer," *Nano Today*, **2**, 18–29 (2007).
79. E. E. Connor, J. Mwamuka, A. Gole, C. J. Murphy, and M. D. Wyatt, "Gold nanoparticles are taken up by human cells but do not cause acute cytotoxicity," *Small*, **1**, 325–327 (2005).
80. C. M. Niemeyer and B. Ceyhan, "DNA-directed functionalization of colloidal gold with proteins," *Angewandte Chemie-International Edition*, **40**, 3685-+ (2001).
81. R. Bhattacharya, C. R. Patra, A. Earl, S. F. Wang, A. Katarya, L. C. Lu, J. N. Kizhakkedathu, M. J. Yaszemski, P. R. Greipp, D. Mukhopadhyay, and P. Mukherjee, "Attaching folic acid on gold nanoparticles using noncovalent interaction via different polyethylene glycol backbones and targeting of cancer cells," *Nanomedicine*, **3**, 224–238 (2007).
82. V. Dixit, J. Van den Bossche, D. M. Sherman, D. H. Thompson, and R. P. Andres, "Synthesis and grafting of thioctic acid-PEG-folate conjugates onto Au nanoparticles for selective targeting of folate

receptor-positive tumor cells," *Bioconjugate Chemistry*, **17**, 603–609 (2006).
83. B. D. Chithrani and W. C. W. Chan, "Elucidating the mechanism of cellular uptake and removal of protein-coated gold nanoparticles of different sizes and shapes," *Nano Letters*, **7**, 1542–1550 (2007).
84. P. H. Yang, X. S. Sun, J. F. Chiu, H. Z. Sun, and Q. Y. He, "Transferrin-mediated gold nanoparticle cellular uptake," *Bioconjugate Chemistry*, **16**, 494–496 (2005).
85. K. Sokolov, D. Nida, M. Descour, A. Lacy, M. Levy, B. Hall, S. Dharmawardhane, A. Ellington, B. Korgel, and R. Richards-Kortum, "Molecular optical imaging of therapeutic targets of cancer," *Advances in Cancer Research*, **96**, 299–344, 2007.
86. G. F. Zheng, F. Patolsky, Y. Cui, W. U. Wang, and C. M. Lieber, "Multiplexed electrical detection of cancer markers with nanowire sensor arrays," *Nature Biotechnology*, **23**, 1294–1301, 2005.
87. R. J. Chen, S. Bangsaruntip, K. A. Drouvalakis, N. W. S. Kam, M. Shim, Y. M. Li, W. Kim, P. J. Utz, and H. J. Dai, "Noncovalent functionalization of carbon nanotubes for highly specific electronic biosensors," *Proceedings of the National Academy of Sciences of the United States of America*, **100**, 4984–4989 (2003).
88. S. D. Caruthers, S. A. Wickline, and G. M. Lanza, "Nanotechnological applications in medicine," *Current Opinion in Biotechnology*, **18**, 26–30 (2007).
89. C. Corot, P. Robert, J. M. Idee, and M. Port, "Recent advances in iron oxide nanocrystal technology for medical imaging," *Advanced Drug Delivery Reviews*, **58**, 1471–1504 (2006).
90. A. K. Gupta and M. Gupta, "Synthesis and surface engineering of iron oxide nanoparticles for biomedical applications," *Biomaterials*, **26**, 3995–4021 (2005).
91. Y. E. L. Koo, G. R. Reddy, M. Bhojani, R. Schneider, M. A. Philbert, A. Rehemtulla, B. D. Ross, and R. Kopelman, "Brain cancer diagnosis and therapy with nanoplatforms," *Advanced Drug Delivery Reviews*, **58**, 1556–1577 (2006).
92. J. K. Raty, T. Liimatainen, M. U. Kaikkonen, O. Grahn, K. J. Airenne, and S. Yla-Herttuala, "Non-invasive Imaging in gene therapy," *Molecular Therapy*, **15**, 1579–1586 (2007).
93. W. J. Rogers, C. H. Meyer, and C. M. Kramer, "Technology insight: *in vivo* cell tracking by use of MRI," *Nature Clinical Practice Cardiovascular Medicine*, **3**, 554–562 (2006).

94. D. L. J. Thorek, A. Chen, J. Czupryna, and A. Tsourkas, "Superparamagnetic iron oxide nanoparticle probes for molecular imaging," *Annals of Biomedical Engineering*, **34**, 23–38 (2006).
95. W. A. Weber, J. Czernin, M. E. Phelps, and H. R. Herschman, "Technology Insight: novel imaging of molecular targets is an emerging area crucial to the development of targeted drugs," *Nature Clinical Practice Oncology*, **5**, 44–54 (2008).
96. Y. M. Huh, Y. W. Jun, H. T. Song, S. Kim, J. S. Choi, J. H. Lee, S. Yoon, K. S. Kim, J. S. Shin, J. S. Suh, and J. Cheon, "In vivo magnetic resonance detection of cancer by using multifunctional magnetic nanocrystals," *Journal of the American Chemical Society*, **127**, 12387–12391 (2005).
97. Y. W. Jun, Y. M. Huh, J. S. Choi, J. H. Lee, H. T. Song, S. Kim, S. Yoon, K. S. Kim, J. S. Shin, J. S. Suh, and J. Cheon, "Nanoscale size effect of magnetic nanocrystals and their utilization for cancer diagnosis via magnetic resonance imaging," *Journal of the American Chemical Society*, **127**, 5732–5733 (2005).
98. J. H. Lee, Y. M. Huh, Y. Jun, J. Seo, J. Jang, H. T. Song, S. Kim, E. J. Cho, H. G. Yoon, J. S. Suh, and J. Cheon, "Artificially engineered magnetic nanoparticles for ultra-sensitive molecular imaging," *Nature Medicine*, **13**, 95–99 (2007).
99. H. T. Song, J. S. Choi, Y. M. Huh, S. Kim, Y. W. Jun, J. S. Suh, and J. Cheon, "Surface modulation of magnetic nanocrystals in the development of highly efficient magnetic resonance probes for intracellular labeling," *Journal of the American Chemical Society*, **127**, 9992–9993 (2005).
100. T. M. Allen and F. J. Martin, "Advantages of liposomal delivery systems for anthracyclines," *Seminars in Oncology*, **31**, 5–15 (2004).
101. L. Cattel, M. Ceruti, and F. Dosio, "From conventional to stealth liposomes a new frontier in cancer chemotherapy," *Tumori*, **89**, 237–249 (2003).
102. J. A. Sparano and E. P. Winer, "Liposomal anthracyclines for breast cancer," *Seminars in Oncology*, **28**, 32–40 (2001).
103. A. Gabizon, H. Shmeeda, and Y. Barenholz, "Pharmacokinetics of pegylated liposomal doxorubicin — Review of animal and human studies," *Clinical Pharmacokinetics*, **42**, 419–436, 2003.
104. N. E. Hynes and H. A. Lane, "ERBB receptors and cancer: The complexity of targeted inhibitors," *Nature Reviews Cancer*, **5**, 341–354 (2005).
105. R. Duncan, M. J. Vicent, F. Greco, and R. I. Nicholson, "Polymer-drug conjugates: Towards a novel approach for the treatment of

131. W. Y. Seow, J. M. Xue, and Y. Y. Yang, "Targeted and intracellular delivery of paclitaxel using multi-functional polymeric micelles," *Biomaterials*, **28**, 1730–1740 (2007).
132. X. L. Wang, R. Jensen, and Z. R. Lu, "A novel environment-sensitive biodegradable polydisulfide with protonatable pendants for nucleic acid delivery," *Journal of Controlled Release*, **120**, 250–258 (2007).
133. F. L. Mi, Y. Y. Wu, Y. L. Chiu, M. C. Chen, H. W. Sung, S. H. Yu, S. S. Shyu, and M. F. Huang, "Synthesis of a novel glycoconjugated chitosan and preparation of its derived nanoparticles for targeting HepG2 cells," *Biomacromolecules*, **8**, 892–898 (2007).
134. T. A. Elbayoumi, S. Pabba, A. Roby, and V. P. Torchilin, "Antinucleosome antibody-modified liposomes and lipid-core micelles for tumor-targeted delivery of therapeutic and diagnostic agents," *Journal of Liposome Research*, **17**, 1–14, 2007.
135. F. Lacoeuille, F. Hindre, F. Moal, J. Roux, C. Passirani, O. Couturier, P. Cales, J. J. Le Jeune, A. Lamprecht, and J. P. Benoit, "In vivo evaluation of lipid nanocapsules as a promising colloidal carrier for paclitaxel," *International Journal of Pharmaceutics*, **344**, 143–149 (2007).
136. J. C. Leroux, "Injectable nanocarriers for biodetoxification," *Nature Nanotechnology*, **2**, 679–684 (2007).
137. T. Lian and R. J. Y. Ho, "Trends and developments in liposome drug delivery systems," *Journal of Pharmaceutical Sciences*, **90**, 667–680 (2001).
138. S. M. Moghimi and S. S. Davis, "Innovations in avoiding particle clearance from blood by kupffer cells — cause for reflection," *Critical Reviews in Therapeutic Drug Carrier Systems*, **11**, 31–59 (1994).
139. S. Y. Cho and H. R. Allcock, "Dendrimers derived from polyphosphazene-poly(propyleneimine) systems: Encapsulation and triggered release of hydrophobic guest molecules," *Macromolecules*, **40**, 3115–3121 (2007).
140. C. M. Paleos, D. Tsiourvas and Z. Sideratou, "Molecular engineering of dendritic polymers and their application as drug and gene delivery systems," *Molecular Pharmaceutics*, **4**, 169–188 (2007).
141. F. P. Seib, A. T. Jones and R. Duncan, "Comparison of the endocytic properties of linear and branched PEIs, and cationic PAMAM dendrimers in B16f10 melanoma cells," *Journal of Controlled Release*, **117**, 291–300 (2007).
142. X. G. Shi, S. H. Wang, S. Meshinchi, M. E. Van Antwerp, X. D. Bi, I. H. Lee and J. R. Baker, "Dendrimer-entrapped gold nanoparticles as a

platform for cancer-cell targeting and Imaging," *Small*, **3**, 1245–1252 (2007).

143. J. F. Kukowska-Latallo, K. A. Candido, Z. Y. Cao, S. S. Nigavekar, I. J. Majoros, T. P. Thomas, L. P. Balogh, M. K. Khan and J. R. Baker, "Nanoparticle targeting of anticancer drug improves therapeutic response in animal model of human epithelial cancer," *Cancer Research*, **65**, 5317–5324 (2005).
144. A. L. Crumbliss, S. C. Perine, J. Stonehuerner, K. R. Tubergen, Junguo Zhao, R. W. Henkens and J. P. O'Daly, "Colloidal gold as a biocompatible immobilization matrix suitable for the fabrication of enzyme electrodes by electrodeposition," *Biotechnology and Bioengineering*, **40**, 483–490 (2004).
145. T. B. Huff, M. N. Hansen, Y. Zhao, J. X. Cheng and A. Wei, "Controlling the cellular uptake of gold nanorods," *Langmuir*, **23**, 1596–1599 (2007).
146. J. Kneipp, H. Kneipp, M. McLaughlin, D. Brown and K. Kneipp, "In vivo molecular probing of cellular compartments with gold nanoparticles and nanoaggregates," *Nano Letters*, **6**, 2225–2231, 2006.
147. R. Shukla, V. Bansal, M. Chaudhary, A. Basu, R. R. Bhonde and M. Sastry, "Biocompatibility of gold nanoparticles and their endocytotic fate inside the cellular compartment: A microscopic overview," *Langmuir*, **21**, 10644–10654 (2005).
148. Paciotti GF, Meyer L, Weinreich D, Goia D, Pavel N, M. RE and a. T. L, "Colloidal gold: a novel nanoparticle vector for tumor directed drug delivery.," *Drug Delivery*, **11**, 169–183 (2004).
149. N. L. Rosi, D. A. Giljohann, C. S. Thaxton, A. K. R. Lytton-Jean, M. S. Han and C. A. Mirkin, "Oligonucleotide-modified gold nanoparticles for intracellular gene regulation," *Science*, **312**, 1027–1030 (2006).
150. T. Neuberger, B. Schopf, H. Hofmann, M. Hofmann and B. von Rechenberg, "Superparamagnetic nanoparticles for biomedical applications: Possibilities and limitations of a new drug delivery system," *Journal of Magnetism and Magnetic Materials*, **293**, 483–496 (2005).
151. D. E. Sosnovik and R. Weissleder, "Emerging concepts in molecular MRI," *Current Opinion in Biotechnology*, **18**, 4–10, 2007.
152. I. Garcia, N. E. Zafeiropoulos, A. Janke, A. Tercjak, A. Eceiza, M. Stamm and I. Mondragon, "Functional ization of iron oxide magnetic nanoparticles with poly(methyl methacrylate) brushes via grafting-from atom transfer radical polymerization," *Journal of Polymer Science Part a-Polymer Chemistry*, **45**, 925–932, 2007.
153. M. Lattuada and T. A. Hatton, "Functionalization of monodisperse magnetic nanoparticles," *Langmuir*, **23**, 2158–2168, 2007.

154. I. H. El-Sayed, X. H. Huang and M. A. El-Sayed, "Selective laser photo-thermal therapy of epithelial carcinoma using anti-EGFR antibody conjugated gold nanoparticles," *Cancer Letters*, **239**, 129–135 (2006).
155. M. Everts, V. Saini, J. L. Leddon, R. J. Kok, M. Stoff-Khalili, M. A. Preuss, C. L. Millican, G. Perkins, J. M. Brown, H. Bagaria, D. E. Nikles, D. T. Johnson, V. P. Zharov and D. T. Curiel, "Covalently linked au nanoparticles to a viral vector: Potential for combined photothermal and gene cancer therapy," *Nano Letters*, **6**, 587–591 (2006).
156. X. H. Huang, I. H. El-Sayed, W. Qian and M. A. El-Sayed, "Cancer cell imaging and photothermal therapy in the near-infrared region by using gold nanorods," *Journal of the American Chemical Society*, **128**, 2115–2120 (2006).
157. W. S. Seo, J. H. Lee, X. M. Sun, Y. Suzuki, D. Mann, Z. Liu, M. Terashima, P. C. Yang, M. V. McConnell, D. G. Nishimura and H. J. Dai, "FeCo/graphitic-shell nanocrystals as advanced magnetic-resonance-imaging and near-infrared agents," *Nature Materials*, **5**, 971–976 (2006).
158. V. P. Zharov, K. E. Mercer, E. N. Galitovskaya and M. S. Smeltzer, "Photothermal nanotherapeutics and nanodiagnostics for selective killing of bacteria targeted with gold nanoparticles," *Biophysical Journal*, **90**, 619–627 (2006).
159. C. P. Barham, R. L. Jones, L. R. Biddlestone, R. H. Hardwick, N. A. Shepherd and H. Barr, "Photothermal laser ablation of Barrett's oesophagus: endoscopic and histological evidence of squamous re-epithelialisation," *Gut*, **41**, 281–284 (1997).
160. J. Y. Chen, D. L. Wang, J. F. Xi, L. Au, A. Siekkinen, A. Warsen, Z. Y. Li, H. Zhang, Y. N. Xia and X. D. Li, "Immuno gold nanocages with tailored optical properties for targeted photothermal destruction of cancer cells," *Nano Letters*, **7**, 1318–1322 (2007).
161. C. Loo, A. Lowery, N. Halas, J. West and R. Drezek, "Immunotargeted nanoshells for integrated cancer imaging and therapy," *Nano Letters*, **5**, 709–711 (2005).
162. R. F. Ismagilov and M. M. Maharbiz, "Can we bufld synthetic, mullficeflullar systems by controfflng developmental signaflng in space and time?," *Current Opinion in Chemical Biology*, **11**, 604–611 (2007).
163. P. S. Dittrich and A. Manz, "Lab-on-a-chip: microfluidics in drug discovery," *Nature Reviews Drug Discovery*, **5**, 210–218 (2006).

164. P. S. Dittrich, K. Tachikawa and A. Manz, "Micro total analysis systems. Latest advancements and trends," *Analytical Chemistry*, **78**, 3887–3907 (2006).
165. J. El-Ali, P. K. Sorger and K. F. Jensen, "Cells on chips," *Nature*, **442**, 403–411 (2006).
166. D. Falconnet, G. Csucs, H. M. Grandin and M. Textor, "Surface engineering approaches to micropattern surfaces for cell-based assays," *Biomaterials*, **27**, 3044–3063 (2006).
167. A. Khademhosseini, R. Langer, J. Borenstein and J. P. Vacanti, "Microscale technologies for tissue engineering and biology," *Proceedings of the National Academy of Sciences of the United States of America*, **103**, 2480–2487 (2006).
168. S. Nagrath, L. V. Sequist, S. Maheswaran, D. W. Bell, D. Irimia, L. Ulkus, M. R. Smith, E. L. Kwak, S. Digumarthy, A. Muzikansky, P. Ryan, U. J. Balis, R. G. Tompkins, D. A. Haber and M. Toner, "Isolation of rare circulating tumour cells in cancer patients by microchip technology," *Nature*, **450**, 1235-U10 (2007).
169. P. Moore, "Cell biology: Ion channels and stem cells," *Nature*, **438**, 699–704 (2005).
170. S. K. W. Dertinger, X. Y. Jiang, Z. Y. Li, V. N. Murthy and G. M. Whitesides, "Gradients of substrate-bound laminin orient axonal specification of neurons," *Proceedings of the National Academy of Sciences of the United States of America*, **99**, 12542–12547 (2002).
171. F. K. Balagadde, L. C. You, C. L. Hansen, F. H. Arnold and S. R. Quake, "Long-term monitoring of bacteria undergoing programmed population control in a microchemostat," *Science*, **309**, 137–140 (2005).
172. G. Martino and S. Pluchino, "The therapeutic potential of neural stem cells," *Nature Reviews Neuroscience*, **7**, 395–406 (2006).
173. S. Temple, "The development of neural stem cells," *Nature*, **414**, 112–117 (2001).
174. E. Fuchs, T. Tumbar and G. Guasch, "Socializing with the neighbors: Stem cells and their niche," *Cell*, **116**, 769–778 (2004).
175. K. A. Moore and I. R. Lemischka, "Stem cells and their niches," *Science*, **311**, 1880–1885 (2006).
176. I. Kratchmarova, B. Blagoev, M. Haack-Sorensen, M. Kassem and M. Mann, "Mechanism of divergent growth factor effects in mesenchymal stem cell differentiation," *Science*, **308**, 1472–1477 (2005).
177. M. P. Lutolf and J. A. Hubbell, "Synthetic biomaterials as instructive extracellular microenvironments for morphogenesis in tissue engineering," *Nature Biotechnology*, **23**, 47–55 (2005).

178. F. M. Watt and B. L. M. Hogan, "Out of Eden: Stem cells and their niches," *Science*, **287**, 1427–1430 (2000).
179. T. P. Fleming, B. Sheth and I. Fesenko, "Cell adhesion in the preimplantation mammalian embryo and its role in trophectoderm differentiation and blastocyst morphogenesis," *Frontiers in Bioscience*, **6**, D1000–D1007 (2001).
180. M. C. Yoder and D. A. Williams, "Matrix molecule interactions with hematopoietic stem-cells," *Experimental Hematology*, **23**, 961–967 (1995).
181. S. Ding and P. G. Schultz, "A role for chemistry in stem cell biology," *Nat Biotech*, **22**, 833–840 (2004).
182. C. J. Flaim, S. Chien and S. N. Bhatia, "An extracellular matrix microarray for probing cellular differentiation," *Nature Methods*, **2**, 119–125 (2005).
183. Y. Soen, A. Mori, T. D. Palmer and P. O. Brown, "Exploring the regulation of human neural precursor cell differentiation using arrays of signaling microenvironments," *Molecular Systems Biology*, (2006).

Chapter Two

Fundamental Mathematical Modeling Techniques for Nano Bio-Systems

Sharon Bewick,
Mingjun Zhang and
William R. Hamel

In this chapter we introduce some of the basic mathematical approaches that can be used to understand and describe biological systems where nanoscale interactions and nanoscale technology are involved. We begin by outlining some of the challenges involved in the modeling of nanoscale systems. We then provide a general framework for choosing a model that is appropriate to the question under consideration. Finally, we survey some common mathematical techniques that are particularly effective when applied to model problems related to nanomedicine and nanomedical phenomena.

2.1 INTRODUCTION

In recent years, there has been considerable excitement over the potential benefits associated with the use of nanotechnology for medical applications. This has led to the growing field of nanomedicine — a subset of nano bio-systems and an active area of research encompassing topics from nanomaterials to nanoelectronic biosensors and molecular nanotechnology. The typical length scale for cells in the human body is two to three orders of magnitude larger than the length scale for nanoparticles and nanodevices. It is because of this difference that nanotechnology offers promising advantages over more traditional medical approaches.

Nanomedicine: *A Systems Engineering Approach* edited by Mingjun Zhang & Ning Xi

www.panstanford.com
978-981-4241-36-6

First, because of their small size, cells will often ingest nanoparticles. As a result, drugs delivered through nanoparticles have improved bioavailability when compared to conventional drugs which are actively cleared from the human body. Second, nanoscale devices are small enough to interact directly with intracellular organelles and proteins. For this reason, it is hoped that they will be capable of specifically targeting and treating damaged cells *in vivo*.[1] Finally, because nanoparticles and nanodevices perform on such a small scale, their effects have an extremely limited range. Therefore, it is widely believed that by using nanotechnology for disease detection, pathogen clearance and genetic damage control, we will be able to selectively direct effector mechanisms, thereby minimizing damage to healthy tissues.

Unfortunately, along with the many advantages offered by nanotechnology, there are also associated disadvantages and complications. Largely, these disadvantages and complications result from our limited understanding of nanoscale interactions, particularly as they relate to biological organisms. While the medical industry hopes to capitalize on the fact that nanoparticles have novel effects on cells and tissue, some of these effects are undoubtedly associated with negative consequences as well. Currently, for example, there is very little understanding of the relationship between nanoscale drug interactions and cell toxicity. Similarly, the long-term effects of nanoparticle accumulation in the body are largely unknown.

Clearly, developing an improved understanding of nanoscale interactions, particularly as they apply to biological systems, is crucial if we are to make further progress in the field of nanomedicine. This, of course, is where mathematical models become essential. Mathematical models can be used to gain insight into the underlying mechanisms governing nanoscale properties and behavior. Furthermore, models can be used to predict consequences associated with nanomedicine, optimize the benefits of a particular nanoscale technique, and guide specific nanoscale experiments to save time and resources.

Since many of the important nanoscale interactions governing nanomedicine involve fundamental physical processes, they can be described, at least partially, through the application of fundamental physical and chemical laws. Therefore, most models that aim to predict the behavior of nanoparticles and nanodevices rely on some combination of conservation laws (mass, energy, momentum), thermodynamic principles, kinetic mechanisms, quantum mechanics, statistical mechanics, classical mechanics, fluid dynamics and electromagnetism. Naturally, applying these physical principles in the context of a biological system can be challenging, and the strength of any mathematical model lies in its ability to focus expressly on the phenomenon being modeled, while still treating the surrounding interactions sufficiently to yield accurate predictions.

Unfortunately, when it comes to modeling in the field of nanomedicine, there is a particularly harsh trade-off between accuracy and insight. This is because nanomedicine, to be effective, requires a high degree of coupling between the physical and chemical processes that occur on the nanometer scale and the biological processes that occur on the micron scale or higher. Obviously, mathematical approaches to understanding nanobiosystems will profit immensely from the development of multi-scale mathematical techniques. Multi-scale modeling, however, is in its infancy, and this is one of the primary challenges faced by mathematicians wishing to explore the effects of nanoparticles or the actions of nanodevices in the human body.

2.2 THE NANOMETER SCALE: CHOOSING LAWS BASED ON THE SYSTEM

The National Physical Laboratory defines nanotechnology as the technology where dimensions and tolerances in the range 0.1–100 nm[2] play a critical role. This definition, however, can be deceptive. While nanoparticles and nanodevices effect processes on the 0.1–100 nm scale, their actions can, and should, have consequences on scales relevant to cells (1–1000 μm), tissues and organs as well. The scale of the phenomenon being modeled is frequently more important to the mathematician than the scale of the effect itself, because the scale of the phenomenon typically determines the physical and chemical laws from which to build the model. If, for instance, one is interested in understanding why a nanoparticle is capable of targeting one cell-type over another, the crucial phenomenon is the electrostatic interactions between the nanoparticle and cell membrane proteins. Since electrostatic interactions depend on charge distributions across contacting surfaces, a quantum mechanical model may be required. In contrast, if one wants to understand the tissue-level effect of a nanoparticle that eliminates one cell-type faster than another, quantum mechanics would add an unnecessary complication, and a simple kinetics model is probably sufficient.

2.2.1 Physical Laws for Phenomena on Scales Significantly Below 1 nm

At scales below $\sim 10^{-10}$ m, the laws of classical physics break down, and phenomena cannot be properly understood without assuming a quantum mechanical description. Technically speaking, quantum mechanical systems can be completely described by the Schrodinger equation:

$$\hat{H}\Psi(\vec{r},t) = i\hbar\frac{\partial\Psi(\vec{r},t)}{\partial t} \tag{2.1}$$

where $\hat{H}$ is the Hamiltonian operator for the N-body system

$$\hat{H} = -\sum_{i}^{N}\frac{\hbar}{2m_i}\nabla_i^2 + V(\vec{r}) \tag{2.2}$$

$\hbar$ is Planck's constant, and Ψ is the N-body wavefunction.

In practice, solutions to the Schrodinger equation are only known exactly for very simple atoms, thus models that depend on the Schrodinger equation almost always require approximations. Frequently, these approximations amount to reducing the N-body problem in the Schrodinger equation to a series of problems that can be solved in terms of single-electron wave functions, ψ_k. This is usually done in two ways. First, according to the Born-Oppenheimer approximation, the atomic nuclei are assumed fixed on time-scales relevant to electron motion, making it possible to solve for the electron wavefunction around static nuclear configurations. This reduces the original N-body problem to an n-electron problem. The resulting n-electron problem is then further simplified by assuming that the entire n-electron wave function can be reasonably approximated as n independent electron wave functions, with each electron responding to the self consistent field of the other $n-1$ electrons.

While many effects important to nanotechnology and nanomedicine can be adequately described through classical mechanics, certain nanoparticles are known to exhibit distinctly quantum behavior. Below 20 nm, for example, gold nanoparticles are effectively quantum dots. As such, they have discrete electron energy levels that can only be understood through quantum mechanics. In terms of nanomedicine, these electron energy levels are important, since they allow for large, size dependent variation in electrical and optical properties.[3] The ability to control and predict these electrical and optical properties has direct applications, particularly with respect to biosensors and biological labeling. Obviously, any mathematical model which attempts to interpret quantum dot behavior in a biological environment, or when bound to a biological molecule, must expressly consider the quantum nature of the phenomenon involved.

2.2.2 *Physical Laws for Phenomena on Scales Near 1 nm*

While classical physics is a valid approximation in the nanometer range, other physical laws, such as those associated with fluid dynamics, are still problematic. For liquid flow through channels with diameters below $\sim$1 nm the assumption that a liquid can be treated as a continuum fails, and the inhomogeneous nature of the liquid's molecular composition becomes an important determinant of liquid behavior. As a result, any attempt to model liquid flow through small channels based on Navier–Stokes equations, or other related physical laws would be doomed to failure.

Along the same lines, diffusion through channels with diameters on the nanometer scale cannot be accurately accounted for using Fick's law. The obvious reason is that diffusion perpendicular to the channel is restricted to scales comparable to the equivalent molecular radius and mean free path of the diffusing molecules. Therefore, the assumption of free Brownian motion in all three directions no longer holds.

In terms of nano bio-systems, examples of highly restricted flow and diffusion include the passage of material through nanoscale channels in silicon chips and the transport of small molecules across pores in the cell membrane. It would be unwise to develop a mathematical description for either of these processes using Navier-Stokes equations or Fick's law. Both phenomena can, however, be modeled with a technique known as molecular dynamics (MD).[4,5] In an MD approach, the macroscopic behavior of a fluid is predicted by simulating a large number of molecules as they move and collide according to fundamental physical laws. Specifically, molecular movement is treated through classic equations of motion while molecular collisions are taken as dependent on electrostatic interactions between the molecules (often approximated as hard-sphere or Lennard-Jones potentials).

Unlike Navier–Stokes equations and Fick's law, the MD approach does not assume effective averaging of fluid properties over space, thus it is a more accurate method for modeling liquids which are significantly non-homogeneous on the length scale of the problem being studied. Unfortunately, while MD methods are accurate, they are also simulation based and, as a consequence, incur high computational costs. Furthermore, the insight gained from an MD simulation is typically situation specific. Therefore, MD models rarely offer a general interpretation, and scenarios which differ slightly from one another must be modeled independently.

2.2.3 Physical Laws for Phenomena on Scales Greater than 1 nm

For length scales beyond the nanometer range, most physical and chemical laws are reasonably valid, at least when applied to physical and chemical systems. Frequently, though, it is tempting to apply physical and chemical laws to explicitly biological systems. Blood, for instance, is often treated as a fluid and modeled according the methods of fluid dynamics. While this is an effective way of building a mathematical description, care must be taken to ensure that all of the assumptions made in deriving the chemical or physical model remain valid in the biological setting. Frequently, the application of physical laws to biological systems is complicated by the fact that the majority of biological components, like macromolecules, organelles and cells, are significantly larger than typical chemical molecules. As a result, length scales that are appropriate for certain physical models are not necessarily appropriate for the corresponding biological models.

As an example, in the last section it was argued that the Navier-Stokes equations were not a suitable description of fluid passing through nanometer channels. This same argument can be applied to blood passing through veins, arteries and capillaries. However, the characteristic length scale changes. Even large solvent molecules are usually less than 1 nm in diameter while the average red blood cell (RBC) is about 7 microns across. Therefore while inhomogeneities become important in a chemical solvent at $\sim 10^{-9}$ m, they become important in blood at $\sim 10^{-5}$ m.

2.2.4 Physical Laws versus Biological Systems

Another challenge presented by biological systems is the fact that, in addition to physical and chemical laws, the system follows certain biological laws associated with processes like cell replication, cell death, cell growth, cell-cell competition, cell-cell cooperation, and cell-line selection. While some of these events are undoubtedly driven by underlying physical and chemical laws, the connections between the physical nature of a system and the biological nature of a system are often obscure. As a result, biological laws, or "rules", are typically incorporated in an *ad hoc* manner according to experimental observation, very basic mechanistic assumptions, or general functional form. For example, populations that exhibit a slow initial growth, followed by a phase of rapid expansion, and then a period of saturation are typically modeled using a Gompertz curve

$$\frac{dN(t)}{dt} = rN(t)\log\left(\frac{K}{N(t)}\right) \tag{2.3}$$

where $N(t)$ is the size of the population at time t, r is the intrinsic growth rate, and K is the number of individuals at equilibrium. Use of the Gompertz curve is largely based on the fact that its general functional form matches growth behavior observed in a wide variety of different populations, both on the cellular and organism level.

An alternate model for population growth, known as the Verhulst equation, has the functional form

$$\frac{dN(t)}{dt} = rN(t)\left(1 - \frac{N(t)}{K}\right) \tag{2.4}$$

where $N(t)$, r and K are defined as before. The Verhulst equation is based on the very general assumptions that

- the rate of reproduction is proportional to the existing population, all else being equal
- the rate of reproduction is proportional to the amount of available resources (represented by the term in brackets on the right-hand side of the equation), all else being equal

A similarly general mathematical description of predator-prey interactions is the Lotka–Volterra model, which relies on a pair of coupled differential equation

$$\frac{dx(t)}{dt} = \alpha x(t) - \beta x(t)y(t) \tag{2.5a}$$

$$\frac{dy(t)}{dt} = \delta x(t)y(t) - \gamma y(t). \tag{2.5b}$$

In the above equations, $x(t)$ and $y(t)$ are the prey and predator populations respectively at time t. The basic assumptions of the model are that

- in the absence of predation, the prey population grows exponentially with a rate constant α
- in the absence of prey, the predator population decays exponentially with a rate constant γ
- the rate of prey consumption is proportional to the number of predator-prey encounters, represented by the term $\beta x(t)y(t)$
- the growth of the predator population is proportional to the rate of prey consumption, represented by the term $\delta x(t)y(t)$

While the Lotka–Volterra model was most famously used to describe the complex interactions between populations of lynx and hare in northern Canada, variants of the Lotka–Volterra model have been applied to cellular biology as well. Encounters between a pathogen and certain types of immune system cells are known to stimulate replication of the immune cell population while at the same time rapidly depleting the pathogen population. This is, effectively, a predator-prey system in which the immune cell functions as the predator, and the pathogen functions as the prey.

2.2.5 Coupling Between Different Scales and Levels of Approximation

Based on the previous discussion, it should be clear that modeling of nanoscale interactions is complicated by the fact that the actions and effects of nanoscale objects bridge the gap between extremely small systems, which are adequately described by quantum mechanics, and very large systems where vast approximations and continuum models are best. Furthermore, while a nanoscale object may *cause* an effect at one level, the effect may actually manifest itself at another, making it practically impossible to strike a balance between including enough information to describe the effect, while at the same time limiting the model to a manageable size.

In certain physical systems, the connections between models at one scale and models at another are very well defined. Statistical mechanics, for instance, provides a bridge between quantum mechanics and thermodynamics. Connections in biological systems are not as obvious. While stochastic methods, and multi-scale modeling techniques can be used as bridges between the properties of small-scale biological events and their manifestations as large-scale biological effects, the highly nonlinear nature of most biological systems makes these bridges less than intuitive.

Naturally, there is no clear-cut distinction between systems which should be modeled at one level of approximation and systems which should be modeled at another, higher or lower, level of approximation. Therefore, when it comes to modeling a particular nanomedical phenomenon, the physical laws chosen, and the degree of approximation accepted are, to some extent, at the modeler's discretion. While length scale can guide decisions, as can previous modeling efforts of similar systems, the final choice of a particular model is often based largely on intuition. This, unfortunately, cannot be avoided, since there is no *a priori* way of knowing that the properties of a thus far unexplained phenomenon stem from behavior occurring on a particular scale.

2.3 ASSUMPTION VERSUS PREDICTION: BASIC APPROACHES FOR MODEL CREATION

Another important factor in determining the best model for a specific nano bio-system involves the desired trade-off between information in the form of experimental properties or relationships, and information in the form of fundamental laws. This leads to a continuum of different modeling approaches, with purely empirical modeling on one end, and purely analytical modeling on the other.

2.3.1 Analytical Modeling

A true analytical model is based entirely on fundamental physical laws. In other words, an analytical model is a model which derives all relationships, and all parameters from first principle considerations. As suggested in Sec. 2.1, models of nano medical systems typically rely on some combination of conservation laws (mass, energy, momentum), thermodynamic principles, kinetic mechanisms, quantum mechanics, statistical mechanics, classical mechanics, fluid dynamics and electromagnetism. While each entry in this list spans a huge number of possible models, many share the same fundamental laws. Of these laws, some of the most important for modeling nano bio-systems are:

Mass conservation: The matter or mass in a closed system cannot be created or destroyed, only rearranged. Strictly speaking, matter can be converted to energy; however, this is rarely of consequence in biological systems.

Momentum conservation: The total momentum in a closed system remains constant.

Energy conservation or the first law of thermodynamics: In any process, the total energy of the universe remains constant

Second law of thermodynamics: the entropy of the universe tends to a maximum, which predicts the direction of a spontaneous process.

Other fundamental laws exist. For instance, in quantum mechanics, there is the Schrodinger equation, in electromagnetism there are the Maxwell equations, and in statistical mechanics there is the postulate of "equal *a priori* probability" which allows definition of the partition function.

Depending on the particular phenomenon being modeled, any number of different laws or combinations of laws could be used. Regardless of the laws employed, however, all analytical models share one property — they do not require input from experimental systems. In other words, there are no parameters that need to be "estimated" or "fit" based on empirical results, and there are no relationships that need to be approximated based on experimental observation. In fact, any parameters that do appear in a purely analytical model are either universal constants or else values that can be estimated from first principles themselves.

In practice, mathematical descriptions of experimental systems are almost never strictly analytical. In thermodynamic models, for instance, heat capacities, heats of formation, absolute entropies and standard electrode potentials are typically tabulated values that have been determined previously through a variety of empirical means. Even though some of these quantities could be derived, at least approximately, from first principles, there would be no obvious advantage to doing so since experimental values are well characterized, while first principle derivations would require huge levels of approximation.

2.3.2 *Empirical Modeling*

On the opposite end of the spectrum from analytical modeling is empirical modeling. Empirical modeling uses simple mathematical functions to approximate relationships that have been determined experimentally. Some common equations used in empirical modeling are listed below:

- Linear functions: $y = ax+b$, which show a straight line relationship between the state variable, y, and the independent variable, x.
- Exponential functions: $y = e^{ax}$ which show exponential growth (from zero) for $a > 0$, and exponential decay (to zero) for $a < 0$.
- Saturation functions: $y = a(1 - e^{bx})$ which grow from zero to saturation for $bx < 0$ and drop rapidly from zero for $bx > 0$.
- Triangular functions:

$$y = \begin{cases} a + bx & x < c \\ e - fx & x \geq c \end{cases}$$

  which consist of two linear functions, one increasing up to c, and another decreasing thereafter.
- The heaviside step function:

$$H[n] = \begin{cases} 0 & n < 0 \\ 1 & n \geq 0 \end{cases}$$

  which is a special, and particularly useful form of the triangular function above, with $a = 0, b = 0, f = 0, c = 0$ and $e = 1$.
- Hill functions:

$$y = \frac{x^n}{\theta^n + x^n}$$

  which were originally devised to describe oxygen binding to hemoglobin, but are now important in characterizing ligand-enzyme interactions. For $n > 1$ binding is positively cooperative, for $n < 1$ binding is negatively cooperative, and for $n = 1$ binding is non-cooperative.

Empirical modeling is often used when the underlying chemical and physical laws are poorly understood, which is frequently true in biological systems. In purely empirical modeling, both the parameters in the model and the functional form of the model itself are determined experimentally.

Consider, for example, the model for pharmacodynamics used by Barbolosi and Iliadis to describe cancer chemotherapy.[6] To account for tumor growth, Barbolosi and Iliadis propose the following equation:

$$\frac{dn(t)}{dt} = \lambda n(t) \ln[\theta/n(t)] - k[c_2(t) - C_{\mathrm{MIN}}]n(t)H[c_2(t) - C_{\mathrm{MIN}}] \quad (2.6)$$

where $n(t)$ is the number of tumor cells at time t, λ and θ are analogous to r and K in the Gompertz equation (Sec. 2.2.4), $c_2(t)$ is the drug concentration at the site of action, k is a parameter describing the reduction in tumor size per unit increase of the drug and C_{MIN} is the therapeutic threshold of the drug below which no tumor cells are killed. The full model includes additional equations describing drug delivery, drug concentration in the plasma, and drug concentration in the tumor tissue. For the purposes of illustrating an empirical model, however, their first equation suffices.

As noted by the authors in the study, the functional form of their equation for tumor growth was largely determined by empirical observations. First the choice of a Gompertz curve to explain cell growth was not motivated by mechanistic assumptions or physical laws but rather, by the fact that the Gompertz curve produces sigmoidal growth consistent with the available data. Second, the use of a Heaviside step function $H(\cdot)$ in the loss term was entirely driven by several studies which have shown that tumor cells are resistant to killing below a minimum drug threshold. Notice how, in this particular description of tumor growth, the functional forms assumed for the growth and death terms and the parameters describing their relative magnitudes (λ, k, θ and C_{MIN}) are fully empirical.

While empirical models are simple, and easily related to experimental systems, the range of applicability for any empirical description is limited by the range of experimental inputs used to develop it. While a system may exhibit certain parameter values or a linear dependence between two variables for a particular range of experimental conditions, there is no guarantee that these relationships will hold in alternate conditions. As a result, empirical models can only be trusted as far as they have been "calibrated" to the biological environment under consideration.

2.3.3 Hybrid Modeling

The large majority of mathematical descriptions used to account for experimental results are neither strictly analytical nor strictly empirical. Rather, most models falls somewhere in between, with certain parameters, or functional forms, chosen based on fundamental laws and assumptions, and other parameters, or functional forms, chosen based on comparison to experiment.

In the field of quantum mechanics, for instance, semi-empirical models are common. As suggested in Sec. 2.2.1, quantum mechanical models often involve simplification of the n-electron problem to a series of n single-electron problems. Frequently, the corresponding single-electron wavefunctions, ψ_k, are found as a linear combinations of basis functions known as atomic orbitals (AO). The process of setting up and solving for these single-electron wavefunctions involves generating and evaluating integrals of the form

$$H_{ij} = \int \varphi_i \hat{H} \varphi_j dv \tag{2.7a}$$

$$S_{ij} = \int \varphi_i \varphi_j dv \tag{2.7b}$$

where the integrals, in each case, are over all of space, φ_i and φ_j are AO basis functions, and $\hat{H}$ is the Hamiltonian for the system. While integrals of this form could be evaluated from first principles, semi-empirical methods actually treat them as parameters. By "calibrating" the H_{ij} and S_{ij} to a variety of experiments, some of the electron–electron correlation effects destroyed by approximating the n-electron problem as n single-electron problems can be regained. Therefore, while the Hamiltonian itself encodes an analytical model (albeit one requiring a significant degree of approximation), the solution often involves input from experiment, and thus empirical techniques.

On the other end of the scale, consider the Verhulst equation. While the functional form of the Verhulst equation is motivated by the underlying fundamental assumption that organisms both reproduce and deplete resources at a fixed per capita rate, the parameters

r and K are typically "fitted" to the data, making the model neither fully analytical nor fully empirical, but rather, a hybrid of both.

2.4 MATHEMATICAL TECHNIQUES

Regardless of the physical laws used, the level of approximation assumed and the degree of experimental "fitting" required, most models involve similar mathematical techniques. Generally speaking, models relate the state of one or more dependent variables to the state of one or more independent variables. Independent variables are those which are either controlled in the experiment, or else change in a predictable manner not associated with the experiment itself. Common examples of independent variables include time, and space (x, y, z). Dependent variables, on the other hand, are those which are *not* controlled in the experiment, but instead, change in response to changes in the independent variables.

There are a limited number of ways in which dependent variables can be related to independent variables, thus the large majority of mathematical models can be divided into four categories based on whether they are continuous or discrete and, separately, based on whether they are stochastic or deterministic. We consider these classifications below.

2.4.1 Discrete Models

Discrete models describe changes in the states of a system over discrete intervals. Typically, the discrete intervals involve steps in time; however, this is not always the case. If x_n represents the state of the system in the nth interval, then a discrete model is a "difference equation", or "updating function" of the form

$$x_{n+1} = g(x_n, x_{n-1}, x_{n-2}, \ldots, n) \tag{2.8}$$

which describes how the state x_{n+1} depends on some, or all of the previous states (also known as state variables), $x_{n-1}, \ldots, x_0$, and the independent variable. Frequently, discrete models are used as approximations for continuous models. In certain situations, however, discrete models are preferable because there is an inherent interval in the problem itself. Seasons, cell-cycles, life-cycles, circadian-rhythms and even periodic medical intervention, for instance, are all associated with discrete time intervals.

A difference equation is said to be of order k if it has the form

$$f(x_{n+k}, x_{n+k-1}, \ldots, x_{n+1}, x_n, n) = 0 \tag{2.9}$$

In other words, the order of a difference equation is the *number* of previous states, k, required to predict the current state x_{n+k}.

Difference equations can be further classified according to their dependence on the independent variable, and their dependence on the state variables. If f does not depend explicitly on the independent variable, the difference equation is said to be autonomous; otherwise it is said to be nonautonomous. Similarly, if all terms in f appear only to the first power of the state variables — in other words, if there are no products $(x_{n+k-i}x_{n+k-j})$, higher powers (x^2_{n+k-i}) or functions $\sin(x_{n+k-j})$ of state variables, f is said to be *linear*; otherwise, it is said to be *nonlinear*. Finally, if every term in f exhibits a dependence on at least one of the state variables, f is said to be homogeneous; otherwise, it is said to be nonhomogeneous. The updating function

$$x_{n+1} = cx_n \tag{2.10}$$

for instance, is autonomous, linear, first order and homogeneous. The classification of a difference equation is important in determining which method should be used to solve it.

2.4.1.1 *Solving a linear difference equation that exhibits limited time-dependence*

While updating functions can be numerically iterated to determine the behavior of a difference equation from one interval to the next, difference equations are particularly useful if closed form solutions can be found. A solution to a difference equation is an expression which relates the state of the function, x_n, to the current interval, n, and the initial conditions without explicit reference to intermediate states, $x_{n-i}, \ldots, x_1$. While solutions can, on occasion, be found by inspection, it is possible to solve certain forms of linear difference equations systematically, as shown in the example below.

Example 2.1. Suppose that a particular cancer is treated by administering chemotherapy once a month. In order to generate a discrete model for the treatment process, we make the following assumptions

(1) Chemotherapy kills a certain fraction, k, of the existing cancer cells rapidly, but is removed from the body (and thus has no

further effect) on a time scale that is short compared to the period of time between treatments.

(2) The cancer drug is only effective at killing newly differentiated cancer cells (cells which have been generated since the previous treatment) — an assumption which comes from experimental observation on a variety of different chemotherapeutics.

(3) During the period between chemotherapy treatments, cancer cells that have not been destroyed by the cancer drug both replicate and die such that, at the end of each time interval, the cancer cell population has increased by the factor $(1 + r)$.

These assumptions are summarized in the difference equation

$$x_{n+2} = x_{n+1} - k(x_{n+1} - x_n) + r[x_{n+1} - k(x_{n+1} - x_n)] \tag{2.11}$$

which simplifies to

$$x_{n+2} = (1-k)(1+r)x_{n+1} + k(1+r)x_n. \tag{2.12}$$

Solution. Notice that this difference equation is linear. In general, a linear kth order difference equation will have k linearly independent solutions. Therefore, since our drug model is an example of a 2nd order difference equation, we expect to find two linearly independent solutions, which we label x^1 and x^2, respectively. In order to solve for x^1 and x^2, we make use of the fact that the terms linear in the state variables have no explicit time dependence. As a result, we can assume that both linearly independent solutions are of the form

$$x_n = \lambda^n \tag{2.13}$$

where λ is a nonzero constant. Substituting Eq. (2.13) into Eq. (2.12) we get

$$\lambda^{n+2} = (1-k)(1+r)\lambda^{n+1} + k(1+r)\lambda^n \tag{2.14}$$

If we divide by λ^n, and move all terms to the left hand side, this becomes

$$\lambda^2 - (1-k)(1+r)\lambda - k(1+r) = 0 \tag{2.15}$$

which is known as the "characteristic equation". Solving the characteristic equation for its roots, or eigenvalues, gives

$$\lambda_{1,2} = \frac{(1-k)(1+r) \pm \sqrt{(1-k)^2(1+r)^2 + 4k(1+r)}}{2} \tag{2.16}$$

Provided that both of the roots in Eq. (2.16) are real and distinct, the two linearly independent solutions to the difference Eq. (2.12) become $x^1 = \lambda_1^n$ and $x^2 = \lambda_2^n$. While it is possible to treat solutions with repeated, or complex conjugate roots, we leave mention of the appropriate linearly independent solutions for these cases to Sec. 2.4.2.1.1.

Returning to our drug model, notice the restriction that the roots be real and distinct is equivalent to the restriction that the discriminants in the equation above for λ_1 and λ_2 be greater than zero. This means that $r > -(\frac{1+k}{1-k})^2$, which is certainly the case in the region of biological interest where spread of the cancer is uncontrolled ($r > 0$).

According to the "superposition principle", any linear combination of solutions to a linear homogeneous difference equation will be a solution itself. Since our original difference equation was homogeneous, we can write a "general solution" as

$$x_n = c_1 x^1 + c_2 x^2 = c_1 \lambda_1^n + c_2 \lambda_2^n. \tag{2.17}$$

Notice that Eq. (2.39) contains two arbitrary constants, c_1 and c_2. These can be determined uniquely for a particular system, provided that two initial conditions, x_0 and x_1, are known.

Suppose, for instance, that medical scans performed both initially, and after the first month of chemotherapy revealed the following information: $x_0 = \alpha$ and $x_1 = \beta$. Substituting $n = 0$ and $x_0 = \alpha$ into the general solution yields an equation in terms of c_1, c_2, the initial condition and the model parameters. Doing the same for $n = 1$ and $x_1 = \beta$ yields a second, similar equation which, together with the first, can be used to solve for c_1 and c_2. This gives

$$c_1 = \frac{(-\Gamma\alpha + R\alpha + 2\beta)}{2R} \quad \text{and} \quad c_2 = -\frac{(-\Gamma\alpha - R\alpha + 2\beta)}{2R} \tag{2.18}$$

where

$$\Gamma = (1-k)(1+r) \quad \text{and} \quad R = \sqrt{(1-k)^2(1+r)^2 + 4k(1+r)}. \tag{2.19}$$

The full solution to the original difference equation in (2.12) can then be constructed using the Eqs. (2.16)–(2.19) and (2.13). The state of the system x_n, can then be determined solely on the basis of the interval number, n, the initial conditions, α and β, and the model parameters, r and k.

While the example above was for a linear, autonomous, homogeneous second order difference equation, the technique can, with some slight modifications, be extended to higher order difference equations, and difference equations which are nonhomogeneous. It cannot, however, be used to solve a nonlinear difference equation, or a difference equation which have time-dependent coefficients in front of the state variables, $x_{n-k-1}, \ldots, x_0$.

2.4.1.2 The analysis of first-order nonlinear difference equations

In general, the analysis of nonlinear difference equations is significantly more complicated than the analysis of linear difference equations. While a complete introduction to nonlinear methods is beyond the scope of this chapter, we introduce two basic concepts which will give the reader some insight into the behavior of nonlinear difference equations.

Much can be learned about a particular difference equation by examination of its behavior around what are termed "fixed point" or "steady-state" solutions. A fixed point is a state of the system, x^*, which remains at x^* for all subsequent iterations of the difference equation. For the general autonomous, first order difference equation

$$x_{n+1} = f(x_n) \tag{2.20}$$

fixed points can be found by setting $x_{n+1} = x_n = x^*$, and then solving

$$x^* = f(x^*) \tag{2.21}$$

for x^*. The following example illustrates how this is done.

Example 2.2. Suppose a particular pharmaceutical is administered weekly to fight against a persistant viral infection. In order to generate a discrete model, we make the following assumptions

(1) The increase in the viral population over the week is r times the viral population at the beginning of the week.
(2) At the end of the week, a blood sample is taken, and the amount of drug administered is chosen as a saturating function, $a(1-e^{bx})$, of viral load.
(3) The fraction of the existing viral population destroyed by the drug is proportional to the amount of drug administered, and the amount of virus in the system. The proportionality constant is taken as d.

(4) As with chemotherapy, the drug's effect can be approximated as instantaneous, since the drug is cleared rapidly from the body, and thus has little impact on viral dynamics after a brief burst of antiviral activity.

These assumptions are summarized in the difference equation

$$x_{n+1} = x_n + rx_n - d(x_n + rx_n)a\left(1 - e^{b(x_n+rx_n)}\right) \tag{2.22}$$

which simplifies to

$$x_{n+1} = da(1+r)\left(\frac{1}{da} - 1 + e^{b(1+r)x_n}\right)x_n \tag{2.23}$$

Solution. Notice that this difference equation is *not* linear as a result of the exponential function. Therefore, a solution to Eq. (2.23) cannot be found using the method introduced in Sec. 2.4.1.1. Rather than look for a solution *per se*, we begin by looking for fixed points. To do so, we substitute $x_{n+1} = x_n = x^*$ into Eq. (2.23), which gives

$$x^* = da(1+r)\left(\frac{1}{da} - 1 + e^{b(1+r)x^*}\right)x^* \tag{2.24}$$

Solving Eq. (2.24) for x^*, we find that there are two fixed points

$$x^* = 0, \quad \frac{\ln\left(\frac{(1+r)(da-1)+1}{(1+r)da}\right)}{b(1+r)}. \tag{2.25}$$

The first fixed point has an obvious interpretation. If the antiviral drug forces the viral population to zero at any point in time, the viral population will never recover, but rather, will remain at zero for all future time points. The second fixed point is less intuitive. It represents an equilibrium reached between viral replication and drug control.

While a system which reaches a fixed point will remain at the fixed point indefinitely, there is no guarantee that a system which is perturbed slightly from a fixed point will return to it. This brings up the issue of stability. Small perturbations away from a *stable* fixed point will decay, allowing the system to regain its original equilibrium state. Small perturbations away from an *unstable* fixed point, however, will increase in magnitude, forcing the system to a new equilibrium position. Many of the techniques used for the stability analysis of difference equations are similar to those used for stability analysis of continuous models. Rather than discuss them here, we focus, instead, on a graphical method known as "cobwebbing", which is unique to discrete models.

Notice that first order difference equations like those in Eq. (2.20) or (2.23) describe the state of the system in the succeeding interval, $n+1$, as a function of the state of the system in the current interval, n. It is therefore possible to plot the state of the system in the succeeding interval versus the state of the system in the current interval. The red line in Fig. 1 is one such plot for Eq. (2.23). The blue line in Fig. 1(a). is a plot of the line $x_{n+1} = x_n$. All intersections between the red line and blue line represent fixed points, since they satisfy both the difference equation and the restriction that x_n does not change from one interval to the next ($x_{n+1} = x_n$). Just as predicted in Eq. (2.25), there is one fixed point at $x^* = 0$, and one fixed point at $x^* \neq 0$. For reference in the discussion that follows, we term these fixed points x^*_{cured} and $x^*_{\text{persistent}}$, respectively.

A plot like the one shown in Fig. 1. offers more than a graphical interpretation of fixed points. It can also be used to visualize the trajectory of the state of the system as it moves through successive intervals. Beginning with the initial state x_0 on the horizontal $\hat{x}_n$-axis, we draw a vertical line from x_0 on the $\hat{x}_n$-axis to the point (x_0, x_1) on the red curve defined by the difference equation. This is the point labeled A in Fig. 1(b). The next iterate will have $x_n = x_1$. Therefore, we locate x_1 along the $\hat{x}_n$-axis by drawing a horizontal line from x_1 on the $\hat{x}_{n+1}$-axis to the point (x_1, x_1) on the blue curve $x_{n+1} = x_n$. This is the point labeled B in Fig. 1(b). We now have x_1 on the $\hat{x}_n$-axis, from which we repeat the process just described, drawing a vertical line from x_1 on the $\hat{x}_n$-axis to the point (x_1, x_2), and then a horizontal line from x_2 on the $\hat{x}_{n+1}$-axis to the point (x_2, x_2). These two steps give C and D in Fig. 1(b). Further iterations complete the sequence of horizontal and vertical lines shown.

The technique described above is known as "cobwebbing". In Fig. 1(b), it is easy to see that when the initial state, x_0, is greater than $x^*_{\text{persistent}}$, successive iterations bring the system back to $x^*_{\text{persistent}}$. Figure 1(c) shows a similar cobwebbing diagram for an initial state, x_0, less than $x^*_{\text{persistent}}$. Again, after repeated iterations, the system similarly arrives at $x^*_{\text{persistent}}$. Since small perturbations away from

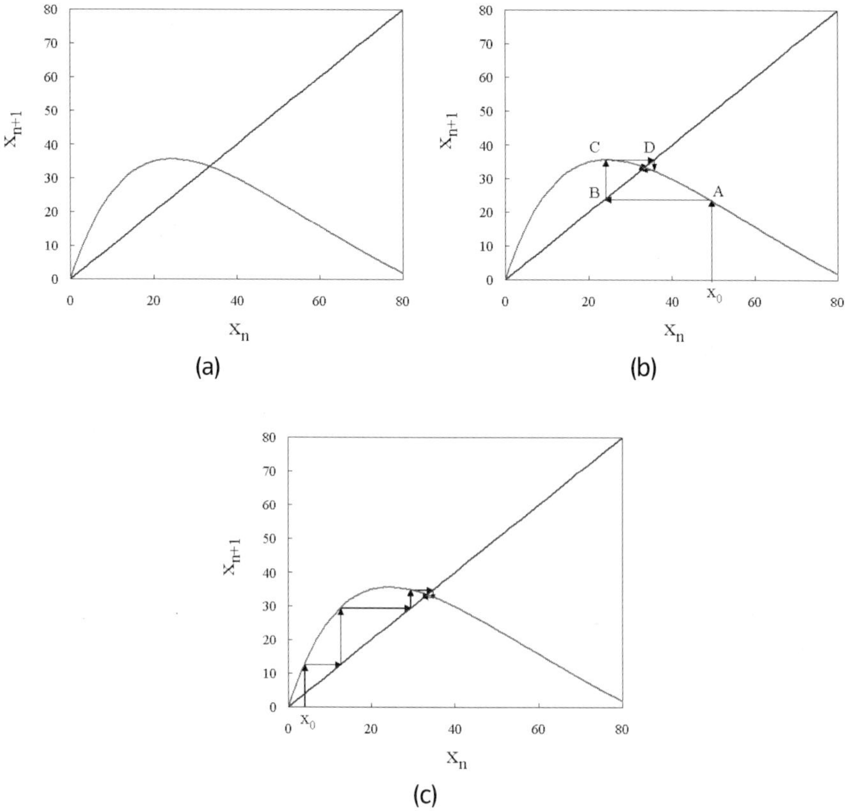

Figure 1. The cobwebbing technique applied to Eq. (2.23) for the parameters $r = 2.9, d = 1.045, a = 1, b = -0.0096$.

$x^*_{\text{persistent}}$ return to $x^*_{\text{persistent}}$, while small perturbations away from x^*_{cured} progress to $x^*_{\text{persistent}}$. $x^*_{\text{persistent}}$ is a stable fixed point, while x^*_{cured} is an unstable fixed point. This, of course, is not good news for the doctor treating the patient, because it indicates that the virus population will eventually settle at a constant nonzero level, at least for the set of parameters chosen in Fig. 1.

2.4.2 Continuous Models

In contrast to discrete models, which view change in the state of the system as the result of a series of *discrete* steps in the independent variable(s), continuous models view change in the state of the system as the result of the *continuous* evolution in the independent variable(s). Continuous models represent both the change in the state of a system and the change in the independent variable(s) using differential equations. When a differential equation has only one independent variable, it is termed an ordinary differential equation (ODE), otherwise, it is termed a partial differential equation (PDE).

2.4.2.1 *The ordinary differential equation*

Ordinary differential equations have the form

$$\sum_{i=0}^{N} a_i(x)\frac{d^i f(x)}{dx^i} = g(f(x)) \tag{2.26}$$

where x is the independent variable, $f(x)$ is the dependent, or state variable, and the differential equation is expressed as a linear combination of derivates of the state variable with coefficients $a_i(x)$. When ODE models are used in biology, time is frequently the independent variable, although systems which can be approximated as one-dimensional may also be modeled using ODEs with a single spatial independent variable. We have already seen examples of ODEs in Sec. 2.2.4. Equations (2.3), (2.4), (2.5a) and (2.5b) are all examples of ODEs.

Biological models may require coupled differential equations when a description of more than one state variable is necessary. This is the case, for instance, in the Lotka-Volterra model from Sec. 2.2.4, where separate ODEs are required for the predator and for the prey. In addition, it is sometimes convenient to rewrite ODEs with nth derivatives as a system of ODEs involving n first derivatives. The general form for a system of coupled differential equations is

$$\begin{aligned} \frac{df_1(x)}{dx} &= g_1(f_1(x), \ldots, f_n(x)) \\ &\vdots \\ \frac{df_n(x)}{dx} &= g_n(f_1(x), \ldots, f_n(x)) \end{aligned} \tag{2.27}$$

The total number of state variables, and thus the total number of individual ODEs in a system of coupled differential equations is referred to as the dimension, or order of the system. Note that since an ODE with an nth derivative can be rewritten as a system of n ODEs with first derivatives, the order of a single ODE corresponds to its highest derivative. For example, the Gompertz curve (2.3) and the Verhulst Eq. (2.4) are one-dimensional, or first order systems, while the Lotka-Volterra model is a two-dimensional, or second order system.

As with difference equations, ODEs can be classified as autonomous or nonautomonous, based on whether or not they exhibit an explicit dependence on the independent variable, x. Dependence on x has not been included in Eq. (2.25), however, because an n-dimensional system of nonautonomous ODEs can be transformed into an $(n + 1)$-dimensional system of autonomous ODEs.

A system of ODEs can also be classified as homogeneous or nonhomogeneous. If the system contains no ODEs with constant terms it is homogeneous, otherwise it is nonhomogeneous. Additionally, ODEs can be linear or nonlinear. If all state variables, $f_i(x)$, appear to the first power only, the ODE is linear, otherwise it is nonlinear. Notice that these definitions are identical to their counterparts for difference equations.

Just as the classification of ODEs is similar to the classification of difference equations, methods for analyzing and solving ODEs are similar to methods for analyzing and solving difference equations. We consider a few possibilities below.

2.4.2.2 Exact solutions for ODEs

Certain classes of ODEs can be solved exactly. As with difference equations, a solution to an ODE provides a direct relationship between the state of the system and the value of the independent variable. Some of the simpler methods for solving ODEs are separation of variables, integration by parts, partial fraction expansions, and series solutions. Entire textbooks are devoted to solving ordinary differential equations, and we cannot possibly hope to cover all of the different techniques here. Instead, we will consider a method that is directly analogous to the one used in Example 2.1, however this time we will apply it to a continuous model, and extend it to a nonhomogeneous ODE.

Example 2.3. Consider a drug which is delivered at a continuous rate γ to the blood stream via IV. In order to be effective, this drug must reach a particular target tissue. If we denote the concentration of the drug in the blood and the concentration of the drug in the tissue b and h, respectively, then we can model the flow of the drug from the IV, through the blood to the tissue according to the basic pharmacokinetic model[7]

$$\dot{b}(t) = \gamma - c_{bt}b(t) \tag{2.28a}$$
$$\dot{h}(t) = c_{bt}b(t) - c_d h(t) \tag{2.28b}$$

where the overdots denote differentiation with respect to time, c_{bt} is the rate constant for passage of the drug from the blood to the target tissue, and c_d is the rate constant for drug decay in the target tissue itself. (We assume negligible drug decay in the blood). This is a second order system of linear ODEs. Notice that, as a result of γ in (2.28a), this system is also nonhomogeneous.

Solution. In order to solve a nonhomogeneous system of differential equations, we begin by finding a "general solution" for the corresponding homogeneous system of equations

$$\dot{b}(t) = -c_{bt}b(t) \tag{2.29a}$$
$$\dot{h}(t) = c_{bt}b(t) - c_d h(t). \tag{2.29b}$$

We do this, as in Example 2.1, by searching for linearly independent solutions with the functional form

$$b(t) = \beta e^{\lambda t} \tag{2.30a}$$
$$h(t) = \Gamma e^{\lambda t}. \tag{2.30b}$$

Substituting Eqs. (2.30a) and (2.30b) into Eqs. (2.29a) and (2.29b) gives

$$\lambda \beta e^{\lambda t} = -c_{bt}\beta e^{\lambda t} \tag{2.31a}$$
$$\lambda \Gamma e^{\lambda t} = c_{bt}\beta e^{\lambda t} - c_d \Gamma e^{\lambda t}. \tag{2.31b}$$

Notice that we can rewrite Eqs. (2.31a) and (2.31b) in matrix form

$$\begin{bmatrix} -c_{bt} - \lambda & 0 \\ c_{bt} & -c_d - \lambda \end{bmatrix} \begin{bmatrix} \beta \\ \Gamma \end{bmatrix} e^{\lambda t} = 0. \tag{2.32}$$

In order to satisfy Eq. (2.32)

$$\det \begin{bmatrix} -c_{bt} - \lambda & 0 \\ c_{bt} & -c_d - \lambda \end{bmatrix} = c_{bt}c_d + (c_{bt} + c_d)\lambda + \lambda^2 = 0 \tag{2.33}$$

which is the characteristic equation for the system of ODEs in Eq. (2.28). As in Example 2.1, the characteristic equation can be solved for λ, giving

$$\lambda_1 = -c_d, \quad \lambda_2 = -c_{bt} \tag{2.34}$$

which are the eigenvalues of the matrix in (2.32). The corresponding eigenvectors are

$$\vec{v}_1 = [0, 1] \quad \text{and} \quad \vec{v}_2 = [\, c_d - c_{bt}, c_{bt}]. \tag{2.35}$$

Since we expect the eigenvalues in Eq. (2.34) to be both real and distinct, we write the two linearly independent solutions in Eq. (2.30) as

$$b(t) = 0, \quad h(t) = e^{-c_d t} \tag{2.36a}$$

and

$$b(t) = (c_d - c_{bt})e^{-c_{bt}t}, \quad h(t) = c_{bt}e^{-c_{bt}t}. \tag{2.36b}$$

Once again, the "general solution" for the homogeneous differential equations can be written as a linear combination of the two linearly independent solutions.

$$\begin{bmatrix} b(t) \\ h(t) \end{bmatrix} = c_1 \begin{bmatrix} 0 \\ 1 \end{bmatrix} e^{\lambda_1 t} + c_2 \begin{bmatrix} c_d - c_{bt} \\ c_{bt} \end{bmatrix} e^{\lambda_2 t}. \tag{2.37a}$$

Had the eigenvalues been identical, though, the correct linearly independent solutions would have given the linear combination

$$\begin{bmatrix} b(t) \\ h(t) \end{bmatrix} = c_1 \begin{bmatrix} 0 \\ 1 \end{bmatrix} e^{\lambda_1 t} + c_2 t \begin{bmatrix} c_d - c_{bt} \\ c_{bt} \end{bmatrix} e^{\lambda_2 t}. \tag{2.37b}$$

Similarly, for eignenvalues with imaginary components, the correct linear combination would have been

$$\begin{bmatrix} b(t) \\ h(t) \end{bmatrix} = rc_1 \begin{bmatrix} 0 \\ 1 \end{bmatrix} \cos(\phi t) + rc_2 \begin{bmatrix} c_d - c_{bt} \\ c_{bt} \end{bmatrix} \sin(\phi t) \tag{2.37c}$$

where

$$\lambda = A + iB, \quad r = \sqrt{A^2 + B^2} \quad \text{and} \quad \phi = \arctan(\mathrm{B/A}). \tag{2.38}$$

Note that Eq. (2.37) is directly analogous to Eq. (2.17). In Example 2.1, however, the original difference equation was homogeneous, while in this example, the original differential equation is nonhomogeneous. We have not, therefore, found the final "general solution" for the inhomogeneous ODE. To solve inhomogeneous differential equations, we search for an additional 'particular solution". Any function free of arbitrary parameters that satisfies a nonhomogeneous differential equation is said to be a "particular solution" of that ODE. Notice, for example, that $b(t) = \gamma/c_{bt}$, and $h(t) = \gamma/c_d$, is a particular solution for Eq. (2.28). The "general solution" for the nonhomogeneous system of ODEs is then the "general solution" for the homogeneous system, plus the "particular solution" for the nonhomogeneous system.

$$\begin{bmatrix} b(t) \\ h(t) \end{bmatrix} = c_1 \begin{bmatrix} 0 \\ 1 \end{bmatrix} e^{\lambda_1 t} + c_2 \begin{bmatrix} c_d - c_{bt} \\ c_{bt} \end{bmatrix} e^{\lambda_2 t} + \begin{bmatrix} \gamma/c_{bt} \\ \gamma/c_d \end{bmatrix}. \tag{2.39}$$

As might be expected, Eq. (2.39) contains two unknowns, c_1 and c_2. These can be found using the initial values $b(0)$ and $t(0)$. Suppose, for instance, that at the beginning of the monitoring period, $b(0) = D$ and $t(0) = 0$. Substitution of these initial values into Eq. (2.73) shows that

$$c_1 = \frac{c_{bt}\left(c_d D - \gamma\right)}{c_d\left(c_d - c_{bt}\right)}, \quad c_2 = \frac{\left(c_{bt} D - \gamma\right)}{c_{bt}\left(c_d - c_{bt}\right)} \tag{2.40}$$

Further substituting Eq. (2.40) back into Eq. (2.37) we find

$$\begin{aligned} b(t) &= \left(D c_{bt} - \gamma\right) e^{-c_{bt} t} + \frac{\gamma}{c_{bt}}, \\ h(t) &= \frac{c_{bt}(c_d D - \gamma)}{c_d\left(c_d - c_{bt}\right)} e^{-c_d t} + \frac{(c_{bt} D - \gamma)}{\left(c_d - c_{bt}\right)} e^{-c_{bt} t} + \frac{\gamma}{c_d} \end{aligned} \tag{2.41}$$

which is the final solution. Notice that expressions for $b(t)$ and $h(t)$ are now explicitly in terms of time, meaning that the time evolution of the system is fully defined.

The method outlined above can be applied to any system of linear differential equations provided the coefficients, a_i, in Eq. (2.26) have no dependence on the independent variable. An alternate technique for solving such systems is known as the Laplace Transform method. The Laplace transform of a function, $f(x), x \in [0, \infty]$ is defined as

$$L(f(x)) = F(S) = \int_0^{\infty} f(x) e^{-Sx} dx \tag{2.42}$$

where S is the Laplace operator. The Laplace transform, L, is useful in that it can convert the differentiation of $f(x)$ to the multiplication of $f(x)$ by S

$$L\left(\frac{d(f(x))}{dx}\right) = SF(S) - f(0) \tag{2.43}$$

which converts a differentiation expression into an algebraic expression. Applying the Laplace transform to the homogeneous system of ODEs in Eq. (2.29) we have

$$SB(S) = -c_{bt}B(S) - b(0) \tag{2.44a}$$

$$SH(S) = c_{bt}B(S) - c_dH(S) - h(0) \tag{2.44b}$$

Equations (2.44a) and (2.44b) are now algebraic, and it is possible to solve for $B(S)$ and $H(S)$ as

$$B(S) = \frac{b(0)}{S + c_{bt}}, \tag{2.45a}$$

$$H(S) = \frac{h(0)}{S + c_d} + \frac{c_{bt}B(S)}{S + c_d} = \frac{h(0)}{S + c_d} + \frac{c_{bt}b(0)}{(S + c_d)\,(S + c_{bt})} \tag{2.45b}$$

$B(S)$ and $H(S)$ can be converted back to b(t) and h(t) by applying an inverse Laplace transform. This gives expressions similar to those in Eq. (2.36).

2.4.2.2.1 Steady states and qualitative analysis

For some linear ODE systems, and certainly for most nonlinear ODE systems, it is impossible to find analytical solutions. As in difference equations, in the absence of a full solution, the behavior of the system can be partially interpreted by analyzing the steady-states, and the behavior of the system near these steady states. For the set of coupled ODEs, in Eq. (2.27), the steady states can be found as

$$\begin{gathered} g_1(f_1^*(x), \ldots, f_n^*(x)) = 0 \\ \vdots \\ g_n(f_1^*(x), \ldots, f_n^*(x)) = 0 \end{gathered} \tag{2.46}$$

where f_i^* is the value of the ith state variable at the fixed point. In the example that follows, we consider the behavior of the steady states in a one-dimensional system.

Example 2.4. Viral population dynamics within a host system can be approximated using the Verhulst equation (2.4), modified by a term which represents viral death[8]

$$\dot{v} = rv\left(1 - \frac{v}{k}\right) - pv \tag{2.47}$$

where r is the viral replication rate, k the viral carrying capacity, and p the viral death rate. We assume that r, k and p are all greater than 0. While Eq. (2.47) has an analytical solution, for the purpose of illustration, we choose to examine its behavior using a qualitative approach instead.

Solution. First, we find the steady states associated with Eq. (2.47). These are

$$0 = rv^*\left(1 - \frac{v^*}{k}\right) - pv^* \tag{2.48}$$

$$v^* = 0, \frac{k(r-p)}{r}. \tag{2.49}$$

The steady states in Eq. (2.49) can be plotted on a line as shown in Figs. 2(a) and 2(b) for $r > p$ and $r < p$, respectively.

In order to determine the behavior of the ODE system when it is not at one of these steady states, we consider the sign of $\dot{v}$ along the line segments between the fixed points. If $\dot{v} > 0$, v increases, while if $\dot{v} < 0$, v decreases. Consider, for example, the three separate line segments marked (i), (ii) and (iii) in Fig. 2(a). The first line segment,

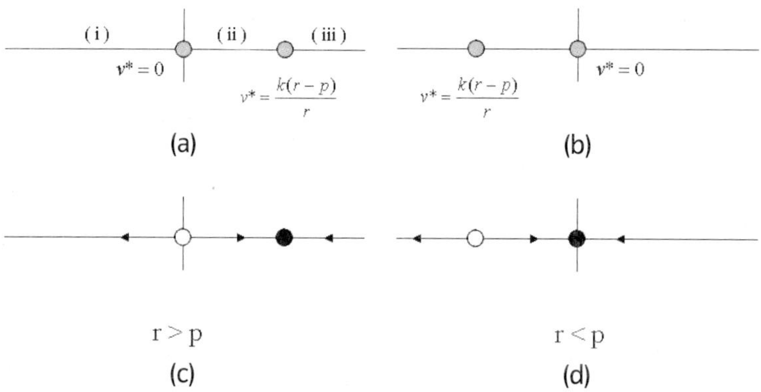

Figure 2. Qualitative analysis of Eq. (2.47).

(i), has $v < 0$. To determine the behavior of v in this region, we substitute $v = 0 - \delta$ into Eq. (2.47)

$$\dot{v} = r(0-\delta)\left(1 - \frac{(0-\delta)}{k}\right) - p(0-\delta) = (r-p)(-\delta) - \frac{r(-\delta)^2}{k}. \quad (2.50)$$

Notice that for $r > p$ we expect $\dot{v} < 0$, meaning that v decreases below $v = 0$.

In line second line segment, marked (ii), we have $0 < v < k(r-p)/r$. This time we substitute $v = 0 + \delta$ into Eq. (2.47)

$$\dot{v} = r(0+\delta)\left(1 - \frac{(0+\delta)}{k}\right) - p(0+\delta) = (r-p)(\delta) - \frac{r(\delta)^2}{k}. \quad (2.51)$$

Equation (2.51) shows that for small $\delta < k(r-p)/r, \dot{v} > 0$, meaning that v increases between $v = 0$ and $v = k(r-p)/r$.

Finally, in the third line segment, marked (iii), we have $v > k(r-p)/r$. We therefore substitute $v = k(r-p)/r + \delta$ into Eq. (2.67)

$$\begin{aligned}\dot{v} &= r\left(\frac{k(r-p)}{r} + \delta\right)\left(1 - \frac{\left(\frac{k(r-p)}{r} + \delta\right)}{k}\right) - p\left(\frac{k(r-p)}{r} + \delta\right) \\ &= -(r-p)(\delta) - \frac{r(\delta)^2}{k} \quad (2.52)\end{aligned}$$

which gives $\dot{v} < 0$, meaning that v decreases above $v = k(r-p)/r$. Putting all of this together we get Fig. 2(c), where the open circle denotes an unstable fixed point, while the closed circle denotes a stable fixed point. A similar analysis for Fig. 2(b). ($r < p$) gives Fig. 2(d).

Careful inspection of Fig. 2 reveals that while we have not obtained an analytical solution to Eq. (2.47), we have, in fact, explained much of the behavior that might be expected. For $r > p$, the system will exhibit two biologically relevant steady states, one at $v = 0$, and another at $v = k(r-p)/r$. The latter of these two steady states will be stable, thus the system will tend to move towards a viral population of $k(r-p)/r$. In contrast, for $r < p$, the system will exhibit only one biologically relevant steady state ($v < 0$ is not biologically relevant). This steady state will be stable, thus for $v > 0$, the virus population will decrease over time until it goes extinct. In terms of medical treatments, then, clearing the virus entirely requires $r < p$, although the steady state population decreases with decreasing r, even for $r > p$.

Interesting behavior can be observed in even fairly simple nonlinear ODEs analyzed using a more descriptive approach like the one illustrated above. In particular, an important aspect that can be elucidated using a qualitative approach is the dependence of the system dynamics on parameter values. In the example above, it was clear that the behavior of the system was strongly influenced by the parameters r, p and k. In general, steady states can appear, disappear or even change stability depending on parameter values (consider, for instance, what happens to the fixed points in Eq. (2.47) for $r = p$). These qualitative changes in the dynamics of the system are known as bifurcations, and can have important consequences as far as the physical systems being modeled are concerned. Figure 3 shows a bifurcation diagram for Eq. (2.47).

In this diagram, we plot the steady state values of the virus population, v^*, as a function of r for fixed p and k. The stability, indicated by the dotted lines (unstable) and the solid lines (stable), follows from our previous arguments. The bifurcation in Fig. 3 is known as a "transcritical bifurcation". Many other types of bifurcations exist,

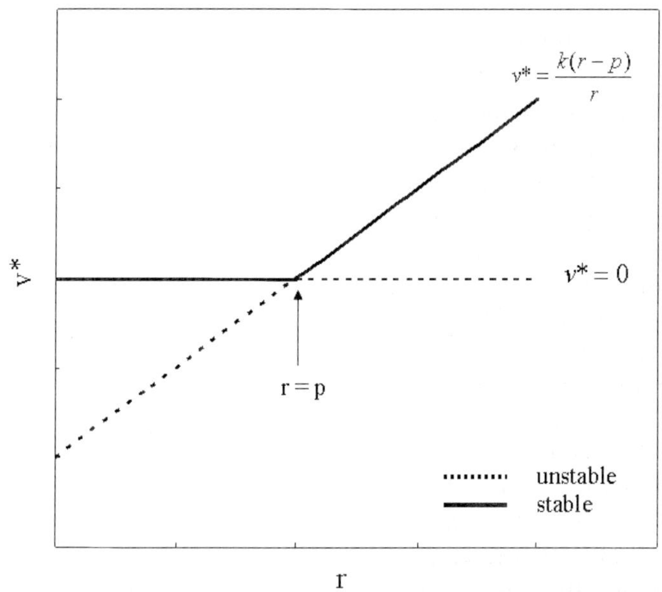

Figure 3. Bifurcation diagram for Eq. (2.47).

and the behavior of higher order ODEs can become extraordinarily complex.

2.4.2.2.2 The partial differential equation

Partial differential equations (PDEs) have the form

$$\sum_{i=0}^{N} a_i(x,y,\ldots)\frac{d^i f(x,y,\ldots)}{dx^i} + \sum_{i=0}^{M} b_i(x,y,\ldots)\frac{d^i f(x,y\ \ldots)}{dy^i} + \cdots$$
$$= g(f(x,y,\ldots)) \qquad (2.53)$$

where x and y are independent variables, $f(x,y,\ldots)$ is the dependent, or state variable, and the differential equation is expressed as a linear combination of derivates of the state variable with coefficients $a_i(x,y,\ldots)$ and $b_i(x,y,\ldots)$. PDE models most often appear in biological systems when spatial dimensions are considered. The spatial movements of a population of cells over a surface, for instance, can be described as a PDE with two independent variables, while diffusion through a chamber can be described by a PDE with three independent variables. Again, we have already seen an example of a PDE in Eq. (2.2). Another common PDE in biological systems comes from the use of Fick's law for passive diffusion

$$J = -D\nabla\varphi(x,y,z) \qquad (2.54)$$

where J is the diffusion flux, D is the coefficient of diffusion, and $\varphi(x,y,x)$ is a function describing the concentration of the diffusing chemical as a function of spatial location.

PDEs are usually quite difficult to solve exactly. Occasionally methods like Backlund transformations, integral transforms, Fourier transforms or separation of variables may yield an analytical solution; however, in many cases, PDEs can only be solved by numerical methods.

2.4.3 *Deterministic Models*

All of the examples that we have thus far encountered have been deterministic models. In deterministic models, the dynamics of the system include no chance events. Said differently, given a deterministic model, and knowledge of the initial state of the system, it is possible to completely determine all future states. In general, there are two reasons for assuming a deterministic model. First, if the model accurately accounts for all of the relevant independent variables, and the independent variables are, themselves, known with certainty, then the evolution of the dependent variables can be known with certainty as well. In biological systems, the large number of independent variables, and the typically poor characterization of their interactions, mean that nanomedical models are not, in general, motivated by determinism in this sense (arguably, neither are models of physical systems, since they suffer from uncertainty at the quantum level).

Purely deterministic models should not, however, be eliminated entirely from biological modeling. Provided that the system under consideration is large enough, small fluctuations caused by either incomplete knowledge of the independent variables themselves, or incomplete characterization of their interactions will, on average, balance out. Therefore, while individual events may, effectively, be random, the behavior of the system as a whole will evolve in a deterministic fashion.

In certain biological systems, however, fluctuations do not balance out. Many biological processes, for instance, rely on a variety of large protein molecules present at very low concentrations. Due to their scarcity, it can hardly be assumed that fluctuations in the concentrations of these chemicals are effectively averaged. In fact, in certain cases, biological organisms have developed methods for amplifying small fluctuations through positive feedback. As a result, rather than averaging out, random events can actually drive the behavior of the entire system. Similar arguments can be made about specific biological processes, like mutation, which rely on rare chance events. Biological systems whose dynamic behavior is highly dependent on random or poorly characterized occurrences cannot be adequately described by deterministic models. Instead, an entirely different mathematical framework, known as stochastic modeling, must be employed.

2.4.4 Stochastic Models

Stochastic models describe the probabilities associated with certain events or outcomes, and the evolution of those probabilities as a function of one or more independent variables. In many biological models, we are not concerned with individual outcomes, *per se*, but rather, with some *function* of the outcomes. As a result, stochastic models often involve what is known as a "random variable". Consider a sample space, S, which is the collection of all possible outcomes for a particular event. We define the random variable, X, as a function which associates a particular real number value to each event. As the independent variable(s) in a system evolve, the random variable evolves as well, generating a family of functions {X(t)}, where t is used to denote the independent variable. This family of functions, $\{X(t)\}$, is called a *stochastic process*.

2.4.4.1 Markov chains

The simplest stochastic process is one in which the probability of a future event is entirely characterized by the current state of the system, and does not depend on any past states. This is analogous to a first order difference equation where $x_{n+1} = f(x_n)$. In our discussion of difference equations, however, we focused on deterministic models. To extend this concept to a stochastic model, we write

$$\Pr\{X(t_{n+1}) = x_1 | X(t_n) = x_2\} \tag{2.55}$$

where Pr is the probability associated with an event, and "|" means "given that". Equation (2.55) thus reads "the probability that X will equal x_1 at t_{n+1}, given that X equals x_2 at t_n". Implicit in Eq. (2.55) is the assumption that the value of the random variable at t_n, $X(t_n)$, depends only on the most recent state of the system, and is thus history-independent. History-independent random processes are known as Markov processes. A Markov chain is a model which tracks the progression of a Markov process as the independent variable changes.

Analysis of Markov processes relies on the "law of conditional probability", which defines the probability of a certain event, given that another, correlated event has already occurred. The law of conditional probability states

$$\Pr\{E_1 | E_2\} = \frac{\Pr\{E_2 \cap E_1\}}{\Pr\{E_1\}} \tag{2.56}$$

where $\cap$ symbolizes the intersection, "and". Equation (2.56) indicates that the probability of observing event 1, E_1, given that event 2, E_2, has already occurred is equal to the probability of observing *both* E_1 and E_2 divided by the probability of observing E_1. Equation (2.56) allows us to define "transition probabilities", which describe the likelihood of a random number with $X = x_i$ "transitioning" to the value x_j in the next t interval,

$$\Pr\{E_j|E_i\} = \Pr\{X(t_{n+1}) = x_j|X(t_n) = x_i\} = p_{ij}. \tag{2.57}$$

For a system where the transition probabilities are constant and there are N possible values for the random variable, there is an $N \times N$ "transition probability matrix" of the form

$$P = \begin{bmatrix} p_{11} & p_{12} & \cdots & p_{1N} \\ p_{21} & p_{22} & \cdots & p_{2N} \\ \vdots & & & \\ p_{N1} & p_{N2} & \cdots & p_{NN} \end{bmatrix}. \tag{2.58}$$

To track the probability of being in each state as the independent variable progresses, we define a "probability vector", $\mathbf{u} = [u_1, u_2, \ldots, u_N]^T$. A Markov model for transitions then takes the form

$$\mathbf{u}_{n+1} = P\mathbf{u}_n \quad \text{with } \mathbf{u}_0 \text{ given.} \tag{2.59}$$

The following example illustrates the application of a Markov Chain model to a biological system.

Example 2.5. Consider a cancer cell which, during the course of a single cell cycle, may undergoing mitosis to produce a replicate daughter cell, fail to replicate but persist, or die and be cleared from the body. If we denote p_m the probability that the cell replicates, and p_d the probability that the cell dies, $1 - p_m - p_d$ is the probability that the cell fails to replicate, but persists. From this information, we can model the probability of having $X(t_n) = c$ cancer cells in the body after n cell cycles.

Solution. The first step in solving this problem is to formulate transition probabilities of the form $\Pr\{X(t_{n+1}) = x_j|X(t_n) = x_i\}$. In light of the information that we have about the system, it is easier to generate transition probabilities using basic probability arguments than it is to try to apply Eq. (2.56). The arguments are as follows. The total increase (or decrease) in cancer cells must equal the number of successful replications, r, minus the number of cell deaths, d

$$x_j - x_i = r - d \tag{2.60}$$

On the other hand, the total number of cell replications and cell deaths in a single cell cycle cannot exceed the number of cells present at the beginning of that cycle. When both of these points are considered

$$\Pr\{X(t_{n+1}) = x_j | X(t_n) = x_i\}$$
$$= \sum_{k=0}^{\min(x_i, x_i - x_j/2)} \binom{x_i}{k} \binom{x_i - k}{x_j - x_i + k} p_d^k p_m^{x_j - x_i + k} (1 - p_m - p_d)^{2x_i - 2k - x_j} \tag{2.61}$$

where the sum is over cell deaths, and the upper bound on the sum reflects the constraint that the total number of births and deaths not exceed the number of cells at the beginning of the cycle.

From Eq. (2.61) we can write a transition probability matrix. For the purposes of illustration, we will limit our tumor size to 5 cells, thus the set of allowable values for the random variable is $X \in \{0, 1, \ldots, 5\}$. To accommodate the limited size of the tumor, we assume that all transitions which would otherwise take the system beyond 5 cells instead return the system to its 5 cell upper limit. The transition matrix for $p_d = 0.2$ and $p_m = 0.3$ is shown in Eq. (2.62).

$$P = \begin{bmatrix} 1 & 0.2 & 0.04 & 0.008 & 0.0016 & 0.00032 \\ 0 & 0.5 & 0.2 & 0.06 & 0.016 & 0.004 \\ 0 & 0.3 & 0.37 & 0.186 & 0.0696 & 0.0224 \\ 0 & 0 & 0.3 & 0.305 & 0.172 & 0.074 \\ 0 & 0 & 0.09 & 0.279 & 0.2641 & 0.1597 \\ 0 & 0 & 0 & 0.162 & 0.4767 & 0.73958 \end{bmatrix}. \tag{2.62}$$

If we consider an initial state with a single cancer cell, $\mathbf{u}_0 = [0,1,0,0,0,0]^T$, we can find the probability of having any number of cancer cells between 0 and 5 after one cell cycle

$$\mathbf{u}_1 = P\mathbf{u}_0 = \begin{bmatrix} 0.2 & 0.5 & 0.3 & 0 & 0 & 0 \end{bmatrix}. \tag{2.63}$$

Similarly, after a second cell cycle, the probability of having any number of cancer cells between 0 and 5 will be

$$\mathbf{u}_2 = P\mathbf{u}_1 = P^2\mathbf{u}_0 = [0.312 \quad 0.31 \quad 0.261 \quad 0.09 \quad 0.027 \quad 0]. \tag{2.64}$$

After t cell cycles, then, the probability of having any number of cancer cells between 0 and 5 will be

$$\mathbf{u}_t = P\mathbf{u}_{t-1} = P^t\mathbf{u}_0. \tag{2.65}$$

While repeated application of the transition probability matrix will show how the probabilities change over repeated cell cycles, an equilibrium state $\mathbf{u}^*$ can be found as well. An equilibrium probability vector is a probability vector that doesn't change with repeated application of the transition probability matrix. In other words, it is a probability vector which has stabilized.

$$\mathbf{u}^* = P\mathbf{u}^* \tag{2.66}$$

From Eq. (2.66), it is clear that $\mathbf{u}^*$ is an eigenvector of the probability transition matrix with eigenvalue $\lambda = 1$. Solving for $\mathbf{u}^*$ gives $\mathbf{u}^* = [1, 0, 0, 0, 0, 0]$, suggesting that the cancer will eventually disappear from the body. This result should make intuitive sense. Since there is always a finite probability that all of the cells present at t_n will be dead by t_{n+1}, the system should, at some point, reach the state $X(t) = 0$. Unlike a system in any of the other states, $X(t) \in \{1, 2,...,5\}$, a system with $X(t) = 0$, is trapped in that state for all future time points. Therefore, all systems will eventually end up with $X(t) = 0$, as reflected in the equilibrium probability vector which suggests $X(t \to \infty) = 0$ with certainty.

2.4.4.2 *The generating function*

Another important concept in stochastic modeling is the "generating function". For any discrete random variable X that assumes values in the natural numbers n with probability p_n, the generating function is defined as

$$g(s) = \sum_{n=0}^{\infty} s^n p_n, \quad 0 \leq s \leq 1. \tag{2.67}$$

All of the information about the random variable X is contained in the generating function. Notice, for instance, that the probability of the random variable taking on any value, n, can be found as

$$p_n = \left.\frac{1}{n!}\frac{d^n g}{ds^n}\right|_{s=0} \tag{2.68}$$

while the mean of the random variable can be found as

$$E(X) = \sum_{n=0}^{\infty} np_n = g'(1). \tag{2.69}$$

Not only are generating functions an attractive method for summarizing information about the random variable, but also, they are a convenient way to deal with sums of independent random variables. If X is one random variable, and Y is another, then the sum of $X + Y$ has the generating function $g(s)h(s)$, where $g(s)$ is the generating function for X and $h(s)$ is the generating function for Y. In the example below, we use this relationship to model Polymerase Chain Reaction (PCR).

Example 2.6. PCR is a technique in molecular biology whereby a small amount of DNA or RNA is taken from a probe and multiplied to levels at which it can be detected. PCR is a necessary step in DNA fingerprinting, and has potential diagnostic applications. For many of these applications, knowing the quantity of nucleic acid present in the original sample can be important. PCR involves incubating DNA or RNA strands with a mixture of primers and nucleotides that allow them to replicate. When a strand replicates, it produces two daughter strands. Strands which do not replicate are destroyed. Replication is, however, a stochastic process, with a probability of doubling, p_d, ranging between 0.6 and 0.8. We can use this information to derive an estimate for the nucleic acid content in the original PCR sample, A_0, based on the nucleic acid content in the sample after t PCR generations.[9]

Solution. The system begins with a certain number of parent nucleic acid strands, $Y_0 = A_0$. All future generations will have Y_t parent strands, where Y_t is a random variable which depends on the number of replicating parent strands in the preceding PCR cycle. We use $h_t(s)$ to denote the generating function for Y_t, noting that

$$h_0(s) = s^{A_0}. \tag{2.70}$$

Since the random variable Y_0 takes on the value A_0 with certainty. To account for the stochastic replication of the parent strands in the tth PCR cycle, we introduce a total of Y_t more independent, identically distributed random variables X_i, each with generating function $g(s)$. For individual X_i

$$\begin{aligned} &\Pr\{X = 2\} = p_2 = p_d, \quad \Pr\{X = 0\} = p_0 = 1 - p_d \quad \text{and} \\ &\Pr\{X = n\} = p_n = 0, \quad n \neq 0, 2 \end{aligned} \tag{2.71}$$

since each parent strand can produce 2 daughter strands with probability p_d, or can be destroyed, with probability $1 - p_d$. Along with Eq. (2.67), this gives

$$g(s) = (1 - p_d) + p_d s^2. \tag{2.72}$$

At the end of each PCR cycle, the "daughter" strands from the current generation become the "parent" strands for the future generation, thus the number of parent strands in the $t + 1$ cycle, Y_{t+1}, can be found by summing the random variables, X_i, corresponding to replication of strands in the t cycle

$$Y_{t+1} = \sum_{i=1}^{Y_t} X_i = \left(g(s)\right)^{Y_t} \tag{2.73}$$

where the second equality makes use of the relationship between summed independent random variables and their generating functions. Equation (2.73) is complicated by the fact that the number of parents in the t generation, Y_t, is a random variable itself. If we denote $q_{tm} = \Pr\{Y_t = m\}$ the probability of having m strands at the beginning of the tth PCR cycle, we can write the generating function for Y_{t+1} as

$$h_{t+1}(s) = \sum_{m=0}^{\infty} \Pr\{Y_t = m\}(g(s))^m = \sum_{m=0}^{\infty} q_{tm}(g(s))^m = h_t(g(s)). \tag{2.74}$$

Notice that

$$\begin{aligned} h_1(s) &= h_0(g(s)) = (g(s))^{A_0} \\ h_2(s) &= h_1(g(s)) = h_0(g(g(s))) = (g(g(s)))^{A_0} \\ &\vdots \\ h_t(s) &= (g^t(s))^{A_0} \end{aligned} \tag{2.75}$$

Therefore, we can then find the mean of Y_t using Eqs. (2.75) and (2.69)

$$E(Y_t) = h_t'(1) = g'(1)h_{t-1}'(1) = R_0 E(Y_{t-1}) \tag{2.76}$$

where $R_0 = g'(1)$. It follows that $E(Y_1) = R_0 A_0$, and thus that

$$E(Y_t) = R_0^t A_0. \tag{2.77}$$

An experimental estimate for R_0 can therefore be found as

$$\hat{R}_0 = \frac{\tilde{Y}_{t+1}}{\tilde{Y}_t} \tag{2.78}$$

where the tildas indicate that Y_t and Y_{t+1} are empirically determined, and thus subject to some variation between PCR experiments. Having an estimate for R_0, an estimate for A_0 is

$$\hat{A}_0 = \frac{\tilde{Y}_t}{\hat{R}_0^t}. \tag{2.79}$$

In this section, we have dealt primarily with discrete random variables. It should be noted, however, that random variables can also take on continuous values. In that case, it is necessary to define a "probability density function", $f(a)$ such that

$$\Pr\{a_1 \le A < a_2\} = \int_{a_1}^{a_2} f(\alpha)d\alpha. \tag{2.80}$$

Readers interested in stochastic modeling are directed to textbooks which cover the topic in more depth and, in particular, deal with the field of stochastic differential equations.

2.5 SUMMARY

The purpose of this chapter has been to introduce the reader to both the difficulties and techniques associated with modeling biological processes where nano-scale interactions are involved. Certainly, we have only had time to deal with the most common modeling methods. Other mathematical formulations and computational routines can be found in the literature. Network models, for instance, are discussed in Chapter 5. Cellular automata are also commonly encountered in biological modeling, particularly when phenomena like pattern formation are involved.

While modeling techniques can be applied to all scales and all types of physical processes, from subatomic particles, to galaxies, modeling nano bio-systems is particularly challenging because the phenomena involved frequently span a wide range of different scales. As a result, it is difficult to strike a balance between a model which fails to capture the detail required for accurate prediction and understanding, and a model which is so detailed that it yields no insight and incurs huge computational costs. Clearly, as the field of multi-scale modeling develops, new methods for integrating mathematical descriptions from a wide range of different scales can be expected. This should have huge benefits in terms of applying a mathematical description to nanomedicine.

Despite the obvious challenges of scale and complexity involved in developing any model which might be applied to a nano bio-system, the benefits of such a model make the effort well worthwhile. At present, our understanding of nanotechnology as it applies to living organisms is very limited. There are questions about the safety of nanomedical approaches, the long term effects of nanoparticles in the human body, and the optimal design of nanotechnology applications. All of these questions must be answered before the true potential of nanomedicine can be realized. While experimental methods for characterization of nano bio-effects will always be important, mathematical modeling will likely play an important role in the understanding, development, and optimization of new nanomedical techniques. In the rest of this volume, specific models for nanomedicine are developed and discussed, many of which rely on methods and mathematical concepts developed in this chapter.

REFERENCES

1. H. B. Frieboes, J. P. Sinek, O. Nalcioglu, J. P. Fruehauf and V. Cristini, "Nanotechnology in Cancer Drug Therapy: A Biocomputational Approach," in *BioMEMS and Biomedical Nanotechnology*, Springer (2006)
2. A. Franks, "Nanotechnology," *J. Phs. E: Sci Instrum*, **20**, 1442–51 (1987).
3. M. C. Daniel and D. Astruc, "Gold Nanoparticles: Assumbly, Supramolecular Chemistry, Quantum-Size-Related Properties, and Applications towards Biology, Catalysis, and Nanotechnology," *Chem Rev*, **104**, 293–346 (2004).
4. M. Ciofalo and M. W. Collins, "Hennessy TR. Modelling nanoscale fluid dynamics and transport in physiological flows," *Med. Eng. Phys.*, **18**, 437–451 (1996).
5. M. Ferrar, "The mathematical engines of nanomedicine," *Small*, **4**, 20–25 (2008).

6. D. Barbolosi and A. Iliadis, "Optimizing drug regimens in cancer chemotherapy: a simulation study using a PK-PD model," *Computers in Biology and Medicine*, **31**, 157–172 (2001).
7. W. Tao and M. Zhang, "Mathematical modeling for medical automation," in *Systems Engineering Approach to Medical Automation*, R. A. Felder, M. Alwan and M. Zhang (eds.), Artech publishing house (2008).
8. D. Wodarz, S. E. Hall, K. Usuku, M. Osame, G.Ogg, A. McMichael, M. Nowak and R. M. Bangham, "Cytotoxic T-cell abundance and virus load in human immunodeficiency virus type 1 and human T-cell leukemia virus type 1," *Proc. R. Soc. Lond. B*, **268**, 1215–1221 (2001).
9. G. Vries, T. Hillen, M. Lewis, J. Müller and B. Schönfisch, "A course in mathematical biology: Quantitative modeling with mathematical and computational methods," *Siam* (2006).

Chapter Three

A Mathematical Formulation of the Central Dogma of Molecular Biology

Rui Gao, Juanyi Yu, Mingjun Zhang, Tzyh-Jong Tarn and Jr-Shin Li

In biology, the synthesis of proteins is an important problem. Essentially, this is a process of transferring genetic information from nucleic acids to polypeptides. In terms of molecular biology, this information flow is named the Central Dogma. But this process results from experiments only and uses character-based expression, such as the expression of DNA and RNA sequences, the characterizations of polarities of amino acids, the descriptions of protein structures, and the processes of information transfer from DNA to RNA and proteins, under molecular biology convention. Even though this character-based model describes the whole process precisely, it becomes a bottleneck for the development of inter-discipline of system science, information science, and molecular biology. Thus, an appropriate and efficient mathematical model will be a good language to help build up the connections among different disciplines and to stimulate the studies of this process. In the light of those advantages, we will develop a mathematical formulation for the Central Dogma.

A mathematical model is helpful in understanding the theoretical aspects of DNA computation, and also useful in applying mathematical tools to solve DNA computation problems. Zhang proposed a simple abstract model of molecular computers in 1996.[1] Based on this work, Zhang *et al.* presented a mathematical formulation of DNA

Nanomedicine: *A Systems Engineering Approach* edited by Mingjun Zhang & Ning Xi

www.panstanford.com
978-981-4241-36-6

computation in 2006.[1,2] According to certain rules, character-based DNA computation is converted into a numerical computation problem. Related propositions about DNA hybridization have also been carefully presented.

Another important modeling method in biological computation is the Hidden Markov Model (HMM). HMM is a statistical model that considers all possible combinations of matches, mismatches, and gaps to generate an alignment of a set of sequences. These models are primarily used for protein sequences to represent protein families or sequence domains, but they are also used to represent patterns in DNA sequences, such as RNA slice junctions. In 1987, Lander and Green used an HMM in the construction of genetic linkage maps.[3] In 1989, Churchill employed an HMM in sequence analysis to produce an HMM that represented a sequence profile (a profile HMM) to analyze sequence composition, and patterns.[4] In the 1990s, HMMs have been widely used to model sequences and proteins broadly. White *et al.* used an HMM to model super families of protein.[5] Asai *et al.* applied an HMM to predict the secondary structure of proteins, and obtain higher prediction rates than previously achieved.[6] HMMs are also used in producing multiple sequence alignments.[7–10] These HMMs generate sequences with various combinations of matches, mismatches, insertions, and deletions, and give these sequences probabilities, depending on the values of the various parameters in the models.

HMMs often provide a multiple sequence analysis as good as, or even better than, other methods such as global alignment and local alignment methods, including profiles and scoring matrices. This approach also has a number of other strong features: it is well grounded in probability theory, no sequence ordering is required, guesses of insertion/deletion penalties are not needed, and experimentally derived information can be used. The disadvantage of using HMMs is that at least 20 sequences (and sometimes even more) are required to accommodate the evolutionary history of the sequences.[11] Another problem with HMM is that the training set has to be quite large (50 or more sequences) to produce a useful model for the sequences. The difficulty in training the HMM residues is that many parameters need to be determined to obtain a suitable model (the amino acid distributions, the number and positions of insertion and deletion states, and the state transition frequencies add up to thousands of parameters). The purpose of the training data is to find

a suitable estimate for all these parameters. When trying to make an alignment of short sequence fragments to produce a profile HMM, this problem is worsened because the amount of data for training the model is even further increased. Furthermore, though a HMM is a powerful modeling tool in terms of protein family modeling and gene finding, it is not good at describing the entire process of biological information transferring, such as the Central Dogma of Molecular Biology.

Based on the modeling methods discussed above, this chapter presents a complete mathematical formulation of molecules involved in the Central Dogma of molecular biology. This chapter is organized as follows. First, a brief introduction of the Central Dogma is presented. Several biological information transfers, including DNA replication, transcription and translation, and special information transfers, are described in Sec. 3.1. In Sec. 3.2, the mathematical formulations of DNA sequences are presented. Three methods are presented to convert character-based DNA sequences into numerical sequences and one method to describe the hybridized DNA double-strand sequence. Next, theoretical propositions about DNA sequence operation and DNA hybridization are provided. Section 3.3 discusses the models of RNA self-hybridized sequences, polarities of amino acids and the secondary structures of proteins based on the existing models of DNA. The method presented here not only models the RNA secondary structures, but also reproduces the original single-strand RNA sequence. Selected models of secondary structures of protein are also illustrated. Based on the models, relevant results about determinations of components of RNA self-hybridized sequences, polarities of amino acids, and secondary structures of proteins are given. In Sec. 3.4, a unified mathematical framework describing the information transfer process of the Central Dogma is presented. We can express a complete biological process by transformation between several matrices, which makes it possible to apply existing or modified control theory to molecular biological systems. Discussions and conclusions are presented at the end of this chapter.

3.1 INTRODUCTION OF THE CENTRAL DOGMA

The Central Dogma describes the flow of genetic information from DNA to RNA to proteins, which forms the backbone of molecular biology. In the first step of this process, DNA replicates itself. Then the DNA sequence in a gene copies its information to the messenger RNA (mRNA), which then passes genetic messages to a ribosome, which "reads" this information and uses it for protein synthesis. This entire process is called general transfers. On the other hand, there also exist special or unknown transfers in the Central Dogma. For example, inverse transcription, RNA replication, and direct transfers from DNA to proteins are examples of special transfers, and prions are the special cases of unknown transfers.

3.1.1 The Central Dogma of Molecular Biology

The Central Dogma of Molecular Biology was proposed by Francis Crick in 1958[12] and then was restated in his *Nature* paper in 1970.[13] The main idea of the Central Dogma is that once "information" has passed into protein it cannot get out again, i.e., information transformation is possible only from nucleic acid to itself, or to proteins, but impossible from proteins to itself, or to nucleic acid. Information here means the character-based sequences of nucleic acid for DNA, RNA, and amino acid for proteins.

The Central Dogma of Molecular Biology describes the frame in which the sequential information transfers among semantophoric biopolymers consecutively. In general, there are three classes of biopolymers in live organisms: DNA, RNA and proteins. Theoretically, there exist nine possible transformations between any two of them. Those transformations are divided into three classes, general transfer, special transfer, and unknown transfer. General transfer happens in usual biological processes; special transfer occurs only where viruses are involved or in the laboratory; while unknown transfer is a theoretical class that has not been detected. The three classes of transfers are shown in Table 1.

Table 1. Three classes of informational transfers proposed by the Central Dogma

General transfer	Special transfer	Unknown transfer
DNA → DNA	RNA → DNA	Protein → DNA
DNA → RNA	RNA → RNA	Protein → RNA
RNA → Protein	DNA → Protein	Protein → Protein

3.1.2 Biological Sequential Information

Biopolymers are molecules made of repetitive units called monomers. Biopolymers, including DNA, RNA and proteins, are linear high polymers; that is, each monomer connects with at least two other monomers. The chain of monomers encodes information efficiently. The information transfer described by the Central Dogma is reliable, deterministic transfer, in which the sequence of biopolymers is used as the construction template of the other biopolymer sequence. Those sequences depend completely on the original biopolymer. The biological sequential information flow of the Central Dogma is shown in Fig. 1.

3.1.3 General Transfer of Biological Sequential Information

As stated above, there are three types of general transfers of biological sequential information: DNA replication, transcription, and translation.

DNA replication. DNA replication refers to the process of self-reproduction. To transfer genetic information from ancestors to their descendants, DNA must be replicated accurately, by dissolving and recomposing the super-helices of protein. In generating process of cell or organism, first the double-strand DNA helix is dissolved by using DNA polymerases and proteins, its main template is then copied.

Transcription. Before the synthesis of a protein begins, the information in DNA regions known as genes is unidirectionally transferred to mRNA strands in a process called transcription. Through this process, one strand of DNA double helix is used as a template to synthesize an mRNA by the RNA polymerase and transcriptor. In eukaryote cells, this mRNA migrates from the nucleus to the

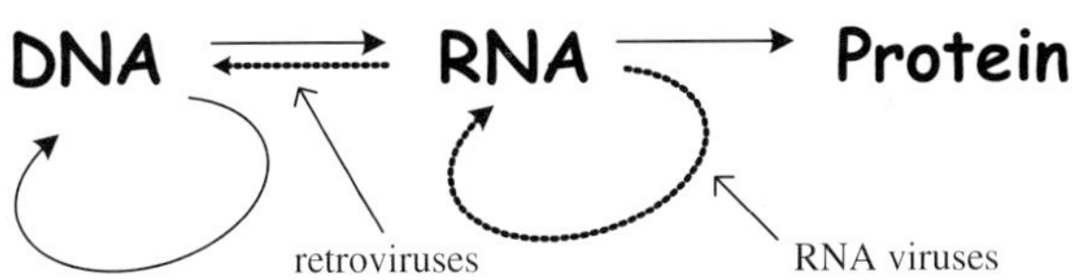

Figure 1. Biological sequential information flow of the Central Dogma. Cited from: http://www.cbs.dtu.dk/staff/dave/DNA_CenDog.html.

cytoplasm and then goes through different types of maturation by alternative splicing, in which the non-coded sequences of mRNA, called introns, are eliminated. Thus, the new sequence generated by rearrangement is different from the original one.

Translation. Translation defined as the process of protein synthesis using mRNA as templates is the last step of information transfer from DNA to proteins. The mature mRNA, from transcription finds its way to a ribosome. Three consecutive nucleotides of the coded mRNA are grouped together to form different codons. In prokaryotic cells, there are no nuclear membranes, so transcription and translation are linked. But in eukaryotic cells, the sites of transcription and translation are usually segregated, so transcription happens in the nucleus, and translation takes place in the cytoplasm. Then mRNA transports out of nucleus into cytoplasm. The ribosome binds to the mRNA from the start codon, which is recognized only by the initiator transfer RNA (tRNA). The ribosome then sequentially matches appropriate codons in mRNA to corresponding tRNA anti-codons, thereby adding correct amino acids in the sequence-encoding gene. Encountering with the stop codon of mRNA, the complete polypeptide chain is released from the ribosome and begins to fold into the correct conformation. This folding process continues until a natal polypeptide becomes a mature protein. The correct folding process is quite complex and may require other proteins. Occasionally, proteins may splice themselves. In this case, the abandoned internal blocks are called "intein".

3.1.4 Special and Unknown Transfers of Biological Sequential Information

Besides the general transfer discussed above, there are two other classes of information transfers, special transfer and unknown transfer. Under certain laboratory conditions or in the present of viruses, genetic information may be transferred from RNA back to DNA and from RNA to itself. The former is called inverse transcription, and the latter is named RNA transcription. Furthermore, in special biological systems, DNA can be used as template to represent protein directly. Though unknown transfers are believed not to happen, there exists an exception, prions.

Inverse Transcription. Inverse transcription regarded as the reverse of transcription is the transfer of information from RNA to DNA. It is believed to occur in the case of reverse transcriptase viruses, such as HIV and in higher eukaryotes, and in the case of retrotransposons. However, it is not the general case in most living organisms.

RNA Replication. RNA replication, the copying of RNA to another RNA, may be possible because of intervention by RNA viruses.

Direct Translation from DNA to Protein. Direct translation from DNA to protein is illustrated in a cell-free system (i.e., in a tube), using extract from *E. Coli* that contains ribosomes, but not intact cells. These cell fragments can represent proteins from foreign DNA templates. Neomycin has been found to enhance this effect.

Prions. Prions are examples of unknown transfer. Prions are proteins that can propagate themselves by changing the conformations of other molecules in the same class of proteins. This change occurs in the process of transmitting genetic information from one generation to the next generation, i.e., protein $\rightarrow$ protein.

3.2 MATHEMATICAL FORMULATIONS OF DNA SEQUENCES

It is well known that within cells of any living species, there is a substance called deoxyribonucleic acid (DNA), which is a double-strand helix of nucleotides carrying the genetic information of the cell. This information is the code used within cells to form proteins and is the building block upon which life is formed. A DNA molecule consists of three types of building blocks, sugar molecules, phosphate groups, and bases. The sugar and phosphate groups are strung together in an alternating fashion, forming the so-called sugar-phosphate backbone of the molecules. Due to the asymmetric structure of their sugar-phosphate backbones, DNA strands are assumed to have an orientation. One end of the strand is usually designated as the 3′ end (referring to the index of the carbon molecule to which the terminal phosphate group is attached), while the other is referred to as the 5′ end. A single-strand DNA consists of a chain of molecules called bases that protrude from the sugar-phosphate backbone. The bases in DNA strands can be partitioned into two groups of elements, known as *purines* and *pyrimidines*. Purines include the bases *adenine* (*A*) and *thymine* (*T*), while pyrimidines include the bases *guanine* (G), and *cytosine* (C). Since the sugar-phosphate backbone of DNA molecules has a fixed structure, at the first level of abstraction, DNA strands can be represented by oriented words over the four-letter alphabet of their bases. And at the second level of abstraction, DNA molecules can be viewed as geometrical structures, more specifically, as two-dimensional shapes. Such shapes arise from the affinity of the bases to bond with each other and form stable folded configurations. Frequently, a simple bonding rule is obtained: G binds to C via three hydrogen bonds, and vice versa, while A binds to T via two hydrogen bonds, and vice versa. These bonding rules are known as *Watson-Crick complementation*.

This section will be divided into three parts. In the first part (Sec. 3.2.1), several mathematical propositions will be proposed for modeling character-based DNA sequences. Then based on the modeling method of transfer matrices, a model of a hybridized double-strand DNA sequence will presented in the second part. The last part will give theoretical results based on the mathematical formulations in the first two parts.

3.2.1 Several Modeling Methods of Character-Based DNA Sequences

This section presents the mathematical formulations of DNA sequences. Based on the formulations, character-based DNA sequences are converted into numerical sequences. There are three methods to describe the modeling processes: the first one uses complex numbers, the second is to use integer numbers, and the last converts DNA sequences into vectors.[2]

(1) Complex Number Representation: Define a function $f(x):\{A, T, G, C\} \to \{1, -1, i, -i\}$ as

$$f(x) = \begin{cases} 1, & x = A; \\ -1, & x = T; \\ i, & x = G; \\ -i, & x = C; \end{cases} \tag{3.1}$$

where x is one of the four nucleotides.

The complementary base of each DNA base x can then be calculated by the following inverse function:

$$\bar{x} = f^{-1}(-f(x)) = \begin{cases} T, & x = A; \\ A, & x = T; \\ C, & x = G; \\ G, & x = C. \end{cases} \tag{3.2}$$

By the definition of Eqs. (3.1) and (3.2), complementary DNA sequences (either numerical or character-based) can be easily obtained, as we only need to specify single-strand DNA segments. The complementary strand can be generated using the above functions in either character-based or numerical format.

(2) Integer Number Representation: DNA bases can be mapped as integer numbers as well. Define a function $f(x) : \{A, T, G, C\} \rightarrow \{0, 1, 2, 3\}$ as

$$f(x) = \begin{cases} 0, & x = A; \\ 1, & x = C; \\ 2, & x = G; \\ 3, & x = T. \end{cases} \tag{3.3}$$

Similarly, the complementary base of x can be determined by the following inverse function

$$\bar{x} = f^{-1}(\{3\} - f(x)) = \begin{cases} T, & x = A; \\ G, & x = C; \\ C, & x = G; \\ A, & x = T; \end{cases} \tag{3.4}$$

where $\{3\}$ represents an appropriate finite length sequence consisting of multiple copies of integer 3. The numerical calculation can then be conducted base by base. For example, the numerical sequence of a DNA segment $X = AGGCAT$ is $f(X) = f(AGGCAT) = 022103$. The complementary segment of X can be easily obtained as $\bar{X} = f^{-1}(\{3\} - f(X)) = f^{-1}(311230) = TCCGTA$.

(3) Vector Representation: In vector space analysis, numerical value based DNA sequences can be expressed as rows of a matrix. Addition of such kinds of matrices can be regarded as a DNA hybridization process. Scalar multiplication produces multiple copies of the sequences in a test tube. Consider the four DNA bases $\{A, T, G, C\}^{\mathrm{T}}$ as a vector, and then any DNA strand $X = x_1 x_2 \cdots x_n, n \in N$ can then be expressed as a vector by a transfer matrix Π as

$$X = \Pi \begin{bmatrix} A \\ T \\ G \\ C \end{bmatrix} = \begin{bmatrix} p_{11} & p_{12} & p_{13} & p_{14} \\ p_{21} & p_{22} & p_{23} & p_{24} \\ \vdots & \vdots & \vdots & \vdots \\ p_{n1} & p_{n2} & p_{n3} & p_{n4} \end{bmatrix} \begin{bmatrix} A \\ T \\ G \\ C \end{bmatrix}, \tag{3.5}$$

where $\sum_{j=1}^{4} p_{ij} = 1, \forall i \in N$. Specifically,

$$p_{i1} = \begin{cases} 1, & x_i = A, \\ 0, & \text{otherwise}, \end{cases}$$
$$p_{i2} = \begin{cases} 1, & x_i = T, \\ 0, & \text{otherwise}, \end{cases}$$
$$p_{i3} = \begin{cases} 1, & x_i = G, \\ 0, & \text{otherwise}, \end{cases} \tag{3.6}$$
$$p_{i4} = \begin{cases} 1, & x_i = C, \\ 0, & \text{otherwise}. \end{cases}$$

In the above definition, each row of the matrix Π represents one DNA base. The complementary sequence of $\bar{X}$ can then be obtained by simply swapping column one with column two, and column three with column four as follows:

$$\bar{X} = \begin{bmatrix} p_{12} & p_{11} & p_{14} & p_{13} \\ p_{22} & p_{21} & p_{24} & p_{23} \\ \vdots & \vdots & \vdots & \vdots \\ p_{n2} & p_{n1} & p_{n4} & p_{n3} \end{bmatrix} \begin{bmatrix} A \\ T \\ G \\ C \end{bmatrix}. \tag{3.7}$$

For example, the transfer matrix of a single-strand DNA $X = ACGTGGATCT$ is Π_1 shown in Eqs. (3.8). The complementary sequence of X is $\bar{X} = TGCACCTAGA$, whose transfer matrix is Π_2, shown in Eqs. (3.8),

$$\Pi_1 = \begin{bmatrix} 1 & 0 & 0 & 0 \\ 0 & 0 & 0 & 1 \\ 0 & 0 & 1 & 0 \\ 0 & 1 & 0 & 0 \\ 0 & 0 & 1 & 0 \\ 0 & 0 & 1 & 0 \\ 1 & 0 & 0 & 0 \\ 0 & 1 & 0 & 0 \\ 0 & 0 & 0 & 1 \\ 0 & 1 & 0 & 0 \end{bmatrix}, \quad \Pi_2 = \begin{bmatrix} 0 & 1 & 0 & 0 \\ 0 & 0 & 1 & 0 \\ 0 & 0 & 0 & 1 \\ 1 & 0 & 0 & 0 \\ 0 & 0 & 0 & 1 \\ 0 & 0 & 0 & 1 \\ 0 & 1 & 0 & 0 \\ 1 & 0 & 0 & 0 \\ 0 & 0 & 1 & 0 \\ 1 & 0 & 0 & 0 \end{bmatrix}. \tag{3.8}$$

The above definitions of Eqs. (3.6) and (3.7) make it possible to define a DNA strand as an $n \times 4$ matrix.

The DNA nucleotides may also be defined as a vector directly. For example, $A = [1 \quad 0]^{\mathrm{T}}, T = [-1 \quad 0]^{\mathrm{T}}, G = [0 \quad 1]^{\mathrm{T}}$, and

$C = [0 \quad -1]^{\mathrm{T}}$. Then a DNA sequence can be expressed as a $2 \times n$ matrix, where n is the number of bases for the DNA sequence. For example, a single-strand DNA sequence $X = GATCCAGT$ can be expressed as

$$\begin{bmatrix} 0 & 1 & -1 & 0 & 0 & 1 & 0 & -1 \\ 1 & 0 & 0 & -1 & -1 & 0 & 1 & 0 \end{bmatrix}. \tag{3.9}$$

In a biological process, mutations often occur.[14] Based on the above definitions (3.6) and (3.7), a stochastic transfer matrix Γ can be defined as follows to reflect this phenomenon,

$$\Gamma = \begin{bmatrix} \rho_{aa} & \rho_{at} & \rho_{ag} & \rho_{ac} \\ \rho_{ta} & \rho_{tt} & \rho_{tg} & \rho_{tc} \\ \rho_{ga} & \rho_{gt} & \rho_{gg} & \rho_{gc} \\ \rho_{ca} & \rho_{ct} & \rho_{cg} & \rho_{cc} \end{bmatrix}, \tag{3.10}$$

where ρ_{ij} represents the probability of transformation from DNA base i to j, where $i, j \in \{a, t, g, c\}$. Obviously, $\rho_{ia} + \rho_{ig} + \rho_{it} + \rho_{ic} = 1$, $\forall i \in \{a, t, g, c\}$. ρ_{ii} is the probability for correct transformation.

The inner product (as the inner product in linear algebra) of a DNA sequence X can then be expressed as XX^{T}, which is a diagonal 2×2 matrix. The first and the last elements in the matrix represent the numbers of bases in X from the set $\{A, T\}$ or $\{G, C\}$, respectively.

Once a DNA sequence is converted into a numerical sequence, many interesting properties can be investigated. In the following section of this chapter, theoretical propositions will be presented.

3.2.2 A Mathematical Model of DNA Sequence Hybridization

If base-pairings occur between two individual DNA strands with opposite directions, the resulting process is referred to as DNA sequence hybridization. In eukaryotic species, the genome itself is organized in terms of two DNA strands hybridized so as to form a double helix that is coiled in the cell's nucleus. Hybridization can be complete or incomplete: in the latter case, only subsets of the bases on the strands bind with each other. For two single-strand DNA sequences, if number of their bases is the same and every base on one DNA sequence is complementary to the correspond- ing base on the other sequence with opposite direction, we call the hybridization is prefect. Otherwise, if there are corresponding

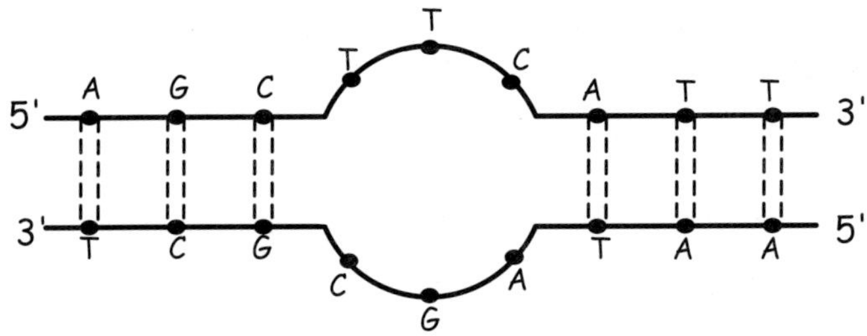

Figure 2. Two imperfectly hybridized DNA strands.

bases on two single DNA strands that aren't complementary to each other, the hybridization is called imperfect hybridization. Perfect hybridization is an ideal case in practice, and is usually used in DNA microarray hybridization. And imperfect hybridization is common. An example of imperfectly hybridized DNA strands is shown in Fig. 2.

According to the transfer matrix introduced above, we will give the models for the double-strand DNA helix. Define two transfer matrices Π_1 and Π_2, and a mathematical model of the DNA helix can be expressed as

$$\mathbf{X} = \Pi_1 \begin{bmatrix} A \\ T \\ G \\ C \end{bmatrix} + \Pi_2 \begin{bmatrix} T \\ A \\ C \\ G \end{bmatrix}, \tag{3.11}$$

where Π_1 and Π_2 are both $n \times 4$ matrices, and n is the length of a single strand of the DNA sequence. Under this definition, we can describe the hybridized DNA helix by concrete transfer matrices. Taking the above imperfectly hybridized DNA sequences as an example, its transfer matrices become

$$\Pi_1 = \begin{bmatrix} 1 & 0 & 0 & 0 \\ 0 & 0 & 1 & 0 \\ 0 & 0 & 0 & 1 \\ 0 & 1 & 0 & 0 \\ 0 & 1 & 0 & 0 \\ 0 & 0 & 0 & 1 \\ 1 & 0 & 0 & 0 \\ 0 & 1 & 0 & 0 \\ 0 & 1 & 0 & 0 \end{bmatrix}, \quad \Pi_2 = \begin{bmatrix} 1 & 0 & 0 & 0 \\ 0 & 0 & 1 & 0 \\ 0 & 0 & 0 & 1 \\ 0 & 0 & 1 & 0 \\ 0 & 0 & 0 & 1 \\ 0 & 1 & 0 & 0 \\ 1 & 0 & 0 & 0 \\ 0 & 1 & 0 & 0 \\ 0 & 1 & 0 & 0 \end{bmatrix}. \tag{3.12}$$

Once a DNA sequence is converted into a numerical sequence or a hybridized DNA helix is expressed as transfer matrices, many interesting properties can be investigated.

3.2.3 Some Theoretical Propositions

By definition (3.1) and by viewing a DNA sequence as a vector in the format of Eqs. (3.9), we get propositions as follows.[2]

Proposition 3.1. *If the base-by-base plus operation of two equal-length numerical value based DNA sequences results in a zero vector, then the two DNA sequences are complementary to each other.*

Proof. Define vectors $X = [x_1 x_2 \cdots x_n]^{\mathrm{T}}$ and $Y = [y_1 y_2 \cdots y_n]^{\mathrm{T}}$ as the two DNA sequences. Under the assumption that $X + Y$ is equal to a zero vector, we can conclude that $x_j = -y_j, j = 1, 2, \ldots, n$, where $x_j, y_j \in \{1, -1, i, -i\}$. By definition (3.1), the sequences are complementary to each other. ■

Proposition 3.2. *If the base-by-base plus operation of all numerical value based DNA sequences in two test tubes results in a zero vector, then the hybridization resulting from mixing the test tubes should be complete. A complete hybridization means all single-strand DNA sequences find their complements.*

Proposition 3.3. *By definition (3.1), if the inner product of two equal-length sequences is not a real number, then the two sequences are not complementary to each other. It can be further claimed that they are not complementary in G and C.*

Proposition 3.4. *By definition (3.1), $\forall X, Y \in R^n$ (n represents the number of DNA bases in the strands), if X and Y have a complete hybridization and $X^T Y = 0$, then X and Y have equal numbers of DNA bases from $\{G, C\}$ and $\{A, T\}$ binding sets. Similarly, if a complete hybridization occurs, but $X^T Y > 0$, it means that there are more bases from the $\{G, C\}$ set than from the $\{A, T\}$ set. Otherwise, if complete hybridization occurs, but $X^T Y < 0$, it means that there are more bases from the $\{A, T\}$ set.*

Proposition 3.5. *Under definition (3.1), $\forall X, Y \in R^n$ (n represents the number of DNA bases in the strands, we have $||X^T Y|| \leq ||X|| + ||Y||$, where $||X^T Y||$ represents the lengths of the DNA strand after hybridization. $||X||$ and $||Y||$ represent the length of single-strand DNA segments. If a* *complete hybridization occurs, it is true that* $||X^T Y|| = n$. *If none of the bases is hybridized, then* $||X^T Y|| = 2n$. *This relationship is similar to the well-known triangle inequality in plane geometry.*

To investigate properties under the formula (3.9) in vector space, we first define equivalent transfer matrices.

Definition 3.6. *Equivalent transfer matrices.* Since each single strand of a double-strand DNA uniquely determines the other strand, a transfer matrix of a single-strand DNA and its complementary are regarded as equivalent to each other. For example, Π_1 in Eq. (3.8), is an equivalent transfer matrix of Π_2 in Eq. (3.8), expressed as $\Pi_1 \Leftrightarrow \Pi_2$, and vice versa. Two DNA sequences are complementary to each other if and only if their transfer matrices are equivalent.

Note that two DNA transfer matrices are equivalent if and only if one matrix is the result of swapping column one with column two, and column three with column four of the other matrix.

Proposition 3.7. *If DNA transfer matrices* $A \Leftrightarrow B$ *and* $B \Leftrightarrow C$, *then* A *is the same as* C.

Definition 3.8. *Similar DNA sequences.* Two equal-length DNA sequences that have less than a certain percent (usually 10% in practice) consecutive bases are regarded as similar sequences. The binding results for similar sequences may be hard to be distinguished using current molecular techniques.

Proposition 3.9. *Necessary condition for similar sequences. Under the formulation (3.9), if two DNA sequences are similar, then the sum of all columns of the transfer matrices is less than a predefined percent of the summed length of a single DNA sequence.*

The above propositions can be used to check similarities and complementary properties of DNA sequences for DNA computation.

Through the above formulation, sequences of DNA can be regarded as numerical value based sequences, so that their base-by-base operation (i.e., hybridization) results in a zero vector.

Moreover, if we regard a hybridized DNA helix as two matrices defined by Eq. (3.11), we can get the following results concerning perfect hybridization and similar DNA sequences.

Proposition 3.10. *If the transfer matrices defined by (3.11) are equal, i.e., $\Pi_1 = \Pi_2$, we can say that the DNA helix is perfectly hybridized or that one strand in the DNA helix is complementary to the other one. And if $\Pi_1 \neq \Pi_2$, then the DNA hybridization is imperfect. Furthermore, we can determine whether two DNA sequences are the similar sequences by the following condition*

$$||\Pi_1 - \Pi_2||_2/2n < 10\%, \tag{3.13}$$

where $\|\cdot\|_2$ denotes 2-Norms, and n is the length of the single-strand DNA sequence.

3.3 MATHEMATICAL FORMULATIONS OF RNA SEQUENCES AND PROTEINS

Like DNA molecules, RNA molecules also consist of sugar molecules, phosphate groups, and bases.[15] Similar to DNA, RNA strands are also assumed to have an orientation, and their bases are of the same type as DNA, with the exception of the base *thymine*(T) being replaced by *uracil*(U). Similar to the Watson–Crick complementation DNA binding rule mentioned above, there also exists a simple rule in binding between the bases of RNA: A binds to U, G binds to C, and vice versa. If the base pairing occurs among bases within a single strand, the process is termed *self-hybridization* or *folding*. Self-hybridization converts a one-dimensional strand into a two- or three-dimensional strand. These kinds of conformations are usually referred to as secondary and tertiary structures of the sequences. Self-hybridization is usually achieved by imperfect binding between bases on the same strand. In Sec. 3.3.1, models of self-hybridized RNA sequences and some related theoretical results will be shown.

Based on the polarity of the side chain, amino acids are categorized into hydrophilic amino acids and hydrophobic amino acids. Polarity is an important property in protein structures and interactions among proteins. The distribution of hydrophilic and hydrophobic amino acids determines the tertiary structure of the protein, and their physical location on the outside structure of the protein influences their quaternary structure. On the basis of triple coded models of amino acids, we will present ways of determining the polarities of amino acids in Sec. 3.3.2.

In molecular biology, one of the most important problems is how to determine the concrete structures of proteins. A common question is, if the amino acid sequence of a protein is given, what is the protein's structure in three-dimensional spaces? Researchers are curious because the structure of a protein offers us a method to understand its biological function. Generally, four distinct aspects of a protein's structure are taken into the consideration: primary structure, secondary structure, tertiary structure, and quaternary structure. The secondary structure is local regular structure commonly found within a protein. Accurate knowledge of the secondary structure will provide a basis for prediction of the tertiary structure and the quaternary structure. A modeling method for protein secondary structures will be proposed in the last part of this section.

3.3.1 Mathematical Models of RNA Sequence Self-Hybridizations and Some Theoretical Results

In general, an RNA self-hybridization sequence is composed of *dangling strands, stems regions,* and *loops.* A self-hybridized RNA sequence is illustrated in Fig. 3.

A dangling strand is a strand of consecutive unpaired bases at either the 3′ or 5′ end of the sugar-phosphate backbone. A stem region is a helix that includes a perfectly matched Watson–Crick complementary subsequence. Loops can be broadly divided into four classes: *hairpin* loops, *internal* loops, *branching* loops, and *bulge* loops. A hairpin loop is an exterior loop connected to one stem region only, an internal loop connects two stems, and a branching loop has

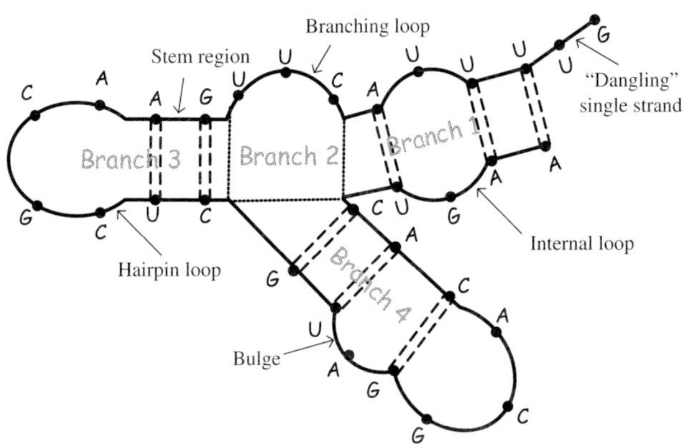

Figure 3. RNA secondary structure.

connections to at least three stems. A bulge is a protrusion between two stems that appears on only one side of the folded structure.

Before formulating an RNA self-hybridized sequence as a mathematical model, we will give *subsection rule* for any secondary structures of RNA.

Subsection Rule 3.11. *By the following rules, we divide an RNA self-hybridized sequence into several branches, which in turn can contain different units.*

(1) *Based on different directions of the sequence, the model is divided into different branches, and every branch is separated into units such as "dangling" single strands, stem regions, and all kinds of loops (except for branching loops);*
(2) *A branching loop usually is defined as an individual branch;*
(3) *The first unit in the first branch of the model may be a "dangling" single strand or a stem region;*
(4) *Every stem region or loop in the secondary structure is an individual unit.*

We demand that the number of bases in every unit should be even, if there are an odd number of bases in a unit, so we will make them an even number by adding a "zero" base. Then we can formulate a mathematical model of a self-hybridized RNA sequence as

$$\sum_j \sum_i \left(\Pi_1^{i,j} \begin{bmatrix} A \\ U \\ G \\ C \end{bmatrix} + \Pi_2^{i,j} \begin{bmatrix} U \\ A \\ C \\ G \end{bmatrix} \right), \tag{3.14}$$

where $\Pi_1^{i,j}$ and $\Pi_2^{i,j}$ are transfer matrices in the ith unit of the jth branch. For example, we construct the model of the RNA sequence in Fig. 3, thus

$$\sum_{i=1}^{4} \left(\Pi_1^{i,1} \begin{bmatrix} A \\ U \\ G \\ C \end{bmatrix} + \Pi_2^{i,1} \begin{bmatrix} U \\ A \\ C \\ G \end{bmatrix} \right) + \left(\Pi_1^{i,2} \begin{bmatrix} A \\ U \\ G \\ C \end{bmatrix} + \Pi_2^{i,2} \begin{bmatrix} U \\ A \\ C \\ G \end{bmatrix} \right)$$
$$+ \sum_{i=1}^{2} \left(\Pi_1^{i,3} \begin{bmatrix} A \\ U \\ G \\ C \end{bmatrix} + \Pi_2^{i,3} \begin{bmatrix} U \\ A \\ C \\ G \end{bmatrix} \right) + \sum_{i=1}^{4} \left(\Pi_1^{i,4} \begin{bmatrix} A \\ U \\ G \\ C \end{bmatrix} + \Pi_2^{i,4} \begin{bmatrix} U \\ A \\ C \\ G \end{bmatrix} \right). \tag{3.15}$$

Here we divide the RNA self-hybridized sequence into four branches (with the dashed lines as shown in Fig. 3), and express all the transfer matrices as follows. In the first branch, $j = 1$, the transfer matrices are

$$\Pi_1^{1,1} = \begin{bmatrix} 0 & 0 & 1 & 0 \\ 0 & 1 & 0 & 0 \end{bmatrix}, \quad \Pi_2^{1,1} = \begin{bmatrix} 0 & 0 & 0 & 0 \\ 0 & 0 & 0 & 0 \end{bmatrix};$$
$$\Pi_1^{2,1} = \begin{bmatrix} 0 & 1 & 0 & 0 \\ 0 & 1 & 0 & 0 \end{bmatrix}, \quad \Pi_2^{2,1} = \begin{bmatrix} 0 & 1 & 0 & 0 \\ 0 & 1 & 0 & 0 \end{bmatrix};$$
$$\Pi_1^{3,1} = \begin{bmatrix} 0 & 1 & 0 & 0 \end{bmatrix}, \quad \Pi_2^{3,1} = \begin{bmatrix} 0 & 0 & 0 & 1 \end{bmatrix};$$
$$\Pi_1^{4,1} = \begin{bmatrix} 1 & 0 & 0 & 0 \end{bmatrix}, \quad \Pi_2^{4,1} = \begin{bmatrix} 1 & 0 & 0 & 0 \end{bmatrix}.$$

In the second branch, $j = 2$, the transfer matrices are

$$\Pi_1^{1,2} = \begin{bmatrix} 0 & 0 & 0 & 1 \\ 0 & 1 & 0 & 0 \\ 0 & 1 & 0 & 0 \end{bmatrix}, \quad \Pi_2^{1,2} = \begin{bmatrix} 0 & 0 & 0 & 0 \\ 0 & 0 & 0 & 0 \\ 0 & 0 & 0 & 0 \end{bmatrix}.$$

In the third branch, $j = 3$, the transfer matrices are

$$\Pi_1^{1,3} = \begin{bmatrix} 0 & 0 & 1 & 0 \\ 1 & 0 & 0 & 0 \end{bmatrix}, \quad \Pi_2^{1,3} = \begin{bmatrix} 0 & 0 & 1 & 0 \\ 1 & 0 & 0 & 0 \end{bmatrix},$$
$$\Pi_1^{2,3} = \begin{bmatrix} 1 & 0 & 0 & 0 \\ 0 & 0 & 0 & 1 \end{bmatrix}, \quad \Pi_2^{2,3} = \begin{bmatrix} 0 & 0 & 1 & 0 \\ 0 & 0 & 0 & 1 \end{bmatrix}.$$

And in the last branch, $j = 4$, we have the transfer matrices

$$\Pi_1^{1,4} = \begin{bmatrix} 0 & 0 & 1 & 0 \\ 0 & 1 & 0 & 0 \end{bmatrix}, \quad \Pi_2^{1,4} = \begin{bmatrix} 0 & 0 & 1 & 0 \\ 0 & 1 & 0 & 0 \end{bmatrix};$$
$$\Pi_1^{2,4} = \begin{bmatrix} 1 & 0 & 0 & 0 \end{bmatrix}, \quad \Pi_2^{2,4} = \begin{bmatrix} 0 & 0 & 0 & 0 \end{bmatrix};$$
$$\Pi_1^{3,4} = \begin{bmatrix} 0 & 0 & 1 & 0 \end{bmatrix}, \quad \Pi_2^{3,4} = \begin{bmatrix} 0 & 0 & 1 & 0 \end{bmatrix};$$
$$\Pi_1^{4,4} = \begin{bmatrix} 0 & 0 & 1 & 0 \\ 0 & 0 & 0 & 1 \end{bmatrix}, \quad \Pi_2^{4,4} = \begin{bmatrix} 0 & 1 & 0 & 0 \\ 0 & 0 & 0 & 0 \end{bmatrix};$$

where we note that in this branch, the second and the fourth units have an added "zero" base.

The propositions, which below follow directly from the above rules and formulations, reveal rules to distinguish dangling single strands, stem regions, and types of loops.

Proposition 3.12. *If the two transfer matrices are the same, i.e., $\Pi_1^{i,j} = \Pi_2^{i,j}$, then the unit is a stem region. If the first unit includes a zero transfer matrix and the other is non-zero, namely, $\Pi_1^{1,1} = 0$ and $\Pi_2^{1,1} \neq 0$, or $\Pi_2^{1,1} = 0$ and $\Pi_1^{1,1} \neq 0$, we can draw the conclusion that there is a dangling single strand existing in the self-hybridized RNA strand.*

Proposition 3.13 helps us to identify the exact type of loops in a branch.

Proposition 3.13. *Suppose that $\Pi_1^{i,j}$ and $\Pi_2^{i,j}$ denote the transfer matrices in the ith unit of the jth branch.*

(1) *If $j \neq 1$, there are more than one units in this branch, and if there is a unit satisfying $\Pi_1^{i,j} = 0$ or $\Pi_2^{i,j} = 0$, but both matrices are not zero at the same time, then the unit denotes a bulge loop.*
(2) *In case (1), if the unit is the only unit in this branch, then the unit is a branching loop.*
(3) *If $j \neq 1$, there are more than one units in this branch, and if the last unit satisfies $\Pi_1^{i,j} \neq \Pi_2^{i,j}$, where both $\Pi_1^{i,j}$ and $\Pi_2^{i,j}$ are not zero matrices, we regard the unit as a hairpin loop.*
(4) *In case (3), if the unit is not the last in the branch, then the unit must be an internal loop.*

Using this proposition, we can identity all units in the example shown in Fig. 3. $(\Pi_1^{1,1}, \Pi_2^{1,1})$ denotes a dangling single strand. There are five stem regions, which are $(\Pi_1^{2,1}, \Pi_2^{2,1})$, $(\Pi_1^{4,1}, \Pi_2^{4,1})$, $(\Pi_1^{1,3}, \Pi_2^{1,3})$, $(\Pi_1^{1,4}, \Pi_2^{1,4})$ and $(\Pi_1^{3,4}, \Pi_2^{3,4})$ respectively. The second branch $(\Pi_1^{1,2}, \Pi_2^{1,2})$ is a branching loop, and $(\Pi_1^{3,1}, \Pi_2^{3,1})$ denotes an internal loop. At last, there are two hairpin loops in this model, $(\Pi_1^{2,3}, \Pi_2^{2,3})$ and $(\Pi_1^{4,4}, \Pi_2^{4,4})$. For the latter loop, we note that a "zero" base is added to make an even number of bases. And last, there is a bulge loop $(\Pi_1^{2,4}, \Pi_2^{2,4})$ in the last branch.

Among the single-strand RNA secondary structures, the most frequently encountered types are hairpins and cruciform RNA folds, as shown in Fig. 4.

Using the modeling method mentioned above, we can obtain mathematical models for these two examples. By the subsection rule, the model has only one branch. To avoid needless complexity, we omit the index j of branches. Then the model for a cruciform RNA

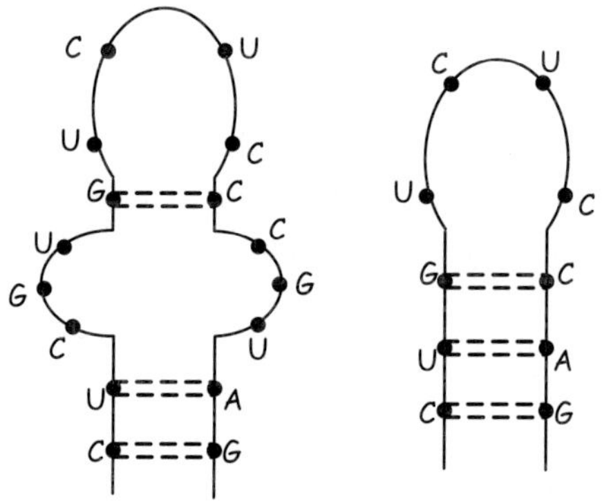

(a) Cruciform RNA fold (b) Hairpin RNA fold

Figure 4. Two RNA secondary structures.

fold is expressed as

$$\sum_{i=1}^{4}\left(\Pi_1^i\begin{bmatrix}A\\U\\G\\C\end{bmatrix}+\Pi_2^i\begin{bmatrix}U\\A\\C\\G\end{bmatrix}\right), \tag{3.16}$$

where

$$\Pi_1^1=\begin{bmatrix}0&0&1&0\\1&0&0&0\end{bmatrix},\quad \Pi_2^1=\begin{bmatrix}0&0&1&0\\1&0&0&0\end{bmatrix};$$

$$\Pi_1^2=\begin{bmatrix}0&1&0&0\\0&0&1&0\\0&0&0&1\end{bmatrix},\quad \Pi_2^2=\begin{bmatrix}0&0&1&0\\0&0&0&1\\1&0&0&0\end{bmatrix};$$

$$\Pi_1^3=\begin{bmatrix}0&0&0&1\end{bmatrix},\quad \Pi_2^3=\begin{bmatrix}0&0&0&1\end{bmatrix};$$

$$\Pi_1^4=\begin{bmatrix}0&0&0&1\\0&1&0&0\end{bmatrix},\quad \Pi_2^4=\begin{bmatrix}1&0&0&0\\0&0&1&0\end{bmatrix}.$$

In this model, we can see that there are two stem regions, (Π_1^1, Π_2^1) and (Π_1^3, Π_2^3); one internal loop, (Π_1^2, Π_2^2); and one hairpin loop, (Π_1^4, Π_2^4).

For the hairpin model shown in Fig. 4, the representation is

$$\sum_{i=1}^{2}\left(\Pi_1^i\begin{bmatrix}A\\U\\G\\C\end{bmatrix}+\Pi_2^i\begin{bmatrix}U\\A\\C\\G\end{bmatrix}\right), \tag{3.17}$$

where

$$\Pi_1^1 = \begin{bmatrix} 0 & 0 & 1 & 0 \\ 1 & 0 & 0 & 0 \\ 0 & 0 & 0 & 1 \end{bmatrix}, \quad \Pi_2^1 = \begin{bmatrix} 0 & 0 & 1 & 0 \\ 1 & 0 & 0 & 0 \\ 0 & 0 & 0 & 1 \end{bmatrix};$$

$$\Pi_1^2 = \begin{bmatrix} 0 & 0 & 0 & 1 \\ 0 & 1 & 0 & 0 \end{bmatrix}, \quad \Pi_2^2 = \begin{bmatrix} 1 & 0 & 0 & 0 \\ 0 & 0 & 1 & 0 \end{bmatrix}.$$

We can see that there are a stem region, denoted by the first unit, and a hairpin loop, denoted by the second unit.

Given the mathematical model of self-hybridized RNA, we can recover the character-based sequence from it. The process is based upon the following proposition.

Proposition 3.14. *From the mathematical model of an RNA self-hybridized sequence, the original single-strand sequence of RNA can be obtained through the following equation,*

$$\begin{aligned} &(\oplus_i \Pi_1^{i,1}) \oplus [\oplus_{j\neq 1}(\oplus_i \Pi_1^{i,j} \oplus \mathrm{inv}(\oplus_i \Pi_2^{i,j}))] \\ &\quad \oplus \mathrm{inv}(\oplus_i \Pi_2^{i,1})[A \quad U \quad G \quad C]^{\mathrm{T}}, \end{aligned} \tag{3.18}$$

where $\oplus$ *denotes the vertical combination of the matrices, namely,* $A \oplus B = [A^{\mathrm{T}} \vdots B^{\mathrm{T}}]^{\mathrm{T}}$. *The function* $inv(\Phi)$ *is defined by*

$$\mathrm{inv}(\Phi) = I_n' \bar{\Phi}, \tag{3.19}$$

where I_n' *is an nth-order square matrix described as*

$$I = \begin{bmatrix} 0 & \cdots & 0 & 1 \\ 0 & \cdots & 1 & 0 \\ \vdots & \iddots & \vdots & \vdots \\ 1 & \cdots & 0 & 0 \end{bmatrix},$$

and $\bar{\Phi}$ *is the equivalent matrix of* Φ, *which follows from Definition 3.6.*

Using Proposition 3.14, we can get the character-based single-strand RNA sequences from the mathematical formulations of the formulas (3.15)–(3.17). For Eq. (3.15), we obtain the original sequence by the above proposition

$$GUUUUACUUGAACGCUCGUAGGCACACUGAA.$$

The other two single-strand RNA sequences described by Eqs. (3.16) and (3.17) are

$$GAUGCCCUCUGUGCUC,$$

and

$$GACCUCUGUC,$$

respectively.

3.3.2 Polarities of Amino Acids in Protein

In the natural evolution of life, we believe that the 20 amino acids are the result of optimization. In biology, we denote an amino acid by a genetic code, a triplet code based on three-letter codons. The complete genetic codes are shown in Fig. 5.

In Fig. 5, the 64 triplets stand for one of the 20 amino acids and for the stop codons. *ATG* coded for *Methionine* is the start codon. Three of the codons, *TAA*, *TAG*, and *TGA*, are the stop codons. Clearly, a given amino acid may be encoded by more than one codon, but a codon can code for only one amino acid. This principle is called *redundancy*. The triplet codon space can be regarded as the binary Cartesian product or Cartesian square on the set $\{A, T, G, C\}$.

The genetic code provides the specificity for protein synthesis. The genetic information in an mRNA molecule can be regarded as a series of non-overlapping, consecutive, three-letter "words". Each

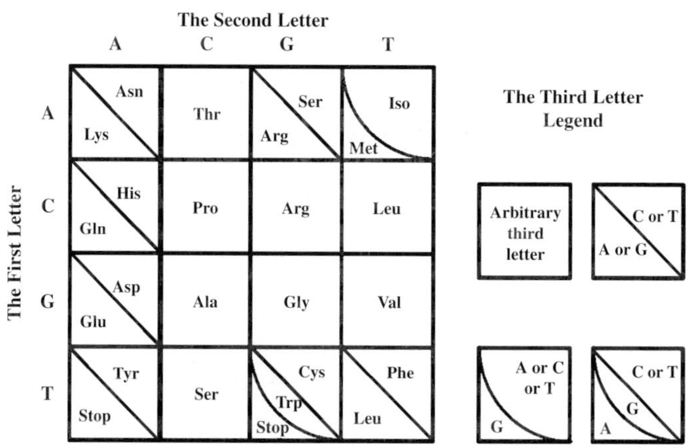

Figure 5. Table of the genetic codes.[2]

sequence of the three nucleotides along the chain specifies a particular amino acid. Each codon is complementary to the corresponding triplet in the DNA molecule from which it is transcribed.

Based on their polarity, the 20 amino acids can be divided into two categories, hydrophilic amino acids, which have polarities, and hydrophobic amino acids, which are non-polar. Non-polar amino acids include *Alanine, Glycine, Isoleucine, Leucine, Methionine, Phenylalanine, Proline, Tryptophan,* and *Valine*. These non-polar acids are also called neutral amino acids. The polar amino acids can be further divided into neutral amino acids, acidic amino acids, and basic amino acids. Neutral amino acids include *Asparagine, Cysteine, Glutamine, Serine, Threonine* and *Tyrosine*. Acidic amino acids are *Aspartic acid* and *Glutamic acid*. Basic amino acids are *Arginine, Histidine* and *Lysine*. From the above mathematical formulation, determining criteria for identifying amino acid polarities can be proposed.

First, define the unit vectors e_i as auxiliary vectors. So $(e_1 e_2 e_3 e_4) = I \in R^{4\times 4}$, where I is the identity matrix. We denote a amino acid by $\Pi[ACGT]^T$, where Π is the transfer matrix. For convenience, we construct the transfer matrix as a square matrix by making up a zero row vector. Namely, we define the "new transfer matrix" $\tilde{\Pi}$ by

$$\tilde{\Pi} = \Pi \oplus [0 \quad 0 \quad 0 \quad 0].$$

Then we can judge the polarities of amino acids through the following propositions.

Proposition 3.15. *An amino acid is polar if its transfer matrix* $\tilde{\Pi}$ *satisfies any one of the following conditions:*

(1) $e_2^{\mathrm{T}} \tilde{\Pi} e_1 = 1$, *or*
(2) $e_2^{\mathrm{T}} \tilde{\Pi} e_2 = 1$ *and* $e_1^{\mathrm{T}} \tilde{\Pi}(e_1 + e_4) = 1$, *or*
(3) $e_1^{\mathrm{T}} \tilde{\Pi}(\sum_{i=1}^{3} e_i) = 1$ *and* $e_2^{\mathrm{T}} \tilde{\Pi} e_3 = 1$.

Otherwise we can call that the amino acid is non-polar.

We can also determine the non-polarity of an amino acid directly.

Proposition 3.16. *An amino acid is non-polar if its transfer matrix* $\tilde{\Pi}$ *satisfies any one of the following conditions:*

(1) $e_2^{\mathrm{T}} \tilde{\Pi} e_4 = 1$, *or*
(2) $e_2^{\mathrm{T}} \tilde{\Pi} e_2 = 1$ *and* $e_1^{\mathrm{T}} \tilde{\Pi}(e_2 + e_3) = 1$, *or*
(3) $e_2^{\mathrm{T}} \tilde{\Pi} e_3 = 1$ *and* $e_1^{\mathrm{T}} \tilde{\Pi} e_4 = 1$.

Furthermore, given the conditions of Proposition 3.15, we can further decide the classification of the polar amino acids by the proposition below.

Proposition 3.17. *The polar amino acid is acidic if its transfer matrix* $\tilde{\Pi}$ *satisfies* $e_1^T \tilde{\Pi} e_3 = 1$ *and* $e_2^T \tilde{\Pi} e_1 = 1$, *or is basic if its transfer matrix* $\tilde{\Pi}$ *satisfies any one of the following conditions:*

(1) $e_1^T \tilde{\Pi} e_1 = 1$, *and* $e_2^T \tilde{\Pi}(e_1 + e_3) = 1$, *and* $e_3^T \tilde{\Pi}(e_1 + e_3) = 1$; *or*
(2) $e_1^T \tilde{\Pi} e_2 = 1$, *and* $e_2^T \tilde{\Pi} e_1 = 1$, *and* $e_3^T \tilde{\Pi}(e_3 + e_4) = 1$; *or*
(3) $e_1^T \tilde{\Pi} e_2 = 1$, *and* $e_2^T \tilde{\Pi} e_3 = 1$.

Otherwise, the polar amino acid is neutral.

Now, we can give the polarity of any amino acid from the above propositions. Here we must note that all propositions are based on triplet codons, which are expressed by transfer matrices. Once the transfer matrix is given, the polarity of an amino acid will be determined.

3.3.3 A Modeling Method for Protein Secondary Structures

Secondary structures in proteins consist of local inter-residue interactions mediated by hydrogen bonds. The most common secondary structures are α helices and β sheets. A common motif in the secondary structure of proteins, the α helix is a right-handed coiled conformation, resembling a spring, in which every backbone $N-H$ group donates a hydrogen bond to the backbone $C=O$ group of the amino acid located four residues earlier. The β sheet, the second form of regular secondary structure in proteins, consists of directional β strands connected laterally by three or more hydrogen bonds, forming a generally twisted, pleated sheet. A β strand is a stretch of amino acids typically 5–10 amino acids long whose peptide backbones are almost fully extended. From the work of Levitt and Chothia,[16] four principal classes of protein structure are recognized, based on the types and arrangements of secondary structural elements. They are all α proteins, all β proteins, α and β proteins, and membrane proteins, respectively. These classes will be described and illustrated. In addition, there are several other classes of proteins recognized in the SCOP or CATH database. Here we only give the mathematical formulations for several basic classes of proteins.[17,18]

We define the structure matrix Ψ and the bases of the proteins $[\alpha \quad \beta \quad loop]^T$, where α denotes an α helix, β denotes a β strand and *loop* denotes regions of protein chains between α helices and β strands. Then the mathematical model of the proteins can be expressed as $\Psi[\alpha \quad \beta \quad loop]^T$.

(1) *All α proteins*: all α proteins comprise a bundle of α helices connected by loops on the surface of the proteins, as shown in Fig. 6.

For the class of protein, the structure matrix Ψ is defined by

$$\Psi = \begin{bmatrix} 1 & 0 & 0 \\ 0 & 0 & 1 \\ 1 & 0 & 0 \\ 0 & 0 & 1 \\ 1 & 0 & 0 \\ 0 & 0 & 1 \end{bmatrix}.$$

We note that in Ψ, all entries in the second column are zeros, which denotes that there is no β sheet in the protein.

(2) *All β proteins*: All β proteins comprise anti-parallel β sheets, usually two sheets in close contact forming a sandwich. Alternatively, a sheet can twist into a barrel with the first and last strands touching. An example of all β proteins is shown in Fig. 7.

The structure matrix Ψ of the all β class proteins is as follows

$$\Psi = \begin{bmatrix} 0 & 0 & 1 \\ 0 & 1 & 0 \\ 0 & 0 & 1 \\ 0 & -1 & 0 \\ 0 & 0 & 1 \\ 0 & 1 & 0 \end{bmatrix},$$

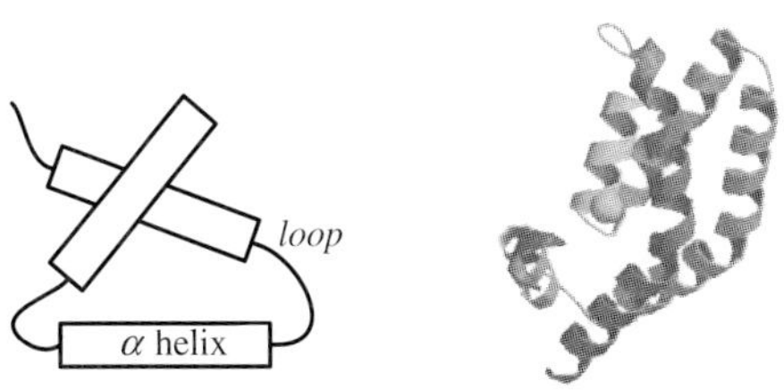

Figure 6. All α proteins: a typical structure and an example (PDB Code: 1a0b).

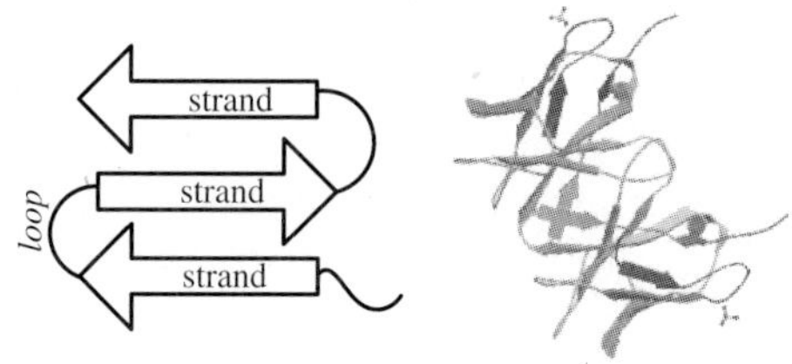

Figure 7. All β proteins: a typical structure and an example (PDB Code: 1cd8).

where –1 in the second column denotes that the corresponding β sheet has a negative direction. In Ψ, it is clear that all entries in the first column are all zeros, which denotes there is no α helix in the protein.

(3) *α and β proteins*: This class of proteins is usually split into $\alpha + \beta$ and α/β, depending on whether the α helices and β strands are segregated in the fold ($\alpha + \beta$) or mixed up and tending to alternate (α/β) . The $\alpha + \beta$ class comprises antiparallel β strands in β sheets and segregated from the α helix. An example of $\alpha + \beta$ proteins is shown in Fig. 8.

The structure matrix Ψ of the $\alpha + \beta$ proteins is

$$\Psi = \begin{bmatrix} 1 & 0 & 0 \\ 0 & 0 & 1 \\ 0 & 1 & 0 \\ 0 & 0 & 1 \\ 0 & -1 & 0 \\ 0 & 0 & 1 \\ 0 & 1 & 0 \end{bmatrix}.$$

From the above expression, we can see that in the protein molecule there are three anti-parallel β sheets.

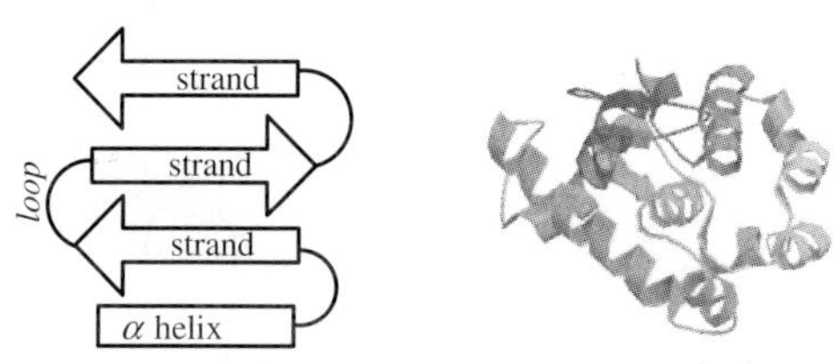

Figure 8. $\alpha + \beta$ proteins: typical structure and an example (PDB Code: 104i).

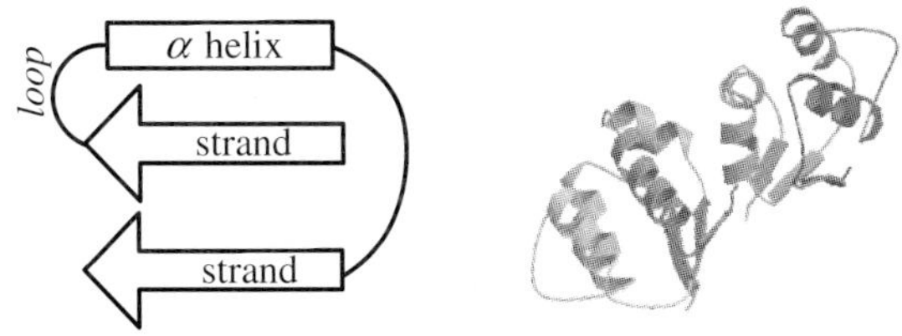

Figure 9. α/β proteins: typical structure and an example (PDB Code: 1a19).

The other class of α and β proteins, α/β proteins, are mainly composed by parallel β sheets with intervening α helices, but many also have mixed β sheets. In addition to forming a sheet in some proteins in this class, as illustrated in Fig. 9, in other proteins parallel β strands may form into a barrel structure surrounded α helices.

The corresponding structure matrix Ψ of the α/β proteins is obtained as follows:

$$\Psi = \begin{bmatrix} 0 & 1 & 0 \\ 0 & 0 & 1 \\ 1 & 0 & 0 \\ 0 & 0 & 1 \\ 0 & 1 & 0 \end{bmatrix}.$$

Comparing with $\alpha+\beta$ proteins, we can see that there are two parallel β strands and an α helix located between the two β strands in the α/β protein molecule.

(4) *Membrane proteins*: In membrane proteins, α helices are of a particular length range and have a high content of hydrophobic amino acids traversing a membrane, features that make this class readily identifiable by scanning a sequence for these hydrophobic regions.

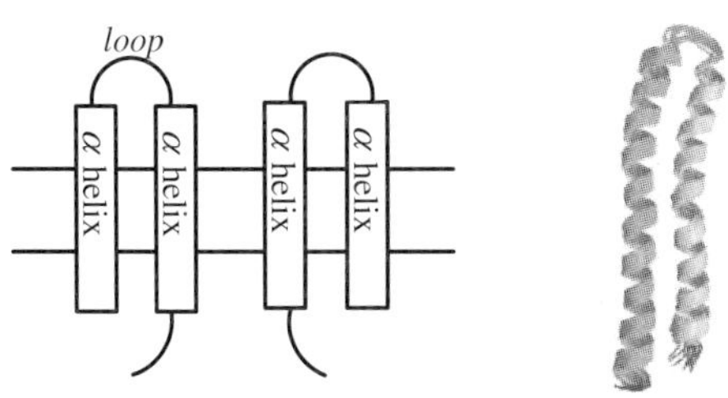

Figure 10. Membrane proteins: typical structure and an example (PDB Code: 1a91).

One model of this class of proteins is

$$(\Psi_1 + \Psi_2)\begin{bmatrix} \alpha \\ \beta \\ \text{loop} \end{bmatrix},$$

$$\text{traverse}(\alpha, \text{membrane}) = 1,$$

where

$$\Psi_1 = \begin{bmatrix} 0 & 0 & 1 \\ 1 & 0 & 0 \\ 0 & 0 & 1 \\ 1 & 0 & 0 \end{bmatrix}, \quad \Psi_2 = \begin{bmatrix} 0 & 0 & 1 \\ 1 & 0 & 0 \\ 0 & 0 & 1 \\ 1 & 0 & 0 \end{bmatrix},$$

and the function traverse(α, membrane) is a Boolean function, which determines whether there are α helices traversing a membrane. If the function is 1, we say that the protein is a membrane protein. If it is 0, there are no α helices traversing a membrane. For convenience, the function will be omitted when it takes zero value.

3.4 A UNIFIED MATHEMATICAL FRAMEWORK FROM DNA TO RNA TO PROTEINS

In nature, the process of information transfer from DNA to RNA, and then to proteins is the process of the synthesis of proteins: the cell gets a message to make certain quantity of specific proteins. The process can be stated simply as follows.

At first, a portion of the DNA double-strand helix unwinds, exposing the genes responsible for the proteins. The nucleotides move along one strand of the exposed gene and form a molecule of mRNA with the help of enzymes, where the base "*U*", instead of the base "*T*", binds with the base "*A*" in RNA. Then the copy of mRNA leaves the nucleus and enters the cytoplasm, passing through the nuclear membranes. An mRNA binds with a ribosome so that it can be decoded by the ribosome. Several specific amino acids, which are represented by triplet codes, are activated by enzymes. A tRNA molecule, which has a specific binding site for a particular amino acid at one end and an anti-codon at the other end, binds with the start codon *AUG* of mRNA. Then tRNA releases its amino acid to link with the amino acid carried by the second tRNA, binding with the subsequent codon along the mRNA strand. The first tRNA molecule now leaves the site. The process repeats itself until a certain tRNA encounters any one of the stop codons, i.e., *UAA*, *UAG*, or *UGA* on the mRNA strand. Next, all the amino acids link in sequence, forming a polypeptide chain. The polypeptide chain is released and folds into its final conformation, and finally a complete protein molecule.

So far in this chapter, we have presented statements for most stages of protein synthesizing. Based on these statements, a unified mathematical framework for this process will be presented in this section and theoretical results will be proposed.

3.4.1 Overall Information Flow and the Unified Framework

The overall information flow in the Central Dogma can be described as a process of transformation of the transfer matrices by the above mathematical models.

Using existing modeling tools, we formulate every stage of the biological information flow separately. Suppose the model of a double-strand DNA helix is

$$\Pi\left(\begin{bmatrix} A \\ T \\ G \\ C \end{bmatrix} + \begin{bmatrix} T \\ A \\ C \\ G \end{bmatrix}\right), \tag{3.20}$$

where Π is the transfer matrix of the DNA helix. When a portion of the helix is unwound, we let the transfer matrix corresponding to this portion be Π_1. Obviously, Π_1 is a submatrix partitioned matrix of Π. An mRNA will transcribe the portion of the single DNA strand. Without loss of generality, we suppose that the template single strand is the first one. Then the expression of the segment of the mRNA becomes

$$\Pi_1'\begin{bmatrix} A & U & G & C \end{bmatrix}^{\mathrm{T}}, \tag{3.21}$$

where Π_1' is the equivalent transfer matrix of Π_1, i.e., $\Pi_1 \Leftrightarrow \Pi_1'$. Next, if the triple code corresponding to the single-strand of mRNA is the start codon, *AUG*, translation begins. The process of translation goes on until the anti-codon of tRNA encounters one of the stop codons, such as *UAA*, *UAG*, or *UGA*. Here we let Π_2' be a partitioned submatrix of Π_1', and then the tRNA sequence with anti-codons is denoted by

$$\Pi_2\begin{bmatrix} A & U & G & C \end{bmatrix}^{\mathrm{T}}, \tag{3.22}$$

where Π_2 is the equivalent matrix of Π_2'. The other end of tRNA has special amino acids, which corresponding to the codons on mRNA. Then we can get a complete polypeptide chain composed of a sequence of special amino acids described as

$$\Pi_2' \begin{bmatrix} A & U & G & C \end{bmatrix}^{\mathrm{T}} = \oplus_{k=1}^{n} \Pi_{2,k}' \begin{bmatrix} A & U & G & C \end{bmatrix}^{\mathrm{T}}, \tag{3.23}$$

where $\Pi_{2,k}'$ denotes the kth partitioned submatrix, which consists of three consecutive row vectors of Π_2'. Define $n = row(\Pi_2')/3$, where $row(\cdot)$ is a function that denotes the number of rows in its parameter matrix, and thus the polypeptide can be represented as $Y = (y_1, y_2, \ldots, y_n)$. In this sequence, each amino acid y_k is denoted by

$$y_k = \Pi_{2,k}' \begin{bmatrix} A & U & G & C \end{bmatrix}^{\mathrm{T}}. \tag{3.24}$$

In biochemical processes, the polypeptide folds into a common motif, such as an α helix and a β strand, in the secondary structure of the protein. Suppose that there are $l$polypeptide chains $Y_1, Y_2, \ldots, Y_l$, which denoted by a vector

$$\mathbf{Y} = \begin{bmatrix} Y_1 & Y_2 & \cdots & Y_l \end{bmatrix}^{\mathrm{T}}.$$

Then the secondary structures of the protein are given as follows. Before the mathematical formulation, we define the unit vectors

$$e_k = [0 \quad 0 \quad \cdots \quad 0 \quad 1 \quad 0 \quad \cdots \quad 0]^{\mathrm{T}} \in R^{l+1}, \quad 1 \le k \le l+1,$$

where 1 is the kth component of the vector. As mentioned in Sec. 3.3.3, the polypeptides folding can be formulated if it obeys one of the four general cases.

(1) If $\mathbf{Y}$ is a set of α helices, namely, Y_i is α helices with $i = 1, 2, \ldots, l$, the secondary structure of a protein belongs to all α proteins, and is denoted by

$$\oplus_{i=1}^{l} \left(e_i^{\mathrm{T}} \oplus e_{l+1}^{\mathrm{T}}\right) \begin{bmatrix} \mathbf{Y} \\ \text{loop} \end{bmatrix}. \tag{3.25}$$

where *loop* denotes regions of protein chains that link all the components of Y together.

(2) If all the components of $\mathbf{Y}$ are β strands, and $\mathbf{Y}$ will fold into all β proteins, the corresponding secondary structure is

$$\oplus_{i=1}^{l} \left(e_i^{\mathrm{T}} \oplus e_{l+1}^{\mathrm{T}}\right) \begin{bmatrix} \mathbf{Y} \\ \text{loop} \end{bmatrix}. \tag{3.26}$$

We note that $\mathbf{Y}$ is different in the two cases, though the above model is formally the same as case (1).

(3) If $\mathbf{Y}$ contains not only α helices but also β strands, the secondary structure will be α and β proteins. Since $\mathbf{Y}$ is only a set of polypeptide chains and there is not an order relation among the components in $\mathbf{Y}$, without loss of generality, we define Y_1 as an α helix and the other components as β strands. Then the secondary structure of $\alpha + \beta$ class is characterized by

$$e_1^{\mathrm{T}} \oplus_{i=2}^{l} \left(e_{l+1}^{\mathrm{T}} \oplus (-1)^i e_i^{\mathrm{T}}\right) \begin{bmatrix} \mathbf{Y} \\ \text{loop} \end{bmatrix}, \tag{3.27}$$

where -1 denotes the negative direction of β strands linked with others. Then through above model, we can conclude that there are anti-parallel β strands in the β sheet. The other class of α and β proteins is the α/β class. For convenience, suppose that Y_2 is α helix and the others are all β strands. The model of the α/β class protein is

$$\oplus_{i=1}^{l} \left(e_i^{\mathrm{T}} \oplus e_{l+1}^{\mathrm{T}}\right) \begin{bmatrix} \mathbf{Y} \\ \text{loop} \end{bmatrix}. \tag{3.28}$$

(4) If $\mathbf{Y}$ folds into a membrane protein, it implies that there are only α helices in $\mathbf{Y}$. The determining function *traverse*$(\cdot,\cdot)$ is necessary, and can be written as

$$\begin{cases} \oplus_{i=1}^{l} \left(e_i^{\mathrm{T}} \oplus e_{l+1}^{\mathrm{T}}\right) \begin{bmatrix} \mathbf{Y} \\ \text{loop} \end{bmatrix}, \\ \text{traverse}(\mathbf{Y}, \text{membrane}) = 1. \end{cases} \tag{3.29}$$

By the above formulation, the process of biological information flow in the Central Dogma can be viewed as a process of transformation of transfer matrices.

3.4.2 Theoretical Results

In this section, we use the mathematical formulations to describe the complete process of biological information transferring from DNA to RNA and to proteins. The biological process is converted into transformations between transfer matrices, as shown in Fig. 11.

Consider the process of translation in the Central Dogma of Molecular Biology. From the previous discussion, the genetic information transferring from mRNA to tRNA begins with the start codon

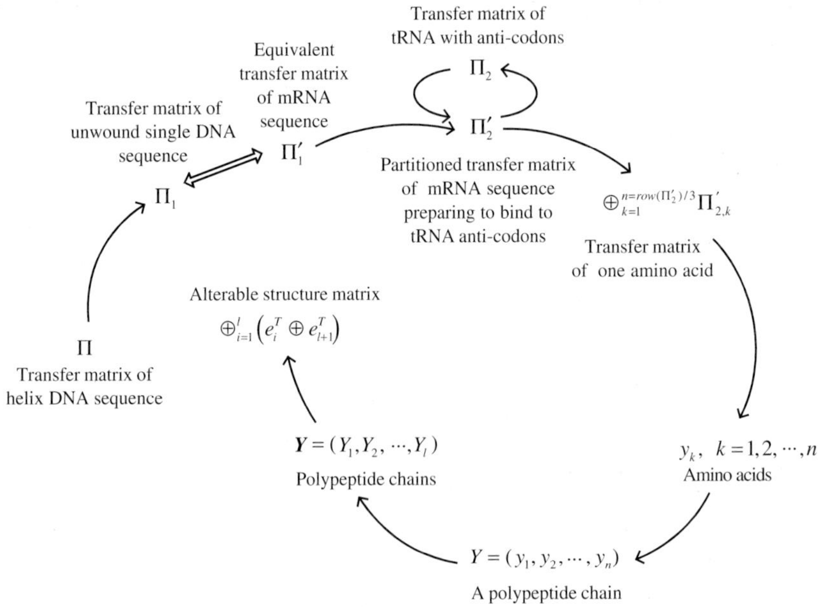

Figure 11. A mathematical formulation of biological information flow.

and finishes when encountering one of the stop codons. The conditions related to start-stop operation in translation, namely, a method to determine the transfer matrix Π_2', are given in the following propositions.

Proposition 3.18. *Based on the mRNA sequence (3.21) transcribed from an unwound DNA sequence, we denote the transfer matrix* Π_1' *by the following formulation*

$$\Pi_1' = \oplus_{j=1}^{n} \Pi_{1,j}'. \tag{3.30}$$

Then the translation begins if there exists $h \in [1, n-2]$, *such that*

$$\oplus_{j=h}^{h+2} \Pi_{1,j}' = \oplus_{i=1}^{3} e_i^{\mathrm{T}}, \tag{3.31}$$

where e_i $(i = 1, 2, 3)$ *are four-dimensional unit vectors defined as in Sec. 3.3.2. If there are several h that can satisfy (3.31), we choose the smallest one and set* $\Pi_{2,1}' = \Pi_{1,h}'$.

Similarly, the determining conditions for the stop codons of the transfer matrix Π_1' are presented.

Proposition 3.19. *The translation ends whenever the codon* $\Pi'_{2,k}$ *of the mRNA which the tRNA encounters satisfies any of the following conditions*

(1) $\Pi'_{2,k} = e_2^{\mathrm{T}} \oplus e_1^{\mathrm{T}} \oplus e_1^{\mathrm{T}}$,
(2) $\Pi'_{2,k} = e_2^{\mathrm{T}} \oplus e_1^{\mathrm{T}} \oplus e_3^{\mathrm{T}}$,
(3) $\Pi'_{2,k} = e_2^{\mathrm{T}} \oplus e_3^{\mathrm{T}} \oplus e_1^{\mathrm{T}}$.

Next, based on the protein model, a method of deciding the classifications of the protein secondary structures will be sketched. For simplicity, suppose that there are two classes of polypeptide chains: one belongs to an α helix and the other belongs to a β strand. **Y** is denoted by $[\alpha\ \beta]^{\mathrm{T}}$, and the unit vectors are defined as e'_1, e'_2, e'_3, such that $(e'_1 e'_2 e'_3) = I \in R^{3\times 3}$. The proposition below follows the model of protein's secondary structures in Sec. 3.3.3.

Proposition 3.20. *For the protein structure matrix* Ψ, *if* $\langle \Psi e'_1, \Psi e'_1 \rangle \neq 0$ *and* $\langle \Psi e'_2, \Psi e'_2 \rangle = 0$, *we say that the structures of proteins are all* α *proteins. If* $\langle \Psi e'_1, \Psi e'_1 \rangle = 0$ *and* $\langle \Psi e'_2, \Psi e'_2 \rangle \neq 0$, *they are all* β *proteins. Furthermore, with the prerequisite of* $\langle \Psi e'_i, \Psi e'_i \rangle \neq 0,\ i = 1, 2$, *we can say that the proteins have* α*/beta structures, if*

$$\langle \Psi e'_2, \Psi e'_2 \otimes (\Psi e'_2) \rangle = \langle \Psi e'_2, \Psi e'_2 \rangle,$$

where $\otimes$ *denotes the Hadamard product of matrices. Otherwise, if*

$$\langle \Psi e'_2, \Psi e'_2 \otimes (\Psi e'_2) \rangle \neq \langle \Psi e'_2, \Psi e'_2 \rangle,$$

the protein structures belong to the $\alpha + \beta$ *class.*

In conclusion, the information flow of the Central Dogma can be precisely described step by step as a transformation among matrices. The introduction of numerical matrices makes it possible to convert a biological process to a mathematical problem. Once the mathematical models are established, theoretical tools of system and computation science can be utilized in the study of proteins synthesis. Furthermore, based on the models, control can be imposed on every step in the Central Dogma, which then makes this biological process controllable. We can get desired protein molecule during the process of control.

3.5 DISCUSSIONS AND CONCLUSIONS

A mathematical formulation of the Central Dogma of Molecular Biology has been proposed in this chapter. Propositions related to the formulation have also been presented. Even though modeling a biological process is still an open problem in molecular biology, numerous studies have suggested that it is an interesting field for systems biology.

Here, the character-based expressions of DNA, RNA and protein molecules are converted into the corresponding transfer matrices, which will help us thoroughly understand the biological process by means of quantitive analysis. Specially, the method proposed here is very useful for us to judge the similarity of DNA sequences, to characterize the process of DNA hybridization and RNA self-hybridization, to decide the polarities of amino acids, and to recover an original RNA sequence from its secondary structure mathematical model . The formulation of secondary structures of proteins is useful to free us from entanglement of biochemical structures of protein molecules, and then to grasp the essential structures quickly. More significantly, this precise formulation satisfies the precondition for system or computer scientists to do researches in the area of biology. Moreover, based on this kind of models, the relevant researches about DNA computation analysis and algorithm design can be further extended.

In addition, the formulation can easily convert a character-based biological information flow problem into a numerical matrix transformation problem. This will allow researchers to build a theoretical framework for the Central Dogma, and analyze biological features when information flows from DNA to RNA, and then to proteins. Some potential problems may be further studied within the proposed mathematical framework, such as how we can move a step further towards understanding the nature of self-hybridization of a single-strand RNA sequence; how we can describe the more complicated structures of proteins, such as tertiary structure and quaternary structure; and how we can correctly control the transfer of useful biological information from nucleic acids to protein macromolecules.

Based on the formulations, we may also introduce advanced control theory into the biological process, which will be helpful in directing protein synthesis.

The goal of this chapter is to start a discussion on mathematical models of the Central Dogma of Molecular Biology. The mathematical formulation given here can be further used to understand the theoretical process by which genetic information flows from DNA to RNA and then to proteins, and can be used for recovering of self-hybridized RNA secondary structures and determining the polarities of special amino acids. The proposed modeling method demonstrates promising potential to better understand biological information transfer and control the synthesis of useful proteins.

REFERENCES

1. M. J. Zhang, "On constructing a molecular computer," in *DNA Based Computers*, ser. Discrete Mathematics and Theoretical Computer Science (DIMACS), R. Lipton and E. Baum, (Eds.), Providence, RI: Amer. Math. Soc., pp. 1–21 (1996).
2. M. Zhang, M. X. Cheng, and T. J. Tarn, "A mathematical formulation of DNA computation," *IEEE Trans. on Nanobioscience*, **5**, 32–40 (2006).
3. E. S. Lander and P. Green, "Construction of multilocus genetic linkage maps in humans," *Proceedings of the National Academy of Sciences of the United States of America*, **84**, 2363–2367 (1987).
4. G. A. Churchill, "Stochastic models for heterogeneous DNA sequences," *Bull Math. Biol.*, **51**, 79–94 (1989).
5. J. V. White, C. M. Stultz and T. F. Smith, "Protein classification by stochastic modeling and optimal filtering of amino-acid sequences," *Math. Bios*, **119**, 35–75 (1994).
6. K. Asai, S. Hayamizu and K. Onizuka, "HMM with protein structure grammar," *Proceedings of the Hawaii International conference on System Sciences*, 783–791 (1993).
7. P. Baldi, Y. Chauvin, T. Hunkapillar, and M. A. McClure, "Hidden Markov models of biological primary sequence information," *Proc. Natl. Acad. Sci.*, **91**, 1059–1063 (1994).
8. A. Krogh, M. Brown, I.S. Mian, K. Sjölander, and D. Haussler, "Hidden Markov models in computational biology: applications to protein modeling," *J. Mol. Biol.*, **235**, 1505–1531 (1994).
9. S. R. Eddy, "Multiple alignment using hidden Markov models," *ISMB*, **3**, 114–120 (1995).
10. S. R. Eddy, "Hidden Markov models," *Curr. Opin. Struct. Biol.*, **6**, 361–365 (1996).

11. G. J. Mitchison and R. M. Durbin, "Tree-based maximal likelihood substitution matrices and hidden Markov models," *J. Mol. Evol.*, **41**, 1139–1151 (1995).
12. F. H. C. Crick, "On protein synthesis," *Symp. Soc. Exp. Biol*, **XII**, 139–163 (1958).
13. F. H. C. Crick, "Central dogma of molecular biology," *Nature*, **227**, 561–563 (1970).
14. R. Durrett, *Probability Models for DNA Sequence Evolution*, Springer, New York (2002).
15. G. Alterovitz and M. F. Ramoni, *System Bioinformatics: An Engineering Case-Based Approach*, Artech House, Inc., Boston (2007).
16. M. Levitt and C. Chothia, "Structural patterns in globular proteins," *Nature*, **261**, 552–558 (1976).
17. D. R. Westhead, J. H. Parish and R. M. Twyman, *Bioinformatics*, BIOS Scientific Publishers Linited, Oxford (2002).
18. D. W. Mount, *Bioinformatics: Sequence and Genome Analysis*, (2nd Edition), Cold Spring Harbor Laboratory Press, Cold Spring Harbor, New York (2004).

Chapter Four

System Approach to Characterize Living *Drosophila* Embryos for Biomedical Investigations

Yantao Shen and Ning Xi

4.1 BACKGROUND

Genetic modification of *Drosophila* embryos has provided us insights that are not only scientifically interesting, but also biologically achievable to cure diseases. The fact that this type of research also has human health care implications was confirmed by the 1995 Nobel Prize award in Physiology or Medicine to Edward B. Lewis, Christiane Nüsslein-Volhard, and Eric F. Wieschaus for their discoveries concerning the genetic control of early embryonic development. In the study, the fruit fly, *Drosophila* melanogaster, has been used as the experimental specimen. This organism is classical in genetics. The principles found in the fruit fly, *Drosophila* melanogaster, apply to higher organisms including man.[1]

The *Drosophila* genome can also provide critical information about human genes that have homophylies with the genes of fruit fly.[2] Of particular interest to medicine, several human diseases are caused by mutations in genes that are analogous to genes found in *Drosophila*. According to the recent research,[3] human and *Drosophila* are similar. About 61% of known human disease genes have a recognizable match in the genetic code of *Drosophila*, and 50% of *Drosophila* protein sequences have mammalian analogues.[3] *Drosophila* is being used as a genetic model for several human diseases including

Nanomedicine: *A Systems Engineering Approach* edited by Mingjun Zhang & Ning Xi

www.panstanford.com
978-981-4241-36-6

neurodegenerative disorders such as Parkinson's, Huntington's, and Alzheimer's diseases.[4]

Drosophila is also being used to study mechanisms underlying immunity, diabetes, and cancer, as well as drug abuse. In addition, scientists can determine the gene function from the loss-of-function phenotype or the overexpression phenotype. Alternately, the function of human genes can be studied by inserting them into *Drosophila* via transposable elements. Recently, Corey Goodman *et al.* have been using *Drosophila* to study the wiring of the human brain and nervous system.[5] The work has led to a better understanding of how human brain develops. Consequently, the combination of the *Drosophila* genome with the well-established genetic tools in the *Drosophila* system will lead to important discoveries for human medicine, and provide a pathway to detecting, developing, treating, and eradicating diseases in humans.[6]

To implement the research on *Drosophila* genome, one of the most important approaches and tasks is the injection of substances that affect the make-up of a cell or an organism. Micro injection such as transgenes, i.e. DNA structures that often consist of a gene and a control component, results in *Drosophila* with new characteristics, since the transgene is integrated into the *Drosophila*'s own DNA. This makes it possible to determine which genes are important for the development of the organism and which organs are affected.[7] However, currently most embryo injections are conducted manually. Geneticists often require at least one year of training to become proficient at the injection skill, and such manual techniques are very time consuming. Furthermore, the average success rate of a manual injection is disappointingly low. The reason for this is that successful micro injection is greatly dependent on moderate injection forces, injection speed and accurate localization.[8] To improve the quality of micro injection, minimizing the damage in vivo caused by current injection methods, a localized, sensorized, and highly efficient micro system should be developed. Although existing micromanipulators can achieve extremely high accuracy in position, the success rate of micro injection is still at the low end due to lack of an effective micro-force sensing and feedback mechanism.

Recently, several developments on micro-force sensing methods used in the characterization of micro injection of different cells or embryos have been reported. Nelson *et al.* introduced their work on the development of a microrobotic cell manipulation system, which employs a multiaxis capacitive force sensor with a tip diameter of 5 μm to characterize the mechanical properties of Mouse Zona Pellucida.[9] In Ref. 10, the mechanical behavior of the Zebrafish embryo chorion is quantified using an unmodeled PVDF force sensor with a resolution of 14.5 μN. The attached injection pipette is 14.6 μm in radius. The measured force is from 100 μN to 800 μN, and an average penetration force of 737 μN is reported in Ref. 10. More recently, Zhang *et al.* presented a micrograting based force sensor integrated with a surface micromachined silicon-nitride probe for penetration and injection into Drosophila embryos.[11] In this work, they found an average penetration force of 52.5 μN $\pm$ 13.2% with a 30μm diameter silicon tip. Comparing with those work, in this chapter, an *in situ* PVDF (Polyvinylidene Fluoride) piezoelectric two-axis micro-force sensing tool with a resolution in the range of sub-μN is presented. The tool is integrated with a glass pipette with an ultra-sharp injecting tip of 1.685 μm in diameter and 2.65° in tip conical angle. In addition, the dynamic model of this tool is developed by incorporating the piezoelectric electro-mechanical relationship into the Bernoulli-Euler beam equation. Using this modeled force sensing tool, the effects of much smaller forces in a minimally invasive area during micro injection of *Drosophila* embryos are explored, that is, the sharp sensing tool can effectively minimize the damage to the embryos as well as further ensure the studies of developmental biology and genetics. Moreover, the micro-force sensing tool can be conveniently integrated into a precision micromanipulator system and is easy to both manufacture and assemble. Without using the complex MEMS technology, the developed sensing system is economical and can be widely used in characterization of mechanical properties of cell or embryos in research labs, and also has the potential for commercialization.

Furthermore, based on the developed PVDF micro-force sensing tool and the event-synchronization[13] of the video and micro-force, a networked human/robot cooperative biomanipulation system is developed. The system can reach heterogenous integration of human and robot functions for achieving reliable, accurate, and efficient biomanipulation or micro injection of cell/embryos. This system can also be applied to single or multiple remote work-cells through LAN or Internet.

This chapter discusses the system approach to measure injection force behavior as well as to characterize mechanical properties of living *Drosophila* embryos using a well modeled *in situ* PVDF (Polyvinylidene Fluoride) piezoelectric micro-force sensing tool with a resolution in the range of sub-μN. The development of such a sensorized biomanipulation tool based on the highly sensitive PVDF film and the minimally invasive pipette injector plays an important role in realizing the system approach. Using this tool, close monitoring of the micro injection and other biomanipulation forces acting on the embryos during the injecting process becomes a reality. In addition, a networked microrobotic biomanipulation platform integrating the micro-force sensing tool can greatly advance operations in micro injection of living *Drosophila* embryos. Several experimental results have clearly demonstrated the effectiveness of the system approach. Those results include the quantitative relationships between the applied force and embryonic structural deformation of embryos in the different stages of embryogenesis, as well as the penetration force behaviors of micro injection.

This chapter is organized as follows. Section 4.2 reviews the modeling, the design, and calibration of both the signal processing unit and the PVDF (Polyvinylidene Fluoride) piezoelectric micro-force sensing tool. The event-based human/robot biomanipulation system is introduced in Sec. 4.3. Section 4.4 demonstrates the experimental results of the system approach to characterize the mechanical properties of living *Drosophila* embryos in different stages. The penetration force behaviors of micro injection are also shown in this section. Finally, the chapter is concluded in Sec. 4.5.

4.2 MICRO-FORCE SENSING TOOL FOR CHARACTERIZATION

4.2.1 Design and Modeling of 2-D Micro-Force Sensing Tool

For effective biomanipulation and micro injection, a self-decoupling two-axis (2-D) PVDF force sensing tool has been developed as shown in Fig. 1.

The 2-D sensing tool is designed based on a serial connecting structure. In each direction, a PVDF composite beam is constructed. Notice that the composite beam is basically a two layer structure whose piezoelectric layer acts as a sensing device bonded to a support beam layer made from polyester. It can also be seen that this structure provides a decoupled force measurement in the Y and Z

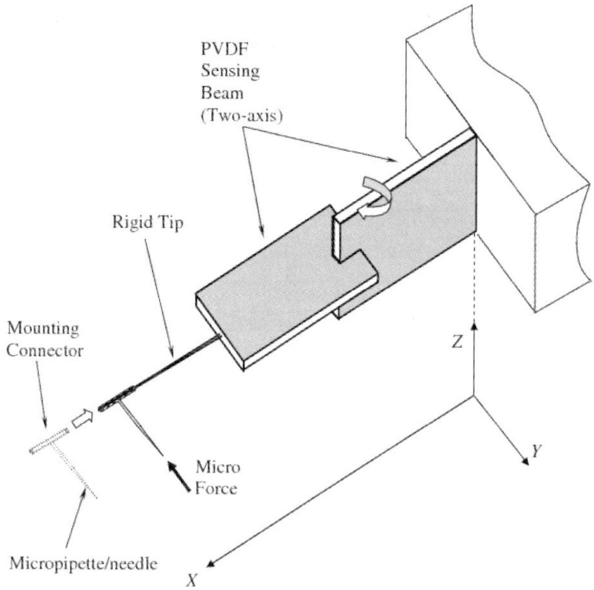

Figure 1. Illustration of the 2-D PVDF micro-force-sensing tool.

axis, which is due to the orthogonal configuration of the two PVDF composite beams', as well as the strong shear forces along the Y and Z axis existing between the two PVDF beams. At the free end of the sensing tool, a rigid steel tip is attached, and a modified micropipette or needle is assembled to the end of the rigid tip.

Notice that, following the configuration shown in Fig. 1, in this work, for the micro injection and mechanical characterization of living *Drosophila* embryos, the only 1-D (along Y axis) micro injection force is measured using the sensing tool.

Subject to the super flexibility of the cantilever based micro-force sensor, development of a dynamic sensing model for achieving accurate micro-force measurement becomes necessary.[14] The dynamic force modeling along one of the force axes (Y axis) is described in detail as follows. To conveniently model the Y axis force sensing, an equivalent and simplified 1-D structure is shown in Fig. 2. In this figure, the deformation of the sensor beam is caused by the applied force acting along the micropipette at the end of the rigid tip.

Following the geometric characteristic of the PVDF sensing layer, since the beam is much wider and longer than the thickness, the strain s_y along the width of the beam can be assumed to

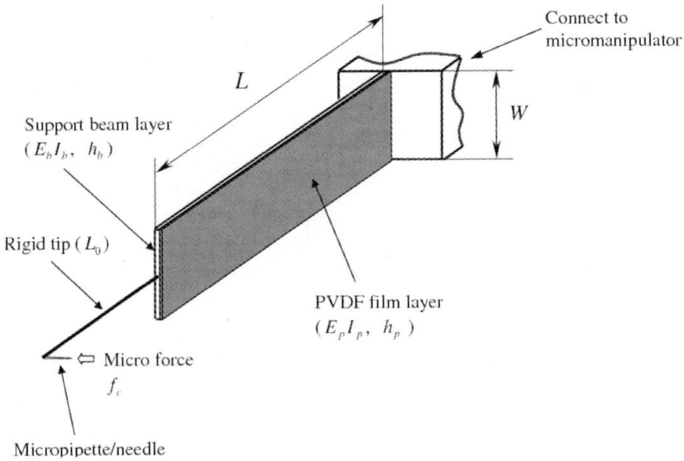

Figure 2. Illustration of the equivalent *Y* axis PVDF force sensing.

be zero.[15] With the above descriptions, based on the piezoelectric transverse effect, the unit piezoelectric equation is: (without considering the pyroelectric effects due to use at stable temperature environment).[12,16]

$$D_3(r,t) = d_{31}\sigma_s(r,t) \tag{4.1}$$

where $D_3(r,t)$ is the normal electric displacement of PVDF film. d_{31} is the transverse piezoelectric coefficient. $\sigma_s(r,t)$ denotes the unit stress of the PVDF sensing layer along the beam length, and $0 \le r \le L$.

The surface area polarization gives a charge $Q_s(t)$ across the PVDF sensing layer surface area S_A $(L \times W)$:

$$\begin{aligned} Q_s(t) &= \int D_3(r,t)dS_A \\ &= \iint_{S_A} D_3(r,t)dydr. \end{aligned} \tag{4.2}$$

Using the mechanics of materials for a cantilever beam,[17] as shown in Fig. 2, the unit stress on the surface of the PVDF sensing layer can be obtained if the external load $f_c(t)$ acts along the micropipette:

$$\sigma_s(r,t) = -cE_p \frac{\partial^2 \omega_s(r,t)}{\partial r^2} \tag{4.3}$$

According to Fig. 2, notice that since two-layer composite beam is considered (the effect of very thin and low Young's modulus electrode layers at the top and bottom surfaces of PVDF layer is omitted). c is the distance between the middle of the PVDF sensing layer and the neutral axis c_n of the composite beam. $\omega_s(r,t)$ is the elastic deflection of the flexible composite beam caused by the micro force $f_c(t)$ along the micropipette. E_p is the Young's modulus of the PVDF sensing layer.

Since the bending charge is the same along the width of PVDF ($s_y = 0$), Eq. (4.2) is rewritten as follows:

$$\begin{aligned} Q_s(t) &= \int_0^L d_{31}\sigma_s(r,t)W dr \\ &= -cE_p d_{31} W \frac{\partial \omega_s(r,t)}{\partial r}\bigg|_0^L. \end{aligned} \tag{4.4}$$

A simplified and effective equivalent circuit model of a capacitor C_P can be used to represent the model of the PVDF sensing layer. And the output voltage $V_s(t)$ of the PVDF sensing layer caused by the micro force can be described by

$$V_s(t) = \frac{Q_s(t)}{C_P}. \tag{4.5}$$

By Laplace transformation, the electrical open-circuit transfer function of the sensing layer is given by

$$V_s(s) = \frac{Q_s(s)}{C_P}. \tag{4.6}$$

To find the dynamic relationship between the sensing output V_s and the micro force f_c acting along the micropipette, a dynamic model of the flexible PVDF sensing beam based on the partial differential equation (PDE) is described firstly. Here the PDE describing the elastic deflection of the flexible composite PVDF sensing beam is a Bernoulli-Euler equation with additional terms due to the external force and moment at the free end of sensing beam as follows:

$$\begin{aligned} & EI\frac{\partial^4 \omega_s(r,t)}{\partial r^4} + \rho A \frac{\partial^2 \omega_s(r,t)}{\partial t^2} \\ &= f_c(t)\delta(r-L) + f_c(t)L_0 \frac{\partial(\delta(r-0)-\delta(r-L))}{\partial r} \end{aligned} \tag{4.7}$$

where E, I, L, L_0 and ρ represent the Young's modulus, inertia moment, length of beam, length of the rigid tip, and linear mass density of the composite beam respectively. We assume that $EI = E_b I_b + E_p I_p$ is the flexural rigidity of the composite sensing beam and $\rho A = \rho_b W h_b + \rho_p W h_p$ is mass per unit length of the sensing beam. It is important to note that E_b and I_b represent the Young's modulus and inertia moment of the polyester layer, and I_P represents the inertia moment of the PVDF sensing layer. ρ_b and h_b represent the mass per unit density and the thickness of the polyester layer, ρ_p and h_p represent the mass per unit density and the thickness of the PVDF sensing layer. $\delta(\cdot)$ denotes the Dirac delta function.

The boundary conditions for the above equation are:

$$\omega_s(0,t) = 0 \tag{4.8}$$

$$EI\frac{\partial \omega_s(0,t)}{\partial r} = 0 \tag{4.9}$$

$$EI\frac{\partial^2 \omega_s(L,t)}{\partial r^2} = f_c L_0 \tag{4.10}$$

$$EI\frac{\partial^3 \omega_s(L,t)}{\partial r^3} = f_c. \tag{4.11}$$

By using the modal analysis method,[18] and assuming that the deformation of the beam has infinite shape modes, the deflection $\omega_s(r,t)$ can be expressed as an infinite series in the following form:

$$\omega_s(r,t) = \sum_{i=1}^{\infty} \Phi_i(r) q_{si}(t) \tag{4.12}$$

where $\Phi_i(r)$ are the eigenfunction satisfying the ordinary differential equation and $q_{si}(t)$ are the modal displacements caused by the micro force.

Using the Lagrange's equation of motion and orthogonality conditions,[18] the differential equation corresponding to each shape mode of the composite beam of the sensing tool is given to be

$$EI\alpha_i^4 q_{si}(t) + \rho A \ddot{q}_{si}(t) = f_c(t)\Phi_i(L) + f_c(t)L_0[\Phi_i'(L) - \Phi_i'(0)] \tag{4.13}$$

where a prime indicates the derivative with respect to position and a dot denotes the derivative with respect to time. α_i are the infinite set of eigenvalues. The natural frequencies ω_i of the composite sensing beam appoximately correspond to the α_i by

$$\omega_i = \alpha_i^2 \sqrt{\frac{EI}{\rho A}}. \tag{4.14}$$

Then by the Laplace transformation of the above equation (4.13), the dynamic relationship between the modal displacements $q_{si}(s)$ and the external micro force is given as

$$q_{si}(s) = \frac{f_c(s)(\Phi_i(L) + L_0[\Phi'(L) - \Phi_i'(0)])}{\rho A(s^2 + \omega_i^2)} \tag{4.15}$$

Recalling Eqs. (4.4) and (4.6), since $\omega_s(r,s) = \sum_{i=1}^{\infty} \Phi_i(r) q_{si}(s)$, by Laplace transform of Eq. (4.4), $Q_s(s)$ can be represented as

$$\begin{aligned} Q_s(s) &= -cE_p d_{31} W \omega_s'(r,s)|_0^L \\ &= -cE_p d_{31} W \sum_{i=1}^{\infty} [\Phi_i'(L) - \Phi_i'(0)] q_{si}(s). \end{aligned} \tag{4.16}$$

Substituting Eq. (4.16) into Eq. (4.6), it has

$$V_s(s) = C_s \sum_{i=1}^{\infty} [\Phi_i'(L) - \Phi_i'(0)] q_{si}(s) \tag{4.17}$$

where $C_s = \frac{-cE_p d_{31} W}{C_p}$.

Subsequently, by combining Eqs. (4.15) and (4.17). The dynamic sensing model that denotes the dynamic relationship between the output voltage V_s of the PVDF sensing layer and the external micro force f_c along the Y axis (micro pipette) is given by

$$\frac{V_s(s)}{f_c(s)} = C_s \sum_{i=1}^{\infty} \left\{ \frac{[\Phi_i'(L) - \Phi_i'(0)]\Phi_i(L)}{\rho A(s^2 + \omega_i^2)} + \frac{L_0[\Phi_i'(L) - \Phi_i'(0)]^2}{\rho A(s^2 + \omega_i^2)} \right\}. \tag{4.18}$$

To achieve the sensing voltage V_s, the PVDF sensing layer is interfaced with the PCI-DAS4020/12 analog/digital & input/output board (Measurement & Computing Co.) in the PC through a developed instrumental amplifier circuit illustrated in Fig. 3. The circuit was constructed using the AD549 ultralow input bias current operational amplifier (Analog Devices Co.) with high input impedance

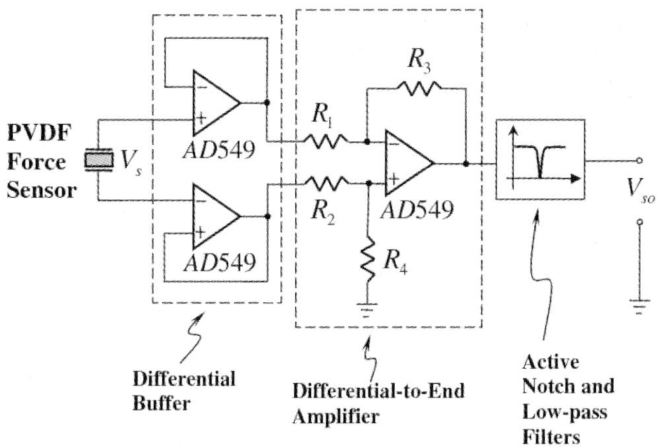

Figure 3. Schematics of the developed electronic interface circuit.

$R_{in} = 10^{13}\,\Omega$ and low bias current 150 fA. Notice that $R_3 = R_4$ and $R_1 = R_2$. The amplifier circuit is used to buffer the open circuit voltage V_s of the PVDF sensing layer, and can convert the high impedance signal generated by the PVDF sensing layer to a low impedance voltage suitable for convenient coupling to the PCI-DAS4020/12 acquisition board. The circuit output V_{so} is an amplified and filtered approximation of the voltage V_s, and can be sampled by the PCI-DAS4020/12 board. The maximum sampling frequency of PCI-DAS4020/12 is 20 MHz with 12-bit AD resolution. The loop time of the force sensing and acquisition is about 60 μs. The transfer function between V_{so} and V_s can be represented as:

$$\frac{V_{so}(s)}{V_s(s)} = \underbrace{\frac{R_3}{R_1}\frac{sR_{in}C_P}{1 + sR_{in}C_p}}_{C_b}. \tag{4.19}$$

To further remove the 60 Hz noise from the data acquisition system, a zero phase notch filter is added in the data collection system.

Finally, by considering the whole sensing system, the global transfer function is found as follows.

$$\frac{V_s(s)}{f_c(s)} = C_b C_s \sum_{i=1}^{\infty} \left\{ \frac{[\Phi_i'(L) - \Phi_i'(0)]\Phi_i(L)}{\rho A(s^2 + \omega_i^2)} + \frac{L_0[\Phi_i'(L) - \Phi_i'(0)]^2}{\rho A(s^2 + \omega_i^2)} \right\}. \tag{4.20}$$

Based on this dynamic equation, the micro force $f_c(t)$ by measuring the output voltage $V_{so}(t)$ from the sensing PVDF is obtained when the initial values $f_c(t_0)$ and $V_{so}(t_0)$ are known.

4.2.2 Calibration and Sensing Performance

As shown in Fig. 4, the developed two-axis PVDF sensing tool for micro injection is demonstrated. In the sensing tool, a fine tipped micropipette with a tip diameter of 1.685 μm and a tip conical angle of 2.65° is attached to the end of the rigid steel tip. The micropipette is made by Omega Dot capillary tubing 30-30-1 of Frederick Haer & Co Inc. The Omega Dot capillary tubing is a unique product which revolutionized micropipette filling. The tiny fiber (100 μm) extruded into the capillary lumen, promotes capillary action and eliminates the need for boiling or vacuum filling of the tip. This tubing is good for single cell recording and micro injection.[19] The multi-layer structure of PVDF composite beam is also shown in a 50X zoom-in microscope picture. There are four layers in the beam. Two thin layers are identical 6 μm thick silver/urethane ink electrode layers with a low Young's modulus of 100 MPa. The other two layers are the 30 μm thick PVDF layer and 125 μm thick Polyester layer, respectively, with Young's moduli of 3×10^9 Pa and 3.8×10^9 Pa. Since the

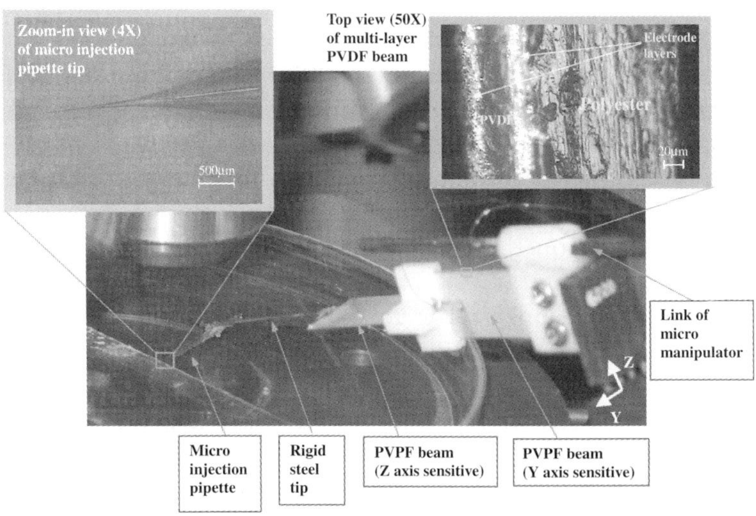

Figure 4. PVDF force sensing tool for biomanipulation.

Young's modulus of the PVDF or Polyester layer is 30 times more than the electrode layers, it is reasonable to neglect the effect of the electrode layers in the model mentioned in page 123 of Sec. 4.2.1. The developed sensing tool has the following dimensions and related parameters:

$$L = 0.018348\,\text{m}; \quad W = 0.010521\,\text{m}; \quad L_0 = 0.05401\,\text{m};$$
$$C_P = 0.88 \times 10^{-9}\text{F}; \quad d_{31} = 23 \times 10^{-12}\text{C/N}; \quad c = 102.5 \times 10^{-6}\,\text{m};$$
$$h_p = 30 \times 10^{-6}\,\text{m}; \quad h_b = 125 \times 10^{-6}\,\text{m};$$
$$E_p = 3 \times 10^9\,\text{N/m}^2; \quad E_b = 3.8 \times 10^9\,\text{N/m}^2;$$
$$\rho_p = 1.78 \times 10^3\,\text{Kg/m}^3; \quad \rho_b = 1.39 \times 10^3\,\text{Kg/m}^3;$$

the amplified gain of the circuit is $K_a = \frac{R_3}{R_1} = 10$. Two sensing relationships are calibrated. That is, the relationship between the sensing output and the deflection of the sensing tip, and the relationship between the micro force and the deflection of the sensing tip. In these calibrations, using a precisely calibrated Mitutoyo 100X microscope (with 50X objective and 2X zoom), and a Sony CCD vision system, the deflections of the sensing tip due to the applied loads are accurately measured with a high resolution of 0.2106 μm/pixels in horizontal axis and 0.2666 μm/pixels in vertical axis on the image plane of the vision system. During the deflection, the related sensing output voltage is recorded. In addition, the micro force information corresponding to the deflection is also obtained assisted by an optically calibrated PVDF high accuracy micro-force sensor. This calibrated force sensor has been successfully used in our previous work.[20] Both the sensing output–deflection and micro force–deflection curves are shown in Fig. 5. In this figure, the star marks (*) represent the calibrated relationship between the sensing output and the deflection of the tip. The blue solid line through the asterisks is its simulation result based on the first shape mode equation derived from Eqs. (4.12), (4.17), and (4.19). Similarly, the triangular marks denote the calibrated relationship between the micro force and the deflection of the tip. The red solid line through the triangular marks is the simulation result of the micro force vs deflection based on the first shape mode equation obtained from Eqs. (4.12) and (4.15). Two calibration results verify the effectiveness of the developed sensing model. In addition, the dynamic force sensing performance is tested. In this test, when the tip deflection due to the sensing tip quickly move in one step to impact a glass substrate is accurately measured,

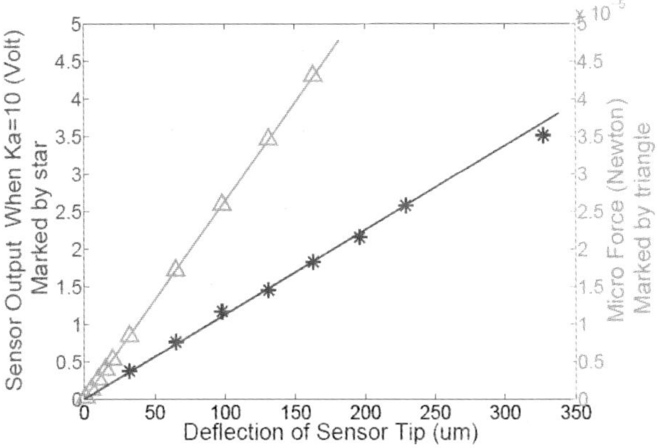

Figure 5. Calibrations: sensing output voltage vs deflection and micro force vs deflection.

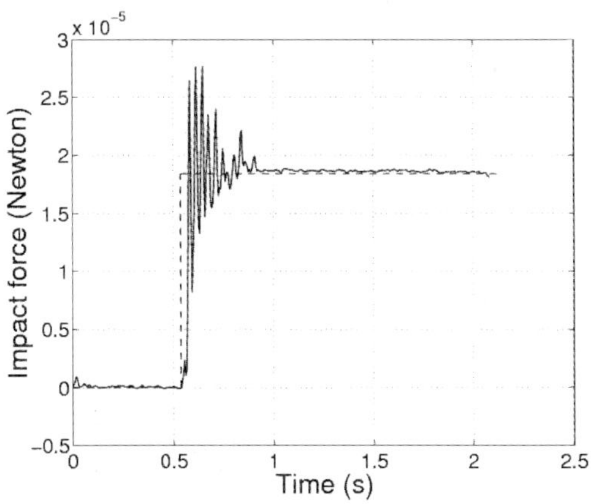

Figure 6. Dynamic force sensing test: step impact and force hold.

then a step force of 18.4 μN can be calibrated using the strain energy method[17,21] of sensing beam. Figure 6 shows this calibrated step force using the dash line. In this figure, the solid line shows an impact force obtained by the sensing tool using Eq. (4.20) in the first shape mode. The result demonstrates the dynamic sensing performance as well as the effectiveness of the developed sensing model. By calibration, the sensitivity of the PVDF sensing tool was estimated to be 40.6 mV/μN $\pm$ 6.5%, the resolution is in the range of sub-μN, which varies with the noise level and the amplified gain. The accuracy is $\pm 6.64\%$ in a full scale (the tip deflection within 250 μm). The spring constant k was estimated to be $0.264\,\text{N/m} \pm 7.1\%$. The linearity was obtained as 6.63% in a full scale. All results verify the performance of the force sensing capability of the sensing tool along the Y axis.

4.3 NETWORKED HUMAN/ROBOT BIOMANIPULATION — A SYSTEM APPROACH

Micro injection requires operations to be performed under a microscope. The visual information on the position of the micro injector and the surrounding workspace can be updated through real-time video feedback to the human operator. By visually observing the injection operations, a human can plan and correct the next operation so as to achieve a reliable and robust injection.[22,23] In this work, both the micro-force and vision as essential action references for an integrated human/robot cooperative system are used. That is, the developed piezoelectric PVDF micro-force sensing tool with a resolution in the range of sub-μN can be integrated with a 3-DOF micromanipulator. The Mitutoyo FS60 optical microscope and a Sony SSC-DC50A CCD Color Video Camera can capture the micro injection process in 30 fps, and feed back the visual information. Notice that, in this system, both the visual and micro-force data streams are transferred or fed back via the network. To ensure the synchronization of two data streams, an event-synchronization method proposed in Ref. 13 is employed. The developed network-based biomanipulation workcell in the Robotics and Automation Laboratory at Michigan State University is shown in Fig. 7.

It consists of a 3-DOF micromanipulator (SIGNATONE Computer Aided Probe Station, a step resolution of 32 nm), a 3-DOF platform, a Mitutoyo FS60 optical microscope and a Sony SSC-DC50A CCD color video camera. The 3-D platform can be controlled to convey the embryos or cells to the working area observed by the microscope. To reduce vibrations, an active vibration isolated table is used during biomanipulation/micro injection. The joystick used in the human/robot cooperative biomanipulation system is a 3-DOF Microsoft SideWinder force feedback joystick. The 3-D movement of the joystick is sent to the 3-D micromanipulator, and the micro forces felt by the PVDF force sensing tool at the front end of micromanipulator are then fed back to operators through the joystick.

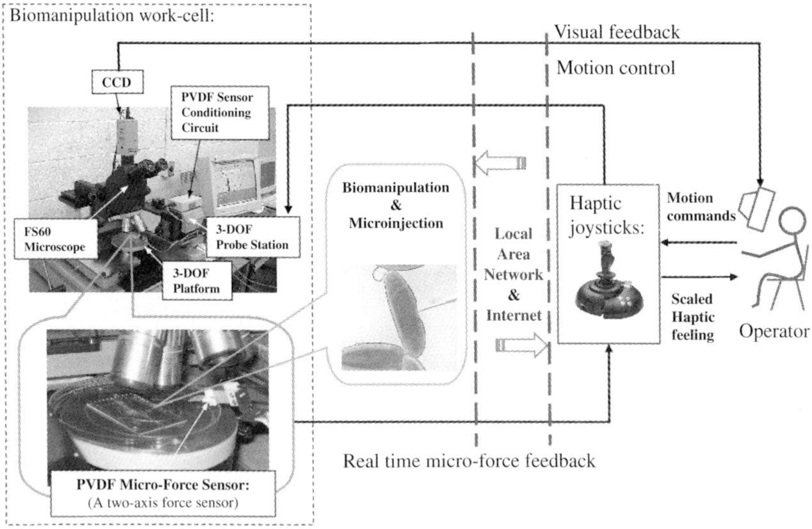

Figure 7. Networked human/robot biomanipulation work-cell at MSU.

4.4 FORCE MEASUREMENT AND MECHANICAL CHARACTERIZATION OF LIVING *DROSOPHILA* EMBRYOS USING SYSTEM APPROACH

4.4.1 *Physical Appearance and Embryonic Stages of Drosophila Embryo*

The *Drosophila* embryo is bilaterally symmetrical, and distinctions between the dorsal and ventral surfaces are indicated by differences in curvature. The dorsal side is flattened while the ventral side is somewhat convex. The dimensions of the embryos are variable, an average length is 500 μm, and the diameter is about 180 μm. In addition, 17 embryonic stages in the living *Drosophila* embryo have been subdivided for a general reference in the embryo research.[24] These stages are defined by prominent features that are easily distinguishable in the living *Drosophila* embryo. For a reference in the experimental section, Table 1 shows the time and events of the 17 stages at room temperature.[24]

4.4.2 *Embryo Preparation*

The *Drosophila* embryos used in the experiments are prepared following a standard procedure. In this work, fresh *Drosophila* embryos

Table 1. Time table of embryogenesis. From Ref. 24.

Stage	Time	Developmental Events
1–4	0:00–2:10 h	Cleavage
5	2:10–2:50 h	Blastoderm
6–7	2:50–3:10 h	Gastrulation
8–11	3:10–7:20 h	Germ band elongation
12–13	7:20–10:20 h	Germ band retraction
14–15	10:20–13:00 h	Head involution and dorsal closure
16–17	13:00–22:00 h	Differentiation

in the early stages (usually stages 1–4, 0–130 minutes after hatching) are collected into a basket, then dechlorinated in 100% bleach for 2 minutes in order to completely remove the outer tough opaque chorion. After the dechlo-rination, the embryos are rinsed thoroughly in 20°C water for 2 minutes to remove all traces of bleach. A soft brush is then used to uniformly transfer embryos to a glass slide with double sticking tape. Finally, embryos on the tape were covered with Halocarbon 700 oil and ready for injection. Notice that, the adhesive force of the tape is strong enough to hold the embryos in place for injection. Each time approximate 50∼100 living embryos were prepared. After the chorion removal mentioned above, those living embryos are still enclosed by an inner homogeneous vitelline membrane, which will be broken through by the sharp pipette during micro injection. Figure 8 shows the 10X microscope image of a sample of fresh *Drosophila* embryos prepared in Dr. Arnosti's Lab of Department of Biochemistry and Molecular Biology at Michigan State University.

4.4.3 Micro Injection Configuration

Using the developed networked human/robot biomanipulation system, the micro injection configuration is shown in Fig. 9.

In this configuration, the embryos are held in place on the sticking tape of a glass slide, which is placed on the 3-DOF platform. The Y axis of force sensing tool is horizontally aligned. This ensures that only a normal injection force is applied along the Y axis of the PVDF force sensing tool. Since the prepared embryos are distributed on

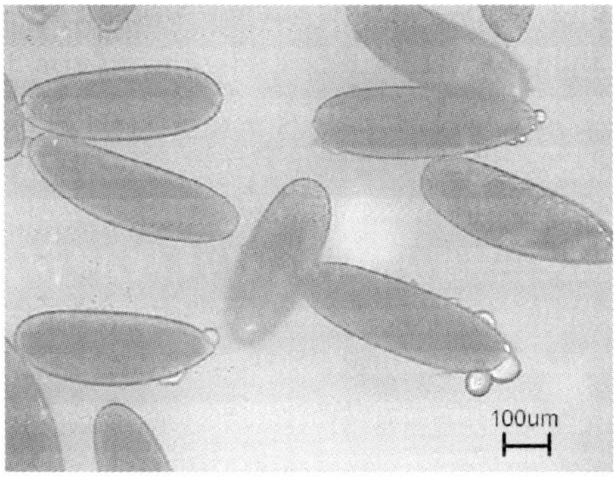

Figure 8. The prepared *Drosophila* embryos for micro injection.

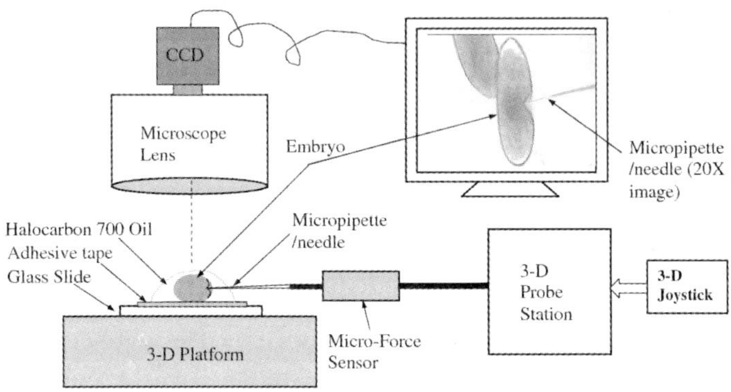

Figure 9. Experimental configuration of micro injection of Drosophila embryos.

the taped glass slide with diverse orientations, the injection angle of the micro pipette on different *Drosophila* embryos, that is, the angle between the incidence direction of the pipette injector and the normal direction of the embryo membrane surface approached, may be different. Based on this configuration, all force measurements and the characterization of mechanical properties of the living embryos were implemented at a stable room temperature of 28°C.

4.4.4 Penetration Force Profile of Micro Injection of Living Embryos

To measure the penetration force profile during micro injection of embryo, the general operation procedures are as follows: first, the

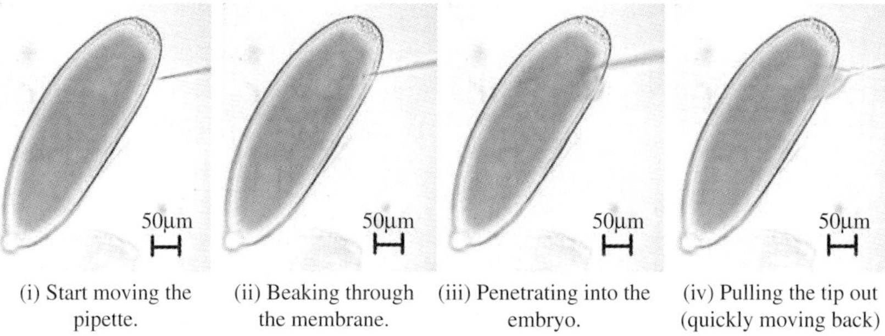

Figure 10. The sequence of penetrating the fresh embryo through one membrane.

operator drives the joystick to move the micropipette tip along the Y axis to approach the embryo for injection. Once the tip penetrates the first membrane and goes inside of the embryo or continues to penetrate the second membrane, in order to stop the pipette tip, the operator quickly moves the micropipette tip back along the *Y* axis.

In Fig. 10, a penetration sequence of the fresh embryo is shown. In this injection, when the pipette tip penetrated the membrane of the embryo, it was moved back. As shown in Fig. 11, the force profile was measured by the developed PVDF sensing tool. The penetration angle is 36.08°, and the speed is about 14.254 μm/s. Notice that,

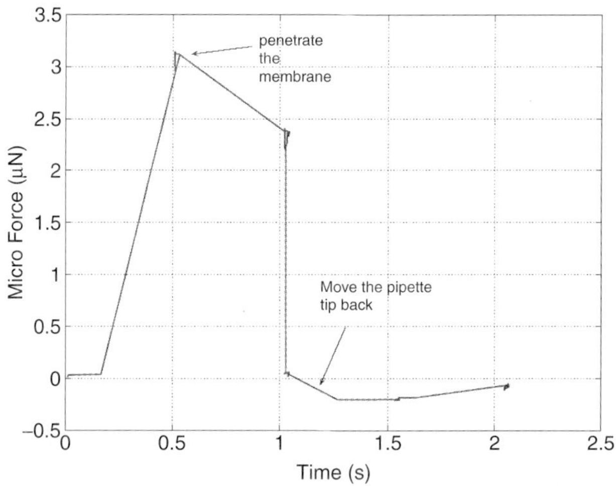

Figure 11. Force profile of penetrating one membrane of fresh embryo.

when the pipette tip is moved back, since the small reverse forces from the edges of broken membrane, the fluid frictions, and the inertial effect, a negative force is shown in the force curve. The measured penetration force is approximately 3.1 μN. The embryo used is in Stage 5 (Blastoderm Stage).

The sequence in Fig. 12 demonstrates penetration of the whole embryo, which means two membranes at the dorsal and ventral sides are punctured. The penetration angle is 17.3°, and the speed is about 58.5 μm/s. Since the first penetration needs to overcome both the internal pressure of closed embryo and the membrane stress, the maximum puncturing force of 7.44 μN appears in the penetration of the first membrane. The penetration force of the second membrane is 3.61 μN. Figure 13 shows the force profile detected by the developed PVDF micro-force sensing tool. Notice that, the embryo used is in Stage 10 (Germ Band Elongation Stage). From the force profile, the penetration force and time of living embryo in Stage 10 are greater and longer than the fresh embryo in Stage 5 demonstrated above. It is reasonable because the plasticity and fiber density of the embryo membrane are both increased when the living embryo becomes gradually mature.[24]

In addition, Fig. 14 shows the complex force profile including 5 contact-release operations and then the penetration of the embryo. Notice that, the speeds of the 5 contacts and 5 releases are diverse. Moreover, when the tip penetrates the first membrane into the embryo, the tip continues to impact the second membrane without piercing it, and then the tip is pulled out. These force behaviors are clearly demonstrated in Fig. 14. The embryo chosen is in Stage 11–12 (Stage of Germ Band Elongation and Retraction). The penetration

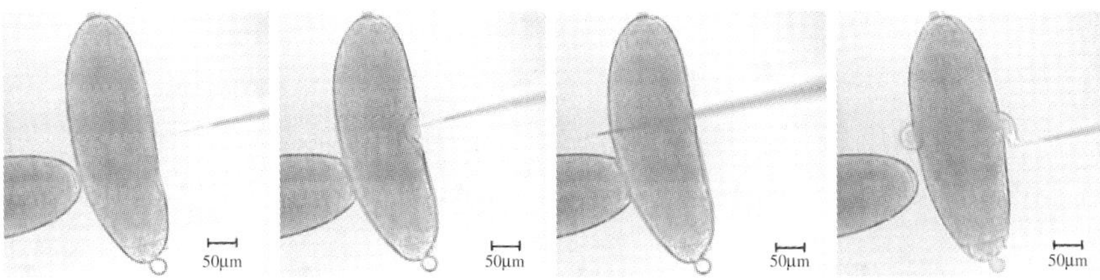

(i) Start moving the pipette. (ii) Beaking through the first membrane. (iii) Penetrating two membranes of embryo. (iv) Pulling the tip out (quickly moving back)

Figure 12. The sequence of penetrating two membranes of living embryo.

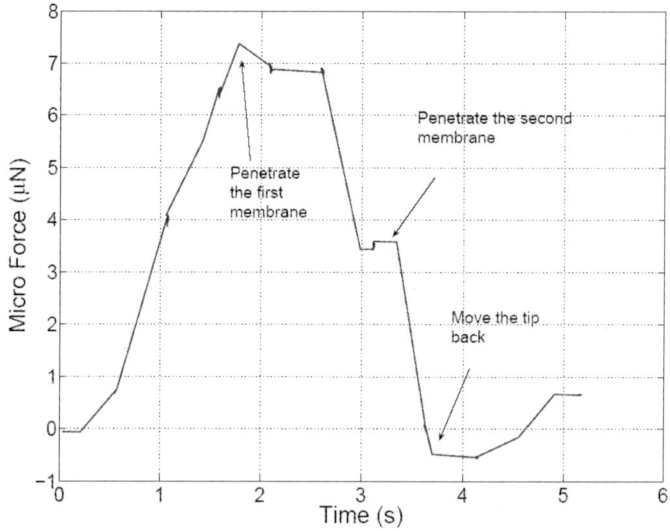

Figure 13. Force profile of penetrating two membranes of living embryo.

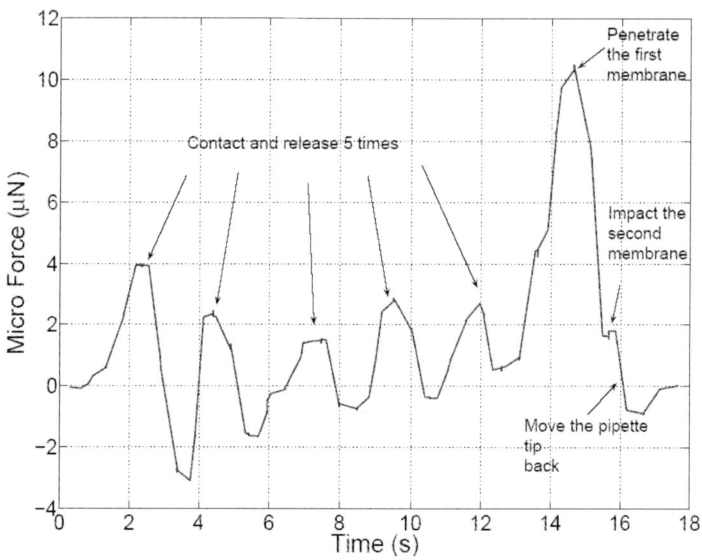

Figure 14. Force profile of complex operation of living embryo.

angle is 43.1°, and the speed is about 44.4 μm/s. The puncturing force is approximately 10.5 μN.

4.4.5 Penetration Forces in Two Directions of Living Embryo Body

The quantitative relationship of penetration forces along two directions of the embryo body is also investigated. Two directions include the anterior-posterior and dorsal-ventral directions, which are illustrated in Fig. 15. Using the prepared embryos between Stage 1 and 5, the penetration forces along two directions are investigated. The preliminary results are shown in Fig. 16. It can be seen that the penetration force along the dorsal-ventral direction is about 3.6 times smaller than the penetrating force along the anterior-posterior direction, in the case of Stage 1–5 embryos are measured.

4.4.6 Quantitative Investigation on Mechanical Properties of Living Embryos at the Different Stages

Using the off-line membrane deformation measurement, which is based on the microscope images with a maximum resolution of 0.702 μm/pixel, and the corresponding micro-force measurement,

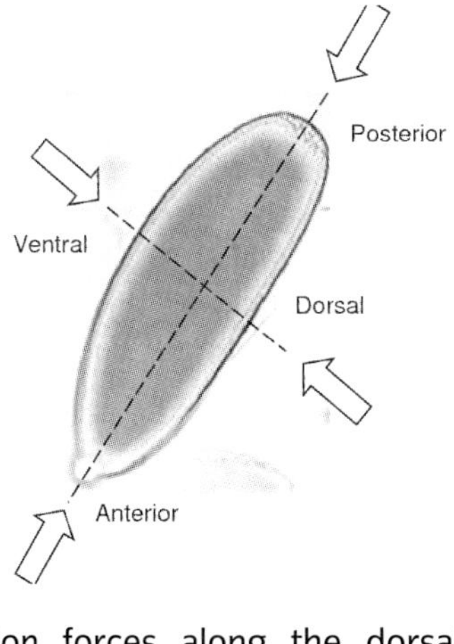

Figure 15. Penetration forces along the dorsal-ventral and anterior-posterior directions.

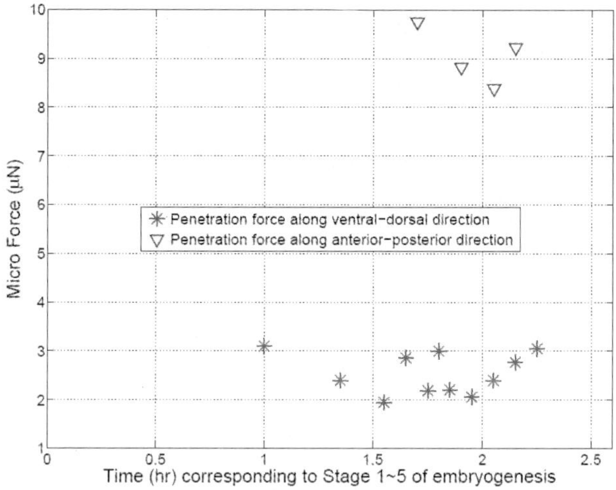

Figure 16. Penetration forces along the dorsal-ventral and anterior-posterior directions.

characterizing and quantifying the differences on mechanical properties of the living *Drosophila* embryos in different stages of embryogenesis are investigated. As a result, the quantitative relationship between the force and the embryomic body deformation is found and established. As shown in Fig. 17, several relationships on force deformation of diverse embryos in the different stages are demonstrated.

It can be seen that bodies of living embryo in the early stages have relatively smaller deformation with respect to the bodies of late stage embryos before being penetrated. In Fig. 17, from the early stages to the late stages, the stiffness of the embryo reduces gradually. This implies that the internal pressure gradually decays, and the plasticity of membrane of living *Drosophila* embryo gradually increase when the embryo becomes mature. In addition, the results in the same stage are repeatable by multiple experiments. These quantitative results will improve the quality of micro injection for the study of the *Drosophila* genetic and developmental projects currently in existence.

4.5 CONCLUSION

This chapter presents the system approach to characterize the force behavior and mechanical properties of living *Drosophila* embryos

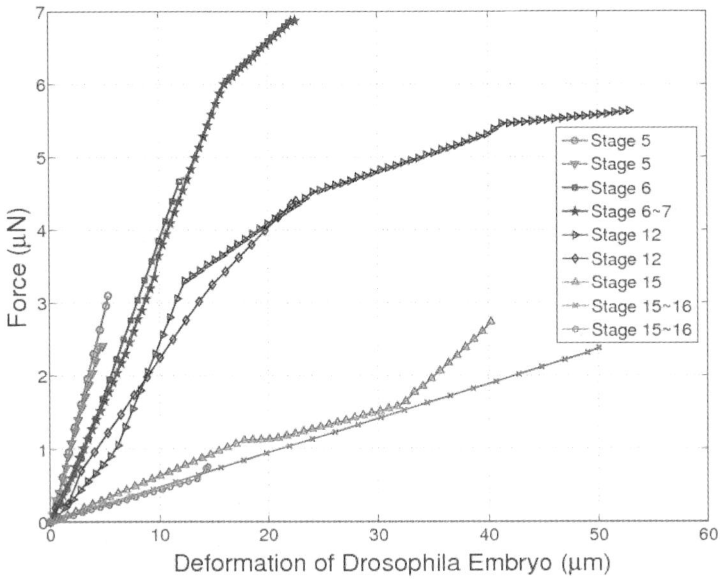

Figure 17. Quantitative relationship between force and deformation of living embryo in different stages.

using an *in situ* PVDF (Polyvinylidene Fluoride) piezoelectric micro-force sensing tool with a resolution in the range of sub-μN. In addition, a networked microrobotic biomanipulation system integrating the developed PVDF micro-force sensing tool has been built up and has greatly advanced operations in biomanipulation and micro injection. Experimental results have verified the effectiveness of the developed system approach. The quantitative relationships between the applied force and embryonic structural deformation for different stages of embryogenesis and penetration force behaviors using an ultra sharp sensing tip were also achieved. Ultimately, the system approach will provide a critical and major step towards the development of automated biomanipuation for minimally invasive injection of living *Drosophila* embryos as well as significant biomedical investigations.

ACKNOWLEDGMENTS

This research work is partially supported under NSF Grants IIS-0713346 and DMI-0500372. The authors would also like to thank

Dr. David N. Arnosti, Professor of Department of Biochemistry at Michigan State University, for his technical advice and help during the process of this research.

REFERENCES

1. Press Release: The 1995 Nobel Prize in Physiology or Medicine. http://nobelprize.org/medicine/laureates/1995/press.html.
2. G. M. Rubin, and E. B. Lewis, "A brief history of Drosophila's contributions to genome research," *Science*, **287**, 2216–2218 (2000).
3. G. M. Rubin *et al.*, "Comparative genomics of the eukaryotes," *Science*, **287**, 2204–2215 (2000).
4. M. B. Feany, and W. W. Bender, "A Drosophila model of Parkinson's disease," *Nature*, **404**, 394–398 (23 March 2000).
5. H. I. Wan, A. DiAntonio, R. D. Fetter, K. Bergstrom, R. Strauss, and C. S. Goodman, "High-wire regulates synaptic growth in Drosophila," *Neuron*, **26**, 313–329 (2002).
6. A quick and simple introduction to Drosophila melanogaster, http://ceolas.org/ VL/fly/intro.html.
7. R. W. Bernstein, and A. Dragland, "Microsurgery on fruit-fly embryos," in *Features, GEMINI 2002/2003*, Web Edition, http://www.ntnu.no/gemini/2002-06e/40-42.htm.
8. Y. Kimura, and R. Yamagimachi, "Intracyto-plasmic sperm injection in the mouse," *Biology of Reproduction*, **52**(4), 709–720 (1995).
9. Y. Sun, K.-T. Wan, K. P. Roberts, J. C. Bischof, and B. J. Nelson, "Mechanical property characterization of mouse zona pellucida," *IEEE Transactions on Nanobioscience*, **2**(4), 279–286 (2003).
10. D.-H. Kim, Y. Sun, S. Yun, B. Kim, C. N. Hwang, S. H. Lee, and B. J. Nelson, "Mechanical property characterization of the zebrafish embryo chorion," *Proceedings of the 26th Annual International Conference of the IEEE EMBS*, pp. 5061–5064 (2004).
11. X. J. Zhang, S. Zappe, R. W. Bernstein, O. Sahin, C.-C. Chen, M. Fish, M. P. Scott, and O. Solgaard, "Micromachined silicon force sensor based on diffractive optical encoders for characterization of micro injection," *Sensors and Actuators A*, **114**, 197–203 (2004).
12. An American national standard: IEEE standard on piezoelectricity, *ANS/IEEE Standard* 176-1987, 1987.
13. I. Elhajj, N. Xi, B. H. Song, M.-M. Yu, W. T. Lo, and Y. H. Liu, "Transparency and synchronization in supermedia enhanced

internet-based teleoperation," *Proceedings of the IEEE International Conference on Robotics and Automation*, pp. 2713–2718 (2002).

14. A. Bicchi, A. Caiti, and D. Prattichizzo, "Dynamic force/torque sensors: Theory and experiments," *Proceedings of the IEEE Conference on Adavnced Robotics*, pp. 727–732 (1997).
15. M. S. Weinberg, "Working equations for piezoelectric actuators and sensors," *IEEE Journal of Microelectromechanical Systems*, **8**(4), 529–533 (1999).
16. Piezo Film Sensors Technical Manual, Internet Version, Measurement Specialties Inc. (August, 1998).
17. W. H. Bowes, L. T. Russell, and G. T. Suter, *Mechanics of Engineering Materials*, John Wiley & Son (1984).
18. L. Meirovitch, *Elements of Vibration Analysis*, New York: McGraw-Hill, Ch. 5, pp. 218 (1975).
19. Data Sheet, *Capillary Tubing*, L022-D pp. 14-15, FHC Inc., website: www.fh-co.com/010300.
20. Y. T. Shen, N. Xi, and W. J. Li, "Force guided assembly of micro mirrors," *Proceedings of the IEEE/RSJ International Conference on Intelligent Robots and Systems*, **3**, 2149–2154 (2003).
21. Y. T. Shen, N. Xi, U. Wejinya, and W. J. Li, "High sensitivity 2-D force sensor for assembly of surface MEMS devices," *Proceedings of 2004 IEEE/RSJ International Conference on Intelligent Robots and Systems*, **4**, 3363–3368 (2004).
22. Y. Zhou, B. J. Nelson, and B. Vikramaditya, "Fusing force and vision feedback for micromanpulation," *Proceedings of the IEEE International Conference on Robotics and Automation*, pp. 1220–1225 (1998).
23. A. Sano, H. Fujimoto, and T. Takai, "Human-centered scaling in micro teleoperation", *Proceedings of the IEEE International Conference on Robotics and Automation*, pp. 380–385 (2001).
24. FlyMove, http://flymove.uni-muenster.de/Stages/StgTabelle.html.

Chapter Five

Learning Signaling Pathway Structures

Karen Sachs and
Solomon Itani

5.1 INTRODUCTION

Three primary types of biomolecular pathways together orchestrate the complex processes required to sustain life: metabolic pathways, in which proteins manufacture cellular building blocks and provide for the energy needs of the cell, genetic regulatory pathways, which control the timing and abundance of de novo synthesis for each of the cell's numerous proteins, from their respective genes, and signaling pathways, which execute a cellular response to environmental cues, often via integration with metabolic and genetic regulatory pathways. The ability to understand the richness of molecular biology — and certainly to attempt to intervene for the sake of desired phenotypic outcomes — depends on successful mapping and analysis of these pathways on a cellular scale. With the advent of increasingly sophisticated measurement and manipulation technologies, and largely as a result of the post genomic age, recent years have seen a massive surge in pathway mapping and subsequent understanding of the intricacies of biological systems.

As informative data becomes available, it is increasingly possible to study biological pathways and processes, at various levels of abstraction. It can be useful to separate analysis efforts into four (somewhat arbitrary) broad levels of increasing detail/decreasing abstraction: compilation of a parts list, characterization and

Nanomedicine: *A Systems Engineering Approach* edited by Mingjun Zhang & Ning Xi

www.panstanford.com
978-981-4241-36-6

identification of function for each component of a parts list, high level models describing the causal structure of pathways, and detailed models including the kinetics and/or specific molecular mechanisms of one or multiple pathways. Although each abstraction level is to some extent self contained, progress at each level feeds into and aids (and is in fact necessary for) progress at the next level of detail; as such, borders between levels are fluid, and the spectrum is in actuality a continuum (Fig. 1).

At the highest level of abstraction, a parts list is defined. Compilation of a parts list for the cell's main players — the proteins — was greatly advanced by the initial sequencing of the human genome;

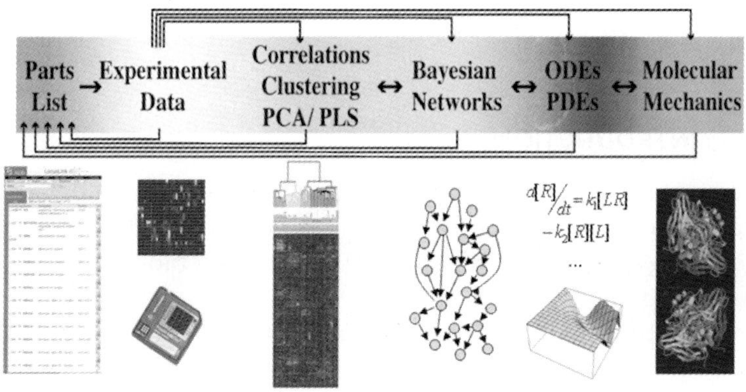

Figure 1. *Continuum of abstraction in modeling and analysis of biomolecular pathways*. At the abstracted end (left side of diagram) is the parts list, crucial for acquisition of experimental data, as well as for more detailed modeling efforts. Somewhat less abstracted are the component characterization approaches, such as those finding correlations among components, clustering techniques, dimensionality reduction techniques like principal component analysis, and methods associating component behavior with phenotypic outcome, like partial least squares regression. Next in the spectrum are pathway structure elucidation approaches, such as Bayesian networks. This chapter focuses on approaches from this level of abstraction. Finally, at the very detailed end are kinetic models showing precise system and even molecular behaviors. The black arrows depict the collaborative and dependent nature of models at different levels of abstractions, models at each level can be used to improve the overall integrated model of a biological system in an iterative manner, by contributing to other model classes both more and less detailed. Figure source: Peter J. Woolf, University of Michigan.

however, the identification and verification of which components exist, and which take part in various biological processes, is an ongoing process and active area of research. A classic example of the functional characterization typical of the next level of detail is gene expression data clustering. In this approach, genes are clustered using unsupervised learning techniques, based on similarity of gene behavior observed in measurements from various experimental conditions and/or timepoints. Genes displaying similar behaviors are assumed to take part in the same or related cellular processes. Various other approaches are used to characterize parts list components.

At the far opposite end of the abstraction continuum lie detailed models, which painstakingly describe pathway dynamics in a mechanism-specific manner using formalisms such as ordinary differential equations. These models can provide tremendous insight into, and predictive ability for, pathways and networks of interest. However, they are difficult to create unless one has extensive detailed knowledge of a pathway, an information state which has not been obtained for most biological processes. Lying in the midrange of the abstraction continuum, and providing the pathway understanding necessary for building detailed models, are high level models which address connectivity of pathway components. These models — which include such formalisms as Boolean networks, fuzzy logic networks and Bayesian networks, attempt to elucidate the pathway structure which describes causal (mechanistic) connections among biomolecules.

Models from across this spectrum have provided intriguing insight into all pathway types comprising cellular systems. In this chapter, we focus on signaling pathways, specifically, their analysis with high level models, striving primarily to elucidate pathway structures. However, because modeling approaches are often agnostic as to biological pathway type, we will also discuss modeling techniques primarily used for genetic regulatory or metabolic pathways. In the remainder of the introduction we describe signaling pathways and their importance in biological systems, and provide a brief introduction to Bayesian networks.

5.1.1 Signaling Pathways

Organism survival depends upon the ability of cells to respond to their changing environment. Single cell organisms must respond to cues indicating the presence of food or toxins, cells in a multicellular organism must react to a multitude of cues from the external environment and from neighboring as well as distant cells. Signaling pathways constitute the primary mechanism by which cells response to cues (Fig. 2). They sense external cues and orchestrate the downstream response, by engaging relevant parts of the cellular machinery to bring about the desired outcome. The extracellular cue may come in many forms — it may be a small molecule, from a nearby sugar source, it may be a hormone secreted from a distant gland and delivered by the bloodstream, or it may be a peptide secreted by, or on the surface of, a nearby cell. Cues are detected by a receptor, a protein that typically spans the cellular membrane of the recipient cell, its external end surveying the extracellular environment for its specific binding partner, called its ligand. In general, each receptor can bind one (or a specific set of) ligands, such that each ligand can trigger a specific process in the recipient cell. The end result of ligand binding is very varied, and can include such effects as cellular motion, proliferation, apoptosis, protein secretion (to signal to other cells), and metabolic alterations.

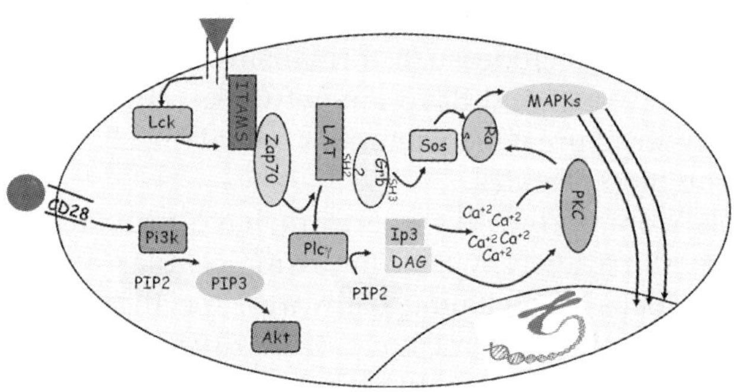

Figure 2. *Diagram of a signaling pathway: T cell signaling*. Extracellular ligands (represented by the green triangle and blue circle) bind to membrane spanning receptor proteins, initiating a cascade of protein modification and phosphorylation events (represented by curved black arrows). While most of the signaling molecules are proteins with enzymatic activity, some are proteins whose main function is to bind other proteins (ITAMS is one example), and others are nonproteins, such as the calcium ions. The end response depicted here is the effect of MAPK members on activity in the nucleus, inducing changes in genetic regulatory pathways. (See page 339 for color illustration.)

Once bound by its ligand, the receptor's internal end becomes modified, enabling it to bind to and/or modify subsequent pathway components. These downstream biomolecules, called signaling molecules, transmit the signal from molecule to molecule in the signaling pathway, each upstream molecule modifying subsequent downstream molecules in the pathway. The modification might be physical (such as a protein cleavage event, which may expose an enzymatic active site), it might be locational (for instance, the transport of a transcription factor into the nucleus, where it can affect gene expression, or recruitment of a protein to the cell membrane, where it can aid in receptor activation), or it might be chemical, most classically, the addition of a phosphate group (PO_4) to a specific protein site. The addition of a phosphate group causes a conformational change in the protein, which often modifies its activity, typically switching it from the "off" conformation to the "on" (or from the "on" to the "off"). Phosphorylation is performed by proteins called kinases, dephosphorylation by phosphatases. Because of the predominance of phosphorylation as a means of protein activation and regulation, signaling pathways often amount to phosphorylation cascades, a series of phosphorylation events that culminate in activation of effector proteins that bring about the desired phenotypic outcome.

The achieve a desired outcome, the signaling pathway often involves other pathway types, such as metabolic and genetic regulatory pathways. The regulation of these different pathways may be tightly intertwined, so their separation into different pathway categories is somewhat artificial. In many cases, for instance, signaling pathways help to regulate metabolism, while metabolic products participate in signaling pathways; similarly, a signaling pathway may culminate in the activation of a set of genes, activating a genetic regulatory pathway, while in turn, one of the activated genes may be responsible for turning off (or further activating) the initiating signaling pathway.

It may seem wasteful and unnecessary to employ dozens of intermediary signaling proteins to convey a signal to a single endpoint. Why not use the receptor to activate an endpoint protein, or at most use a few intermediate secondary messengers? While a few such examples exist, they are the exception rather than the rule. The multistep nature of signaling pathways enables multistage regulation of pathways and ultimately of cellular behavior. In many cases, an initial weak signal is amplified at each step, leading to an eventual thousands-fold amplification. Having many points of control provides the tight regulation necessary to enable the signaling pathway to adapt to various inputs and environments, and to remain robust in the face of heterogeneous conditions. Regulation schemes such as feedback and feedforward loops bestow characteristics like dampening or amplification of a response, and oscillatory responses, which amplify and then fade through repeated cycles. These behaviors can only be achieved with multistep pathways.

The end result of an initiating signal depends on the apparatus triggered by the internal end of the receptor it binds. This apparatus — or sequence of signaling events — may differ from cell to cell, such that the same stimulus can lead to very different responses in different cells. Furthermore, the cell often responds to many varied cues at once. Various pathways may share signaling pathway components or may regulate each other at one or more points, a process known as "crosstalk" among pathways. In essence we can think of all the pathways in the cell as performing the role of a grand processor, taking in multiple cues and integrating them into a final decision, or set of decisions, regarding cell outcome. The end result may be difficult to predict. For instance, a cell may "decide" to apoptose (a process known as programmed cell death, by which the cell, often in response to damage, intentionally self destructs) or to proliferate based on initial cues that are tipped slightly in favor of one behavior or another. A disregulated signaling pathway may lead to too little apoptosis, resulting in accumulation of damaged cells, potential precursors to cancer. Too much apoptosis, on the other hand, underlies neurodegenerative disease. As the orchestrating processes and regulating core of the cell, signaling pathways lead to remarkably robust responses when functioning correctly, but to disasterous endpoints when disregulated. We study them to understand their complex functions, and in the hopes of enabling external controlling interventions, when they become necessary in disease states.

5.1.2 Bayesian Networks

In this chapter, we use a modeling framework called Bayesian networks in order to model signaling pathways. Bayesian networks come from a family of models called graphical models, a family of flexible and interpretable models, in which probabilistic relationships among variables are represented in a graph. In our context, the Bayesian network represents relationships among variables in a signaling pathway, where the variables can represent signaling molecules, small molecules, lipids, or any biologically relevant molecule.

Bayesian networks can uncover statistical relationships among variables from a set of data. Revealed relationships are not restricted to pairwise or linear functions and, in fact, can be arbitrarily complex. Because statistical relationships may imply a physical or functional connection, we can use Bayesian networks to refine existing knowledge or uncover potential relationships in signaling pathways. When interventional data is available (i.e. data in which specific biomolecules have been manipulated or perturbed), we can begin to add a causal interpretation to our model.

In this work, to model signaling pathways, we analyze single cell data from flow cytometry, using the Bayesian network approach. The single cell data is (relatively) abundant but noisy, and the underlying biological processes are noisy as well, so a probabilistic approach is particularly suitable in this domain. The probabilistic nature of the Bayesian network enables it to extract signals from noisy data and to naturally handle uncertainty that arises in the modeling of biological processes.

The probabilistic approach determines dependencies and conditional independencies among variables; for this reason, it is able to include edges that represent meaningful relationships, but exclude those edges that are not necessary, leading to a relatively sparse representation of the underlying dependencies. Thus, given sufficient data, a Bayesian network can provide a first order map of a signaling pathway, and serve as an in silico generator of testable hypotheses.

5.2 RELATED WORK

A number of experimental and computational approaches are used to characterize signaling pathways, using approaches that are quite different (but complementary) to the approach presented here. Additionally, biomolecular pathways other than signaling pathways have been targets of mathematical modeling. In this section, we review signaling pathway characterization approaches which rely on data other than quantification of pathway members, as well as alternate modeling approaches which are in use.

5.2.1 *Protein–Protein Interaction Data*

A. Binding information

Most of the work aimed at elucidating signaling pathways fits the category of elucidation of protein–protein interactions and protein–protein interaction networks. A subset of such interactions (binding events) constitute signaling pathways. The first large scale experimental efforts were presented in 2000.[1–3] These employed yeast two-hybrid screens, which are designed to detect both transient and stable interactions. Assays of protein co-immunoprecipitation coupled to mass spectrometric identification of proteins have also been used; however, these focus primarily on finding components of protein complexes, as they are more suitable for detection of more stable interactions.[4,5] Although these methods suffer from high false-positive and false-negative rates, these networks can be thought of as a noisy super-set, containing some of the interactions present in signaling pathways (other true interactions detected include, e.g. protein association for complex formation). Von Mering *et al.* were among the first to address the task of increasing accuracy of these noisy datasources, proposing to use the intersection of high-throughput experiments.[6] This work resulted in a low false positive rate, but a high false negative rate, finding just 3% of known interactions. Technological improvements have helped to increase the confidence of protein–protein interaction datasets.[7]

More recently, indirect biological data has been integrated with data from the highthroughput interaction screens. In these studies, weak signals from various sources are merged, aiding in detection of real interactions. Such studies use coexpression, Gene Ontology (GO) annotations, localization, transcription factor binding data, and/or sequence information, in addition to high throughput protein–protein interaction data.[8–15] The various information sources are used as input to a classifier, which classifies interactions as true or false. Classification methods include algorithms from data mining, pattern recognition, and grammatical inference.[16,17] Decision trees, random forests, logistic regression, *k*-nearest neighbor, kernel method, naive Bayes, and other scoring functions[18] have been used. Aside from the classifier used, these studies differ in the dataset they use for training and testing their classifier, in the specific features (data types/sources) that they integrate, and in the encoding of the data (whether similar types of experiments are grouped together and merged or summarized, or each used as a separate experiment). Linding *et al.*[19] introduced a scheme to improve the performance of these methods by using contextual factors.

Although these studies are primarily data-driven, a few groups[20–24] have utilized a data-independent approach, predicting interactions based on sequence motifs. More recently, Lu *et al.*[25] and Singh *et al.*[20] have introduced structure based methods, first predicting structures via homology modeling and, using these structures, predicting the likelihood of interaction based on energy considerations. Narayanan *et al.*[26] present a scheme for detecting functionally similar proteins between two networks, with guarantees on efficiency and correctness. More recently, Singh *et al.*[27] developed an elegant algorithm for the global alignment of multiple protein-protein interaction networks, which is based on the following idea: A protein A1 (from network 1) is a good match for protein A2 (in network 2) if the A1's neighbors are good matches for the A2's. Using this algorithm, they discover functional orthologs from five species: yeast, fly, worm, mouse and human. Such orthologs enable transfer of binding information among species. In general the data-independent approaches incorporate data when available (though they have the advantage of being useable even in extremely data poor domains). With various sophisticated computational approaches, these studies have helped greatly improve the accuracy of interaction prediction in an otherwise noisy compendium of possible interactions. Such studies are focused on general protein interactions, rather than those specific to signaling pathways. Information gleaned from these approaches may be useful to incorporate into our approach: first defining a set of protein–to–protein connection possibilities based on the interaction data supplemented by various other datsets, then increasing the likelihood of potential graph arcs in the Bayesian network according to their likelihood of interaction. Such prior knowledge over graph substructures is straightforward to incorporate in the Bayesian network formalism, and constitutes an interesting direction for future work.

B. Cell signaling interactions

A sub-category of protein interaction prediction is that of signaling interaction prediction, a field that is more closely related to the work presented here. Programs such as Scansite use predicted or known modular signaling domains (e.g. a kinase domain) and protein sequence motifs to predict specific signaling interactions (either a specific modification, such as phosphorylation or dephosphorylation, or a binding event).[28,29] Modular domains are distinct and large enough to be recognized directly from protein sequences. The sequence motifs with which they interact, however, are more difficult to detect (due to their small size). These have been identified primarily via experimental binding information from oriented peptide library screening[30–32] and phage display experiments,[33] combined with information from biochemical characterization. More recently, a purely computational approach has been described for this task, in which protein–protein interaction data is used to identify a small set of binding partners for a particular domain.[34] Because the search space is greatly constrained by the binding information, the signals from the motifs' short sequences are detectable.

5.2.2 Models of Molecular Pathways

Other mathematical modeling approaches are used to model molecular pathways in biology, at different levels of detail. These methods are not necessarily applied to signaling pathways — several are applied to genetic regulatory pathways instead — but the modeling approaches are generalizable to any variables of interest for which data is available. We note in fact that our work in Bayesian network modeling of signaling pathway is a direct extension of earlier work applying Bayesian networks to genetic regulatory pathways.[9,10]

The set of possible candidate models that can be used for molecular pathways is large, and so we will focus on models that have proved most effective and useful. We start with the simplest of those models, Boolean Networks, which have been used to model genetic transcription.

5.2.2.1 Boolean networks

The simplest graphical model of a signaling pathway would be to use a graph where the nodes represent the proteins and an edge from protein A to B means that the amount of active protein A directly or indirectly affects the amount of active protein B. There are two types of edges $\rightarrow$ and $-|$. An edge $\rightarrow$ from A to B means that A activates B, an $-|$ edge means that A deactivates B (Fig. 3). If there is an arrow from A to B (activating or deactivating) A is called a "parent" of B.

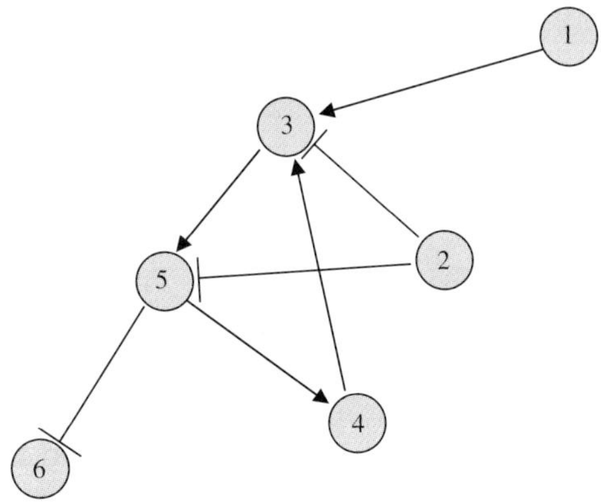

Figure 3. A Boolean network.

The more analytical way of interpreting these graphs is "Boolean Networks". In this description of the network, a protein A is either "on" ($x_A = 1$) or "off" ($x_A = 0$) depending on whether the measured amount of active A is higher than a certain threshold (p_A). Similar to the graphs above, the parents of a certain protein can be activating or deactivating. The assumption here is that protein A is on if and only if at least one of its activating parents is "on" and all of its deactivating parents are off.

Let P_+^A denote the activating parents of A, and P_-^A the deactivating parents. In Boolean Networks, A is on if one or more of P_+^A are active and all of P_-^A are inactive. This can be expressed as follows:

$$x_A = \left(\sum_{B \in P+(A)} x_B \right) \left(\prod_{C \in P-(A)} (1 - x_C) \right).$$

So the equation for the activity of proteins 3 and 5 in Fig. 3 is:

$$\begin{aligned} x_3 &= (x_1 + x_4)(1 - x_2), \\ x_5 &= x_3(1 - x_2). \end{aligned}$$

Remember that all of the values in the equation above are Boolean (1 or 0). These equations can be written for all of the proteins and solved to study the evolution and the steady state(s) of the system.

Boolean Networks have received a lot of attention because of their mathematical simplicity and the abundance of low resolution (high noise) data (that has to be quantized into two levels). Boolean networks offer a great amount of simplicity in that it is rule based. The main two problems with this model are that it is too strict and not representative enough, and that it is not robust in the presence of noise.

Note that this model might give erroneous predictions, that is, under certain rates of different reactions. This is mainly due to the on/off assumption of the relationship between a protein and its parents. In reality, some of the deactivating parents of a certain protein might be active, but that protein will not be totally deactivated (if the activating parents are much more potent than the deactivating one.)

A way to overcome these problems was introduced in Ref. 35. The basic idea is this: Instead of having one graph, a whole family of graphs is used, with corresponding probabilities $\mathbf{c}_i$ that represent how "confident" we are in a certain graph i.

Therefore, one can think of the evolution of the network in two steps:

(1) Randomly choose a graph according to the given probabilities.
(2) Calculate the values the variables take given using the chosen graph (just like in a regular Boolean Network).

This generalization is very helpful in structure learning and makes the model much more robust. Additional advantages are that it allows a notion of "influence" and "sensitivity" of a certain variable to another. The "sensitivity" of a variable X_i to variable X_j in probabilistic Bayesian Networks can be thought of as the probability that X_i changes if X_j does.[35] The influence of a variable is defined in a similar way, and both of those notions can be generalized to collections of variables, so we can think of the sensitivity of a variable to a given set of variables and its influence on another.

5.2.2.2 *Ordinary and partial differential equations models*

A more detailed description of the interaction between proteins can be in the form of a dynamical system. In this description, the amounts of active proteins represent the "states" of the system, and they are related by a set of ordinary differential equations. The choice of equations depends on the level of accuracy/simplicity required from the model, the availability of data to fix the parameters of the model, and assumptions (or prior knowledge) about how the proteins interact.

A simple first-order rate equation for the amount of protein A would look like:

$$\frac{dx_A(t)}{dt} = \alpha_A f(P_A^+(t), P_A^-(t)) - \gamma_A x_A(t).$$

This equation simply means that the rate of change in the active form of protein A is a function of the parents of A, and that there is degradation in that form of A (which is proportional to the amount of protein A that was already active.) Note that there is also a natural dependence on the amount of inactive A, but it is hidden here for notational convenience.

A class of ordinary differential equations that has been extensively used to model is the mass kinetic differential equation model. This model is based on the physiochemical properties of the chemical reactions that activate the proteins. Therefore, the parents of a certain variable A will be collected in groups, where the parents in each group react with each other to produce A (or activate it). On the other hand, the reverse chemical reactions are also taken into account, and so the variables that A would react with (and be deactivated) are also collected in small groups. The effect of each group (parents or reactants) on the rate of change in A is a product of the concentrations of all of the elements in the group. Thus, the differential equations have the following form:

$$\frac{dm_A(t)}{dt} = \sum_i k_i^A \prod_{B_j \in \pi_i(A)} m_{B_j} - \sum_i l_i^A \prod_{C_j \in \sigma_i(A)} m_{C_j}.$$

Here, k_i^A and $l_i A$ correspond to the rates of the reactions of the particular group, and the m's represent the concentrations of the variables.

Differential equations offer a dynamic system representation of the signaling pathway, and thus they can be very detail and expressive. The downside to this modeling method is that it requires a lot of information about the variables, their connectivity, their reactions and the corresponding rates. Thus it can usually only be applied to systems that have been studied extensively, and usually uses a previously known connectivity graph (what variables affect what) that is usually determined by another computational method (such as Bayesian Networks).

Important generalizations of ordinary differential equation models are Partial Differential Equation models, which study the effect of the location (along with time) on the signaling pathway. Because of the existence of gradients in concentrations of different molecules in the cell, the location of a certain protein affects its rate of change.[36] This kind of models is very detailed and descriptive; but similar to ODE's, it is hard to identify all of the important proteins/reactants in a certain system, and harder to collect enough information and data to get a very detailed model.

5.2.2.3 Dynamic bayesian networks

Dynamic Bayesian Networks are a class of graphical models that targets the dynamics of the modeled system. Informally, they are a generalization of Hidden Markov Models and Kalman Filter models, and can be thought of as an extension to BN's that can handle dynamical data and temporal models. Just like Bayesian Networks, DBN's are DAGs and therefore inference and structure learning are relatively easy.

Formally, a DBN for the sequence of variables $X_t = (X_t^1, X_t^2, \ldots, X_t^N)$ is a pair $(B_0, B \rightarrow)$, where B_0 is a BN that models the prior $P(X_0)$, and $B \rightarrow$ is a two-slice temporal BN that details $P(X_t|X_{t-1})$ by a DAG and a set of conditional probabilities distributions (CPDs). Using this model, $P(X_t|X_{t-1})$ is determined as:

$$P(X_t|X_{t-1}) = \prod_{i=1}^{N} P(X_t^i|\pi(X_t^i).$$

The parents of X_t^i, $\pi(X_t^i)$ are determined from the two-slice temporal BN and can be in the slice at time t or that at time $t-1$.

For example, given B_0 and $B \rightarrow$ as below, then

$$P(X_0) = P(X_0^1)P(X_0^2|X_0^1)P(X_0^3|X_0^1)P(X_0^4|X_0^3),$$

and,

$$P(X_t|X_{t-1}) = P(X_t^1)P(X_t^2|X_t^1, X_{t-1}^2)P(X_t^3|X_t^2)P(X_t^4|X_{t-1}^3).$$

We first note that Boolean Networks are a special case of Dynamic Bayesian Networks where each variable can take only two values and the conditional probabilities are deterministic Boolean functions.

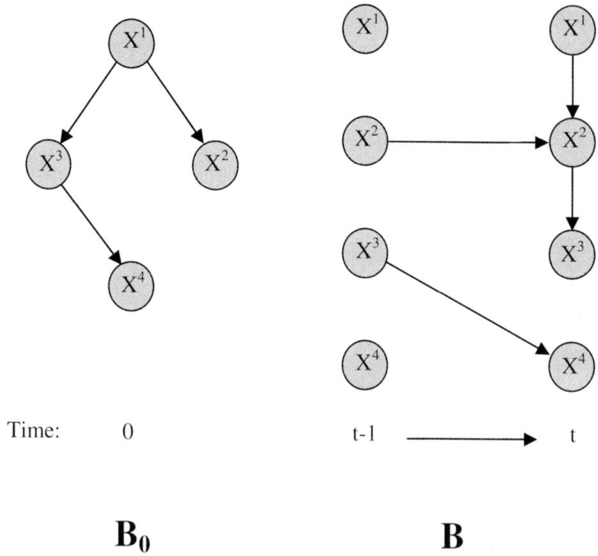

Figure 4. A DBN's decomposition.

Dynamic Bayesian networks are very descriptive, relatively simple, and are very useful for inference and activity prediction. The main challenge with using Dynamic Bayesian Networks is that they require data from several time points.[37] Also, the time step size has to be chosen carefully, or the validity of the model is jeopardized. Similar to the case of Boolean Networks, a family of graphs and their corresponding probabilities can be used instead of one graph. The probabilistic Boolean Network model is a special case of this, just like Boolean Networks is a special case of Dynamic Bayesian Networks.

5.3 BAYESIAN NETWORK TUTORIAL

In this section, we present a Bayesian network tutorial, intended to serve as a basis for understanding Bayesian networks in their general context, with a focus on use in the analysis of signaling pathways. As it is not a comprehensive description, readers intending to pursue the use of Bayesian networks are referred to more in-depth references.[38–40] In this tutorial, we assume only knowledge of basic concepts from probability, including such concepts as Bayes rule, marginalization, and conditional independencies.

5.3.1 Model Semantics

In a Bayesian network, probabilistic relationships are represented by a qualitative description — a graph (G), and a quantitative description — an underlying joint probability distribution. In the graph, the nodes represent variables (in our case, these are biomolecules, usually signaling molecules) and the edges represent dependencies (more precisely, the lack of edges indicate a conditional independency). The graph must be a DAG — a directed acyclic graph. By directed we mean that the edges must be single-headed arrows, originating from one node (the parent node) and ending in another (called the child node). Acyclic indicates that the graph must not include directed cycles, so it should not be possible to follow a path from any node back to itself. (This constraint is a serious limitation in the biological domain, a point which we discuss later.)

For each variable, a conditional probability distribution (CPD) quantitatively describes the form and magnitude of its dependence on its parent(s). This conditional probability distribution is described by a vector of its parameters, θ. The CPD must be consistent with the conditional independencies implied by G. In general, variables in a Bayesian network may be continuous or discrete, and joint probability distributions may take on any form that specifies a valid probability distribution. However, in this discussion, we handle only discrete variables, and multinomial distributions. When discrete variables are used, each variable may take on one of a finite set of states. (E.g. a protein variable may be in state low, medium or high, corresponding to protein abundance.) In general, a Bayesian network represents the joint probability distribution for a finite set $X = \{X_1, \ldots, X_n\}$ of random variables where each variable X_i may take on a value x_i from the domain $\text{Val}X_i$. In our notation, we use capital letters, such as X, Y, Z, for variable names and lowercase letters x, y, z to denote specific values taken by those variables. Sets of variables are denoted by boldface capital letters $\mathbf{X}$, $\mathbf{Y}$, $\mathbf{Z}$, and assignments of values to the variables in these sets are denoted by boldface lowercase letters $\mathbf{x}$, $\mathbf{y}$, $\mathbf{z}$. We denote the parents of X_i in G as $\text{Par}X_i$. θ describes the CPD, $P(X_i|\text{Par}X_i)$, for each variable X_i in X. For illustrative purposes, consider a toy example in which a Bayesian network represents the joint probability distribution between a *cold* virus (c), an *allergy* attack (a), the presence of *pollen* (p), and the occurrence of *sneezing* (s). We may represent the dependencies among these variables in a graph such

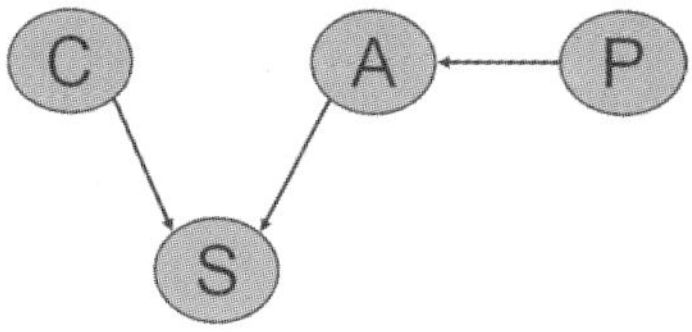

Figure 5. A Bayesian network structure for the variables pollen, allergy, cold and sneezing.

as (Fig. 5). In this graph, *allergy* is the child of *pollen*, *sneezing* is the child of *cold* and *allergy*, and the nodes *cold* and *pollen* have no parents (such nodes are called root nodes). Assume each variable can take on the value 0 ("absent") or 1 ("present").

While the graph appears to represent variable dependencies, its primary purpose is actually to encode conditional independencies, critical for their ability to confer a compact representation to a joint probability distribution. In our toy example, *sneezing* is dependent upon *cold* and *allergy*. *Sneezing* is dependent upon *pollen* as well; however, when the value of *allergy* attack is known, *sneezing* and *pollen* become independent. If we already know that an *allergy* attack is occurring ($a = 1$) (or not occurring — $a = 0$) knowing something about the presence of *pollen* will not help us determine the value of *sneeze*. Therefore, *sneezing* is conditionally independent of *pollen* given *allergy*.

Formally, We say that X is *conditionally independent* of Y given Z if

$$P(X|Y,Z) = P(X|Z)$$

and we denote this statement by $(X \perp Y|Z)$.

The graph G encodes the *Markov Assumptions*: Each variable X_i is independent of its non-descendants, given its parents in G.

$$\forall X_i (X_i \perp \text{NonDescendants}_{X_i} | \text{Pa}X_i)$$

As a consequence of the Markov assumption, the joint probability distribution over the variables represented by the Bayesian network can be factored into a product over variables, where each term is local conditional probability distribution of that variable, conditioned on its parent variables:

$$P(X_1, \ldots, X_n) = \prod_{i=1}^{n} P(X_i|\text{Pa}_{X_i}) \tag{5.1}$$

This is called the chain rule for Bayesian networks, and it follows directly from the chain rule of probabilities, which states that the joint probability of independent entities is the product of their individual probabilities.

A key advantage of the Bayesian network is its compact representation of the joint probability distribution. With no independence assumptions, the joint probability distribution over the variables pollen, allergy, cold and sneeze is $P(P, C, A, S) = P(P)P(C|P)P(A|C, P)P(S|C, P, A)$.[a] For binary variables, this is $1 + 2 + 4 + 8 = 15$ parameters. Employing the conditional independencies, the joint probability distribution becomes $P(P, C, A, S) = P(P)P(C)P(A|P)P(S|A, C)$, or $1+1+2+4 = 8$ parameters. The more sparse a graph structure, the fewer edges it contains and the more conditional independencies it encodes, yielding a greater savings in parameters.

What about the joint probability distribution representation? In the case of the multinomial distributions, each CPD can be presented in a conditional probability table (CPT), in which each row in the table corresponds to a specific joint assignment $\mathbf{pa}_{X_i}$ to $\mathbf{Pa}_{X_i}$, and specifies the probability vector for X_i conditioned on pa_{X_i}. If we assume these variables are binary (each can be in one of two states, either 0 or 1), then in order to specify all the parameters of the CPD for variable X_i with parent(s) Pa_{X_i}, each CPT must have 2^k entries, where k is the number of parents. In general, for a variable X_i with k parents, if each variable takes on one of s states, the number of entries in the CPT will be $s^k * (s - 1)$.

Returning to our toy example, we will assume that the parameters of the CPTs are known:

$P(p = 0)$	$P(p = 1)$
0.7	0.3

$P(c = 0)$	$P(c = 1)$
0.9	0.1

p	$P(a = 0)$	$P(a = 1)$
$p = 0$	0.999	0.001
$p = 1$	0.4	0.6

[a] Or, equivalently, $P(A, C, P, S) = P(A)P(C|A)P(P|A, C)P(S|A, C, P)$, or any other order. If this representation of the joint probability distribution is not familiar, derive it by starting with the more familiar $P(A, B) = P(A|B)P(B)$, then expanding to more variables.

a	c	$P(s=0)$	$P(s=1)$
$a=0$	$c=0$	0.99	0.01
$a=0$	$c=1$	0.4	0.6
$a=1$	$c=0$	0.5	0.5
$a=1$	$c=1$	0.1	0.9

These indicate that, for instance, the probability of having a cold is 0.1 (this is the "general" or a priori probability, not taking into account any information), the probability of pollen being present is 0.3, the probability of having an allergy attack when pollen is present is 0.6 (the person is only somewhat allergic), and the probability of sneezing when the person has a cold but no allergy attack is 0.6. While the graph reveals the conditional independencies, the CPTs demonstrate the strength of dependencies. For instance, although both allergy and cold can cause sneezing, cold alone has a stronger affect than allergy alone ($P(s = 1|c = 1, a = 0) = 0.6$ while $P(s = 1|c = 0, a = 1) = 0.5$). Notice that, as expected, each row in the CPT sums to 1, so the second column of probabilities is redundant. If we consider only the first column, we can count exactly 8 parameters that are specified for this Bayesian network.

5.3.2 Inference

The most common task for which Bayesian networks are used is inference — reasoning from factual knowledge or evidence. This is a "classic" use of Bayesian networks which is applicable whenever (often incomplete) information must be used to evaluate a probabilistic system (examples include assessment of the likelihood of an accident by a car insurance company or a doctor's assessment of the likelihood of a particular patient diagnosis). In an inference task, we wish to know what is the value of a particular node, but we do not have access to that information. Instead, we use the Bayesian network to get an answer in the form "variable X_i is 0 with probability y". Usually we have *evidence* that we take into account. Evidence takes the form of assignments to some other variables. For example, if we wish to know if a person is sneezing, and we know that the person is having an allergy attack, we can use this information to assess $P(\text{sneezing} = 1|\text{allergy} = 1)$. We say that the variable allergy is instantiated, and we call this information evidence. When evidence is available, we may reason about a cause of the instantiated variable, or we may have evidence on the cause, and instead reason about the effect. An example of the former would be $P(p = 1|s = 1)$; $P(s = 1|c = 1)$ is an example of the latter.

To clarify these concepts, lets consider a number of examples. Say we want to know $P(c = 0)$ — the probability that the person does not have a cold. With no available evidence, we need to sum the joint probability distribution over all possible values of the other variables:

$$\begin{aligned}
P(c = 0) = & \sum_{\text{pollen,allergy,sneeze}} P(p)P(a|p)P(c)P(s|a, c) \\
= & P(p = 0)P(a = 0|p = 0)P(c = 0)P(s = 0|a = 0, c = 0) \\
& + P(p = 0)P(a = 0|p = 0)P(c = 0)P(s = 1|a = 0, c = 0) \\
& + P(p = 0)P(a = 1|p = 0)P(c = 0)P(s = 0|a = 1, c = 0) \\
& + P(p = 0)P(a = 1|p = 0)P(c = 0)P(s = 1|a = 1, c = 0) \\
& + P(p = 1)P(a = 0|p = 1)P(c = 0)P(s = 0|a = 0, c = 0) \\
& + P(p = 1)P(a = 0|p = 1)P(c = 0)P(s = 1|a = 0, c = 0) \\
& + P(p = 1)P(a = 1|p = 1)P(c = 0)P(s = 0|a = 1, c = 0) \\
& + P(p = 1)P(a = 1|p = 1)P(c = 0)P(s = 1|a = 1, c = 0) \\
= & 0.7 * 0.999 * 0.9 * 0.99 + 0.7 * 0.999 * 0.9 * 0.01 \\
& + 0.7 * 0.001 * 0.9 * 0.5 + 0.7 * 0.001 * 0.9 * 0.05 \\
& + 0.3 * 0.4 * 0.9 * 0.99 + 0.3 * 0.4 * 0.9 * 0.01 \\
& + 0.3 * 0.6 * 0.9 * 0.5 + 0.3 * 0.6 * 0.9 * 0.05 \\
= & 0.9
\end{aligned}$$

This result is, as expected, the same as the a priori probability of $c = 0$, since we did not take any evidence into consideration. In the case of cold, it is not a very useful calculation. But we can use the same approach to calculate an overall probability for variables that are not root as well. For example, lets say that we want to know what is the overall probability of a person sneezing, when the allergy, pollen, and cold state are not known. We can do a similar calculation:

$$\begin{aligned}
& + 0.3 * 0.4 * 0.1 * 0.6 + 0.3 * 0.4 * 0.9 * 0.01 + 0.3 * 0.6 * 0.1 * 0.9 \\
& + 0.3 * 0.6 * 0.9 * 0.5 = 0.15411
\end{aligned}$$

So the overall probability of sneezing is fairly low; not surprising, since sneezing depends on two variables which occur at low probability. Usually an inference task involves evidence. Lets consider an example in which one or more variables are instantiated. For instance, perhaps we notice that the person is sneezing ($s = 1$), and we wonder if this is due to a cold or an allergy attack. We might examine the CPTs and see that cold alone is more likely than allergy alone to cause sneezing (since $P(s = 1|c = 1) = 0.6$ while $P(s = 1|a = 1) = 0.5$). Can we guess that a cold is likely the cause of the sneezing? We cannot, because we must also take into account the different a priori probabilities of both cold and allergy. If we do the calculation:

$$
\begin{aligned}
&P(c = 1|s = 1)\\
&\quad = \sum_{\text{pollen,allergy}} P(p)P(a|p)P(c)P(s|a,c)/P(s = 1)\\
&\quad = [P(p = 0)P(a = 1|p = 0)P(c = 1)P(s = 1|a = 1, c = 1)\\
&\qquad + P(p = 0)P(a = 0|p = 0)P(c = 1)P(s = 1|a = 0, c = 1)\\
&\qquad + P(p = 1)P(a = 1|p = 1)P(c = 1)P(s = 1|a = 1, c = 1)\\
&\qquad + P(p = 1)P(a = 0|p = 1)P(c = 1)P(s = 1|a = 0, c = 1)]/P(s = 1)\\
&\quad = 0.7 * 0.001 * 0.1 * 0.9 + 0.7 * 0.999 * 0.1 * 0.6\\
&\qquad + 0.3 * 0.6 * 0.1 * 0.9 + 0.3 * 0.4 * 0.1 * 0.6/0.15411\\
&\quad = 0.065421/0.15411\\
&\quad = 0.42451
\end{aligned}
$$

$$
\begin{aligned}
&P(a = 1|s = 1)\\
&\quad = \sum_{\text{pollen,cold}} P(p)P(a|p)P(c)P(s|a,c)/P(s = 1)\\
&\quad = [P(p = 0)P(a = 1|p = 0)P(c = 0)P(s = 1|a = 1, c = 0)\\
&\qquad + P(p = 0)P(a = 1|p = 0)P(c = 1)P(s = 1|a = 1, c = 1)\\
&\qquad + P(p = 1)P(a = 1|p = 1)P(c = 0)P(s = 1|a = 1, c = 0)\\
&\qquad + P(p = 1)P(a = 1|p = 1)P(c = 1)P(s = 1|a = 1, c = 1)]/P(s = 1)\\
&\quad = 0.7 * 0.001 * 0.9 * 0.5 + 0.7 * 0.001 * 0.1 * 0.9\\
&\qquad + 0.3 * 0.6 * 0.9 * 0.5 + 0.3 * 0.6 * 0.1 * 0.9\\
&\quad = 0.097578/0.15411\\
&\quad = 0.63317
\end{aligned}
$$

So given only the information that sneezing is occurring, the more likely cause is actually an allergy attack. Notice that when we perform inference with evidence, we must rule out all possibilities in which the evidence is refuted — in this case, we cannot take into consideration any possibilities in which $s = 0$, since we are given that $s = 1$. The probability distribution over the space of possibilities must sum to 1. Since we are constraining the space of possibilities, in order to maintain the constraint that the probabilities sum to 1, we must renormalize our probability distribution; we do this by dividing by the overall probability of the evidence (in this case, $P(s = 1)$, which we already calculated above). This is an application of Bayes rule: $P(a|s) = P(a, s)/P(s)$. It is this use of Bayes rule for inference tasks that gives Bayesian networks their name. Note also that the probability of cold increased when this evidence was taken into account (we say that our belief in $c = 1$ was increased), because the presence of sneezing made a cold more likely (i.e. $P(c = 1|s = 1) > P(c = 1)$). Although we have not calculated the *a priori* probability of allergy, it is safe to assume we would similarly find that $P(a = 1|s = 1) > P(a = 1)$.

Finally, consider the inference task in which we know the person is sneezing and we also know the person is having an allergy attack. What happens to the probability of cold? How does it compare to the probability of a cold, given only that the person is sneezing, but with no information regarding allergy? If we do the calculation:

$$
\begin{aligned}
&P(c = 1|a = 1, s = 1) \\
&\quad = \sum_{\text{pollen}} P(p)P(a|p)P(c)P(s|a, c)/P(s = 1, a = 1) \\
&\quad = \sum_{\text{pollen}} P(p)P(a|p)P(c)P(s|a, c) \Big/ \sum_{\text{pollen,cold}} P(p)P(a|p)P(c)P(s|a, c) \\
&\quad = [P(p = 0)P(a = 1|p = 0)P(c = 1)P(s = 1|a = 1, c = 1) \\
&\qquad + P(p = 1)P(a = 1|p = 1)P(c = 1)P(s = 1|a = 1, c = 1)] \\
&\qquad\quad /[P(p = 0)P(a = 1|p = 0)P(c = 0)P(s = 1|a = 1, c = 0) \\
&\qquad + P(p = 0)P(a = 1|p = 0)P(c = 1)P(s = 1|a = 1, c = 1) \\
&\qquad + P(p = 1)P(a = 1|p = 1)P(c = 0)P(s = 1|a = 1, c = 0) \\
&\qquad + P(p = 1)P(a = 1|p = 1)P(c = 1)P(s = 1|a = 1, c = 1) \\
&\quad = [0.7 * 0.001 * 0.1 * 0.9 + 0.3 * 0.6 * 0.1 * 0.9] \\
&\qquad\quad /[0.7 * 0.001 * 0.9 * 0.5 + 0.7 * 0.001 * 0.1 * 0.9
\end{aligned}
$$

$$+ 0.3 * 0.6 * 0.9 * 0.5 + 0.3 * 0.6 * 0.1 * 0.9]$$
$$= 0.016263/0.097578$$
$$= 0.16667$$

We see that the presence of an allergy attack reduces our belief that the person has a cold. So our belief in $c = 1$ increases when we know $s = 1$, but then is reduced when we also discover that $a = 1$. This is because the presence of sneezing makes us think the person may have a cold that is the cause of the sneezing. However, when we discover that the person has an allergy attack, we reason that the allergy attack may be the cause of the sneezing, and so we become less convinced that the person has a cold. This process is called *explaining away* (because the allergy attack explains away the sneezing), and it is one way in which a Bayesian network is able to reason.

5.3.3 Structure Learning

There are many applications that utilize Bayesian networks. In an area of interest, the dependency structure may be determined by experts in the domain, the parameters estimated, and the Bayesian network used to determine the probability of unknown events (e.g. for an insurance company) or of a particular diagnosis (in a medical application). In these applications, the values of a subset of the nodes are known, and this information is used to find probability distributions for the unknown nodes. In our case however, the graph structure itself, G, is not known, and in fact our goal is the elucidation of this structure from the experimental data. We find G by searching for a structure that is consistent with the statistical (in)dependencies present in a dataset. Finding the graph structure that may have generated (is most likely to have generated) the observed data is called *structure learning*.

Our strategy for structure learning is as follows: First, we define a Bayesian score which indicates, for a given graph structure, how well the structure reflects the dependencies (and conditional independencies) present in the data. This score allows us to assess individual structures. Armed with this ability, we can now search over possible model structures, until we find the best one (i.e. the one with the highest score), or more accurately, we find a set of high-scoring model structures. Finally, we take our set of high scoring models and average them, in order to avoid overfitting and remain consistently Bayesian in our approach.

5.3.3.1 Bayesian score

To formulate the Bayesian score, we start with a probability metric:

$$\text{score}_{\mathcal{B}}(\mathcal{G} : \mathcal{D})\alpha P(\mathcal{G}|\mathcal{D}) = \frac{P(\mathcal{D}|\mathcal{G})P(\mathcal{G})}{P(\mathcal{D})}$$

This score expresses the posterior probability of the graph given the data, using the formulation of Bayes rule. The marginal likelihood, $P(D)$, will be the same for any structure considered. Neglecting $P(D)$ and taking the log, we get:

$$\text{score}_{\mathcal{B}}(\mathcal{G} : \mathcal{D}) = \log P(\mathcal{D}|\mathcal{G}) + \log P(\mathcal{G}) \tag{5.2}$$

$P(G)$ is the structural prior, expressing the a priori probability of each graph structure. Typically, we employ a uniform prior, so we are not biased towards one structure or another. However, when we have constraints, we can incorporate them. For example, if our model is dynamic, we may have a probability of zero for any graph in which a node representing a later timepoint is influencing an earlier timepoint. Additionally, one could use the prior to incorporate biological knowledge.

Because our dataset is a noisy and finite sample from the true distribution, we are uncertain about the value of our parameters. Therefore, we average over all possible parameter assignments when we calculate $P(D|G)$:

$$P(\mathcal{D}|\mathcal{G}) = \int P(\mathcal{D}|\mathcal{G}, \theta)P(\theta|\mathcal{G})d\theta \tag{5.3}$$

$P(\theta|D)$ in Eq. (5.3) is the parameter prior, in our case it is Dirichlet, specified as:

$$P(\theta) = \text{Dirichlet}(\alpha_{x^1|\mathrm{u}}, \ldots, \alpha_{x^K|\mathrm{u}}) \sim \prod_j \theta_{x^j|\mathrm{u}}^{\alpha_{x^j|\mathrm{u}-1}}$$

where $\alpha_{x^1|\mathrm{u}}, \ldots, \alpha_{x^K|\mathrm{u}}$ are the hyperparameters of the Dirichlet, one corresponding to each $x^j \in \text{Val}(X)$, with $\mathrm{u} \in \text{Val}(\text{Pa}_{X_i})$. The Dirichlet is the *conjugate prior* to the multinomial, ensuring that the posterior expressed in the score will also be distributed as Dirichlet:

$$\text{Dirichlet}(\alpha_{x^1|\mathrm{u}} + M[x^1, \mathrm{u}], \ldots, \alpha_{x^K|\mathrm{u}} + M[x^K, \mathrm{u}])$$

where the *sufficient statistics*, denoted $M[x, u]$, is the number of instances in which $X = x$ and $U = u$ in the dataset. Extracting the sufficient statistics from the data amounts to simply counting, for each variable X_i, how many times it is found in each of its configurations, while its parents in the graph are in each of their possible configurations.

The hyperparamters of the prior are incorporated as *pseudocounts*, corresponding to imaginary datapoints that smooth the posterior by conferring a (small but) nonzero probability to events that do not appear in the dataset. This is important in a data limited domain, where the data may not give a comprehensive reflection of the underlying distribution. For example, if a particular child-parent configuration does not appear in the data (say, protein $B =$ high while protein $A =$ low for the graph $A \rightarrow B$), if we believe this configuration is unobserved yet possible, we can use the prior to express our belief that this configuration has a nonzero probability. The strength of the prior is controlled by the magnitude of the hyperparameters, corresponding to variable number of pseudocounts. Aside from this smoothing effect, the prior is useful for incorporating a complexity penalty, which ensures that the Bayesian score will prefer simpler models, unless a more complex model is supported by a sufficiently large dataset.

The complexity penalty follows from the fact that the Bayesian score integrates over all possible parameters. A more complex model will have more parameters, leading to integration over a larger-dimensional space. For the parameter prior, $P(\theta|G)$, this means that every point in the density function is smaller (since the distribution must integrate to 1). For $P(D|G, \theta)$, the distribution must be strongly peaked over the true parameters to overcome the increase in parameter dimensionality. This can happen (assuming the true underlying graph is indeed complex) when the dataset is large. Otherwise, there is not sufficient data to learn a complex distribution (even if it is the true one), and the Bayesian score will select a simpler graph structure. Because of this complexity penalty, the Bayesian approach avoids overfitting — a process by which we fit our model to noise in the data, rather than a true signal. This is a significant strength, particularly in a data-limited domain.

The integral in Eq. (5.1) can in general be a hard to solve; however, due in part to the decomposability of the Bayesian network and the Dirichlet prior, this integral has a closed form solution. Assuming

Dirichlet priors with hyperparamters $\{\alpha_{X_i^j|u}\}$, the Bayesian score can be expressed as:

$$\text{score}_{\mathcal{B}}(\mathcal{G} : \mathcal{D}) = \sum_i \log \prod_{u \in \text{Val}(\text{Pa}_{X_i})} \frac{\Gamma(\alpha_{x_i|u})}{\Gamma(\alpha_{x_i|u} + M[u])} \times \prod_{x_i^j \in \text{Val}(X_i)} \frac{\Gamma(\alpha_{x_i^j|u} + M[x_i^j, u])}{\Gamma(\alpha_{x_i^j|u})}$$

where Γ is the Gamma function ($\Gamma(n) = (n-1)!$ if n is a natural number) and

$$\alpha_{x_i|u} = \sum_{j \in \text{Val}(X_i)} \alpha_{x_i^j|u}.$$

Note that the score sums over each variable separately, so the contribution of each variable to the overall score can be considered individually (a feature crucial for efficiency of scoring). Furthermore, each local contribution depends only on the sufficient statistics — we need only know the counts from the data and the prior hyperparamters to calculate the score.

5.3.3.2 Searching the space of possible graph structures

For our next step, we have to consider possible graph structures and assess them using the Bayesian score, until we find one (or ones) that score well. We cannot exhaustively examine every graph, because the number of possible graph structures is super-exponential in the number of variables. Instead, we employ a heuristic search to find high scoring models, such as a greedy random search. We begin with a random graph. Then, at each iteration, we select at random an edge to add, delete, or reverse (while keeping within necessary constraints of the graph, such as acyclicity). We score the new graph structure, and keep the change if we find that the score improves. Otherwise, we revert to the previous structure. We then go to the next iteration, and again make a random change, repeating the process. This procedure can get stuck in a local maximum — because we change only one edge at a time, we may encounter a situation in which any single edge change yields a lower score, but a, say, two edge change may lead to a better score. We can never get to this two edge difference, because the first edge change would be rejected. Thus, we are stuck in a score that is locally good (better than any structure that is one edge different) but not very good overall. We avoid this problem in two ways. First, we allow a "bad" edge change (one that leads to a lower score) with some probability. As a second measure, we also repeat the search multiple times, each time starting from a different point in the search space (a different random graph). This is called random restart. Each search iteration requires a calculation of the new model score, but the Bayesian score conveniently decomposes such that each variable's score contribution can be considered in isolation. Therefore, it is only necessary to update the sufficient statistics for any variable which has gained (or lost) a parent as a result of the new edge change.

5.3.3.3 Model averaging

At the end of our search procedure, we have a collection of high-scoring models, one for each random restart. We could simply select the highest scoring model from this collection. However, because our data is noisy and of limited size, we have uncertainty with respect to our dataset and we are concerned about choosing the one highest scoring model that happens to best reflect possibly spurious signals in this dataset. In other words, we are concerned about overfitting. For this reason, rather than select the single highest scoring model, we consider the collection of all high-scoring models, and choose those features that are common to many of them. Each edge is assigned a confidence score based on a weighted average: the sum of the model scores of those models containing that edge, divided by the sum of all the model scores.

5.3.4 Model Properties

So far, we have considered Bayesian networks in their general context, and discussed how to learn the Bayesian network structure from data. In this section, we will delve a bit more into the model structure, to examine how different structures can be extracted from data, even when they represent similar dependencies, what kinds of dependencies are represented, and when a Bayesian network structure can be interpreted as a causal structure.

5.3.4.1 Dependencies and independencies in the graph structure

Explaining Away in Substructures. We have touched on this topic before briefly; here, we revisit it for a more thorough treatment. Returning to our toy example (and neglecting, for a moment, the cold variable), we have the structure $p \rightarrow a \rightarrow s$ (such a structure is sometimes called a chain). In this structure, pollen and allergy are dependent, as are allergy and sneeze. pollen and sneeze are dependent as well: if we know that there is pollen in the air, our belief about the possibility of sneezing increases. However, pollen and sneezing are independent given allergy: If we know the person has no allergic reaction, then the absence or presence of pollen does not change our belief in the possibility of sneezing. Thus, allergy renders pollen and sneeze independent.

Consider this from the structure learning perspective. Examining the variables a, p, s, we may note that all the variables are correlated. We may connect p to a and a to s, but why should we stop there? Afterall, we see that p and s are statistically dependent, so we could also connect p to s. The correct Bayesian network structure will not connect p to s, however, because the dependence between p and a, and between a and s, explains away the dependence between p and s.

How will the conditional independence implied by the ability of a to explain away s's dependence on p manifest itself in the Bayesian score? Consider the CPTs for this structure: In the correct structure, for the variable s, the CPT will specify $P(s|a)$. If we also connect p to s, the CPT must now specify $P(s|a, p)$, so it will contain twice as many parameters. Recall that the Bayesian score penalizes complexity. For the score to allow the extra parameters, the parameters must be more peaked (as they would be if pollen did in fact contribute additional information towards predicting sneeze, for example, if pollen also affected sneezing in some other way, other than by causing an allergy attack). If p affects s only via its affect on a, the complexity penalty will result in the more connected model scoring more poorly; thus, it will ensure that the structure is appropriately sparse.

What other structures encode conditional independencies? Consider the structure $A \leftarrow B \rightarrow C$, sometimes called a fork, in which A and C depend upon B. We have not seen such a structure before, so let us devise an example: say, puddles, rain and umbrellas. In the model puddles $\leftarrow$ rain $\rightarrow$ umbrellas, we see that puddles and umbrellas depend on rain. Because they share a parent, we expect puddles and umbrellas to also be dependent: seeing people with umbrellas might lead us to suspect that there are puddles on the ground, and seeing puddles on the ground may increase our belief that people will be carrying umbrellas (the puddles may have an alternate cause, such as a sprinkler nearby, which does not induce people to carry umbrellas, so while our belief in umbrellas is increased, it is not necessarily certain). What happens when we know that it is raining? As in the chain structure, knowing the value of B (rain) renders A (puddles) and C (umbrellas) independent, because if we know it is raining, then the presence of umbrellas no longer informs us as to the possibility of puddles. In other words, the dependence of each variable on rain *explains* away their dependence on each other, rendering them conditionally independent. As before, the Bayesian score will prefer this structure to one in which an additional edge connects puddles to umbrellas directly.

Let us examine one final graph structure, an important one called a v-structure, which has this configuration: $A \rightarrow B \leftarrow C$ (with no edge between A and C). We saw such a structure in our toy example: cold $\rightarrow$ sneeze $\leftarrow$ allergy. The v-structure is quite unique because in a v-structure, in contrast to the other structures we have seen, two otherwise independent variables may become *dependent*. Consider cold and allergy. Both affect sneeze, but this will not confer a dependence between them. In fact, they are completely independent: the presence of a cold virus does not affect the possibility of an allergy attack and vice versa (to confirm this, return to Sec. 5.3.2 and calculate $P(\text{cold}|\text{allergy})$ and $P(\text{allergy}|\text{cold})$; these will be equal to $P(\text{cold})$ and $P(\text{allergy})$, respectively). What happens if we know the person is sneezing? Suddenly, allergy and cold become dependent; because we know one must be the cause of the sneezing, knowing the value of one helps us to determine the value of the other. We saw an example of this in Sec. 5.3.2: If sneeze $= 1$ and cold $= 1$, our belief in allergy $= 1$ decreases; similarly, if sneeze $= 1$ and allergy $= 1$, our belief in cold $= 1$ decreases; one cause of sneezing explains away the other cause.

5.3.4.2 Selection of model structures

The task of finding the correct model structure relies on assessment of conditional independencies in the domain. Consider a biological domain, in which three kinases affect each other: kinase A phosphorylates kinase B, which then phosphorylates kinase C. From this description, the accurate underlying Bayesian network structure is $A \to B \to C$. The data shows high correlation between A and B, B and C, and A and C. So how can we decide how to connect them? $A \to B \to C$ is one valid structure, but what about $B \to A \to C$? What about $A \to C \to B$? What about a fully connected model (an edge between each two variables)? From the previous sections, we know that although all these models reflect true statistical dependencies in the domain, only $A \to B \to C$ depicts the conditional independency "A and C are independent given B". This will manifest in the data in the following way: the amount of active A predicts the level of active C, as does the level of B. However, B's level is more informative; moreover, *if B is considered, the information from A provides no additional information about C.*

Equivalence Classes. In spite of their usefulness, conditional independencies are often insufficient information for selection of a unique model, because the same set of conditional independencies can often be mapped to multiple models. In the kinase example above, the models $C \to B \to A$ and $A \to B \to C$ depict the conditional independencies in the domain equally as well as the true model ($A \to B \to C$). This is an example of an *equivalence class*. Models with the same set of conditional independencies form an equivalence class. Such models will always receive the same Bayesian score and, therefore, are indistinguishable for a Bayesian network using only observational data. The simplest such example is the class consisting of $A \to B$ and $B \to A$. Because of equivalence classes, when we perform structure learning using observational data, we search for the best equivalence class rather than the best model. Typically, this means we can find a graph with only some of the edges directed (called a PDAG, partially directed graph).

Interventional Data. Equivalence classes are a big problem in practice, but there is a (conceptually) easy way to get around them: using interventional data. Interventional data refers to data in which we intervene experimentally in the biological system, by perturbing the values of individual molecules in defined ways. For example, we may knock out a protein (thus set it equal to zero), we may use RNAi, or, as in this chapter, we may use pharmacological activators and inhibitors. (This is in contrast to observational data, in which the system may be generally stimulated, but no specific variable is forced on or off) The power of interventional data is easy to understand. Assume we have two variables, A and B. A and B are highly correlated, so if we attempt structure learning, we will end up with the equivalence class consisting of $A \rightarrow B$ and $B \rightarrow A$. Now assume we have interventional data in which A is set to zero. In the observational data, generally A and B were both 0 or both 1. Now A is forced to be 0, and we see that B is also zero. This makes us suspect that A might be affecting B, i.e. $A \rightarrow B$ is the correct model. If we also have interventional data on B, and we see that when B is set to zero, A is sometimes 0 and sometimes 1, we can select the model $A \rightarrow B$ with high confidence, because the data demonstrate that B is not affecting A.

How do we translate our intuition regarding interventions into a Bayesian score that incorporates interventions? When we assess a graph structure with the Bayesian score, we are assessing if the dependencies in the data are well represented by those in the graph — in other words, does the child variable (in each case) show a statistical dependence on its parent(s) in the structure. When dealing with interventions, we have a slightly different situation: In those datapoints with intervention, the perturbed variable cannot be dependent on its parent(s), because its value has been externally set (by the experimentor). Therefore, for those datapoints only, we sever the ties between the perturbed variable and its parent(s). The remainder of the data is treated as before. Recall that the Bayesian score sums over the contribution of each variable given its parents, and that this contribution amounts to counts extracted from the data (plus hyperparameters from the prior). Severing the ties of the perturbed variable from its parents amounts to skipping the datapoints with intervention when tabulating the counts for the variable that has been perturbed (only when it is the child). Note that failure to do this would weaken the dependence of the variable on its parents, since they cannot affect the variable's value in those datapoints.

The intuition behind this alteration of the Bayesian score is as follows. Consider again a two variable dataset, with well correlated variables, yielding the equivalence class $A \rightarrow B$ and $B \rightarrow A$. Assuming the true model is $A \rightarrow B$, a perturbation on A would result in interventional data in which the B variable is fixed according to A's value. Thus, when $A \rightarrow B$ is scored, it will score well, as A will appear

5.4.1 Flow Cytometry

In this section, we focus on learning signaling pathway structure by examining single cell data acquired using a technique called flow cytometry. In flow cytometry, molecules of interest in or on cells are bound by antibodies attached to a fluorophore. Cells thus stained pass in a thin stream of fluid through the beam of a laser or series of lasers, and absorb light, causing them to fluoresce. The emitted light is detected and photomultiplier tubes convert this light to electric signals, which are recorded by the flow cytometer, providing a readout of fluorescence, and, therefore, of the abundance of the detected molecules.[45]

Flow cytometry was classically used to measure cell surface markers in order to distinguish functionally distinct cellular subpopulations. More recently, methods have been developed to detect intracellular epitopes (such as e.g. signaling molecules) in order to characterize the cellular response to various conditions and ligands.[46,47] In intracellular flow cytometry, an antibody is often raised to the phosphorylated form of the molecule, under the assumption that this is the active form (or at least the form of interest). However, it is equally appropriate to use antibodies specific to any other form of interest, such as a molecule phosphorylated on an alternate site, or a cleaved form of a molecule (such as caspase 3). To stain intracellular epitopes, it is necessary to fix and permeabilize the cells, in order to allow the antibodies to penetrate the plasma membrane and bind their targets. The procedure involves acquiring a biological sample (this can be cells from a cell line, or, as in this work, primary cells from a human or animal donor), the cells are (typically) treated with various stimuli, then fixed using a cross-linking agent (such as formaldehyde), permeabilized with a detergent or with alcohol (such as Triton or saponin, methanol or ethanol, respectively), then stained with antibodies that are each conjugated to a different fluorophore, and analyzed with the flow cytometer. The flow cytometer determines the relative abundance of each fluorophore in each cell, providing a relative measure of the signaling protein abundance.[48]

A number of issues arise when dealing with intracellular molecules, which mostly tend to lead to false negative results with respect to the presence of a protein (or at least to decrease its apparent abundance). An antibody may not be able to bind its target antigen due to lack of antigen accessibility, if the epitope phosphosite happens to be buried in a protein–protein interaction, or if the protein exists in a cellular compartment that is not permeabilized by standard methods. Phosphoepitope stability is a concern, particularly with respect to treatment of samples prior to fixation. It is necessary to optimize protocols for different specific applications, including, importantly, careful selection of the antibodies. Antibodies that work well in Western blots may not work in the nondenatured, fixation condition of cells in a flow cytometer, and in general control experiments must be carried out to ensure that an antibody is working in a flow context.[48,50]

Another major issue in mutlicolor flow cytometry is fluorophore selection. As discussed above, each antibody is conjugated (directly or indirectly) to a distinct fluorophore. Larger fluorophores may physically interfere with an antibody's binding characteristics or permeability, and thus can be considered another contributor to potential false negative results. In order to quantify each antigen, it is necessary to detect the emission of each antibody separately from the others. It is also necessary to ensure that each fluorophore's absorption spectrum is included in the range of laser light used in the flow cytometer. To accomplish this, it is often necessary to incorporate multiple lasers. A 3-laser flow cytometer may be able to handle as many as twelve distinct fluorophores, hitting a different portion of the excitation spectrum of each (it is not necessary to include the highest point of the excitation spectrum). Although purchasing additional lasers can be expensive, once purchased, it is straightforward to ensure that the absorption spectrum of all the colors (up to the limit possible) is covered. Separating the signal from multiple overlapping emission spectra can be more difficult.[50]

Separation of signals from multiple dyes requires compensation, an adjustment of the measured fluorescence values based on the amount of spillover from one color to another (the amount of emission spectral overlap). We measure fluoresence emissions by selecting an optical filter for each color's detector, that only transmits certain wavelengths of light, thus creating a channel for that color. Although a color may be primarily green (measured in the green channel), it may also have some component of its emission spectrum that is yellow; thus, it spills over into the yellow channel. This will tend to inflate the value reported for the yellow dye. Furthermore, the yellow dye may also spill over into the green channel, causing a similar (but not equivalent in magnitude) inflation effect. For a given fluorophore, the proportion that will be emitted in each channel will always be the same for a particular instrument/instrument setting. For example, green may emit 75% into the green channel and 25% into the yellow, while yellow emits 90% into the yellow channel and 10% into the green. Therefore, by determining this channel ratio for each fluorophore, and measuring the signal in each channel, it is possible to determine how much must be subtracted for correct compensation. In this example, the true green and yellow values are determined by solving the equations $MG = TG + 0.1TY$, $MY = TY + 0.25TG$ for TG and TY, where MG and MT denote the measured values of green and yellow, respectively, and TG, TY denote the true values. In principle this could be expanded to an n-color system. In matrix notation, we solve for T in the equation $M = C \times T$, where M is an $n \times 1$ vector of the measured values in each channel, T is an $n \times 1$ vector of the true values of each color for each channel, and C is an $n \times n$ matrix containing the percent spillover from each color into each other channel. C is determined by measuring each color alone, and determining how much of the total signal spills over into other channels. Since C is known, the system of n equations and n unknowns can be solved for T.

In principle, this approach can be used to include a large number of colors, but in practice, "crowding" the colors in the emission spectrum leads to noisy data. This is because there is error in each measurement (from each channel), and, since compensation involves using measurements from multiple channels to determine a single true abundance, the noise for each color is additive noise from all the channels that contribute to it. This is particularly a problem because some dyes are far brighter than others, so some true measurements are much larger than others. (This can also happen if certain molecules are in far greater abundance or certain antibodies more efficient) When subtracting out the overlapping effect of a bright dye out of a dim channel, even a small measurement error in the bright dye can affect the dim measurement significantly. Therefore, scaling up to increased number of colors must be done carefully. When controls indicate that indeed a bright signal is causing increased spread in the variance of a dimmer signal, it may be preferable to reduce the number of colors in an experiment, rather than contend with noisy data.[51]

5.4.2 *Structure Learning with Multidimensional Flow Cytometry Data*

In a proof of principle study, we applied Bayesian network analysis to multivariate flow cytometry data, in order to learn influence connections among signaling molecules involved in T-cell signaling. Briefly, primary human CD4+ T-cells were treated with a number of stimuli and inhibitors; subsequently, eleven signaling molecules were measured in thousands of cells, and the data were analyzed with a Bayesian network structure learning algorithm to assess the signaling pathway structure. This work has been previously published,[52] therefore, the description here will be abbreviated.

Data were collected after a series of stimulatory cues and inhibitory interventions, with cell reactions stopped at 15 minutes post-stimulation by fixation, to profile the effects of each condition on the intracellular signaling networks of human primary naive CD4+ T-cells, downstream of CD3, CD28, and LFA-1 activation.

We made flow cytometry measurements of 11 phosphorylated proteins and phospholipids (Raf phosphorylated at position S259, mitogen activated protein kinase Erk1 and Erk2 phosphorylated at T202 and Y204, p38 MAPK phosphorylated at T180 and Y182, JNK phosphorylated at T183 and Y185, AKT phosphorylated at S473, Mek1 and Mek2 phosphorylated at S217 and S221 (both isoforms of the protein are recognized by the same antibody), phosphorylation of PKA substrates (CREB, PKA, CAMKII, CASPASE 10, CASPASE 2) containing a consensus phosphorylation motif, phosphorylation of PLC on Y783, phosphorylation of PKC on S660, and phosphor-inositol 4,5 bisphosphate [PIP2] and phosphoinositol 3,4,5 triphosphate [PIP3]). Each independent sample in this dataset consists of quantitative amounts of each of the 11 phosphorylated molecules, simultaneously measured from single cells. In most cases, this reflects the activation state of the kinases monitored, or in the cases of PIP3 and PIP2 the levels of these secondary messenger molecules in primary cells, under the condition measured. Nine stimulatory or inhibitory interventional conditions were used. The complete datasets were analyzed with the Bayesian network structure inference algorithm as detailed below. (See in Fig. 6).

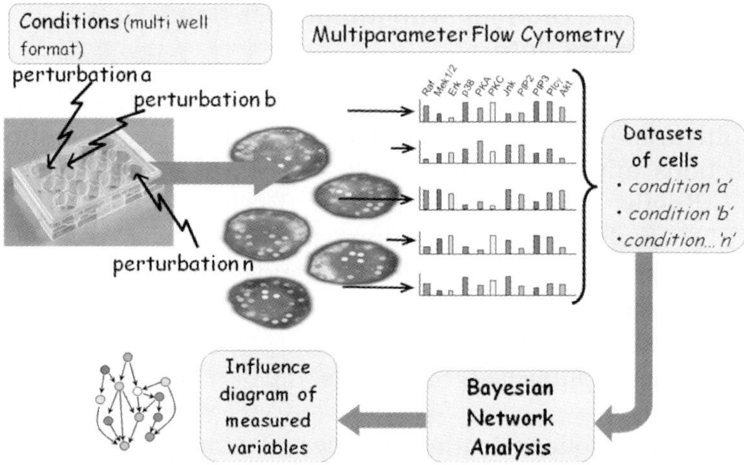

Figure 6. Overview of analysis. Primary human CD4+ T-cells were subjected to a number of stimulatory and/or inhibitory conditions in a multiwell format. Cells were fixed, permeabilized and stained, then profiled using multiparameter flow cytometry. The dataset of protein measurements was subjected to Bayesian network structure learning, yielding an influence diagram.

5.4.2.1 A human primary T-Cell signaling causality map

The resulting de novo causal network model was inferred with 17 high confidence causal arcs between various components. To evaluate the validity of this model, we compared the model arcs — and absent potential arcs-with those described in the literature. Arcs were categorized as: (i) "expected," for connections well-established in the literature, that have been demonstrated under numerous conditions in multiple model systems; (ii) "reported," for connections that are not well known, but for which we were able to find at least one literature citation; (iii) "unexplained," indicates that though the arc was inferred from our model, no previous literature reports were found; and (iv) "missing" indicates an expected connection that our Bayesian network analysis failed to find. Of the 17 arcs in our model, 14 were expected, 16 were either expected or reported, 1 was not previously reported (unexplained), and 4 were missed (Fig. 7).[53–57] Table 2 enumerates the probable paths of influence corresponding to model arcs determined by surveying published reports.

Several of the known connections from our model are direct enzyme-substrate relationships (Fig. 7): (PKA to Raf, Raf to Mek,

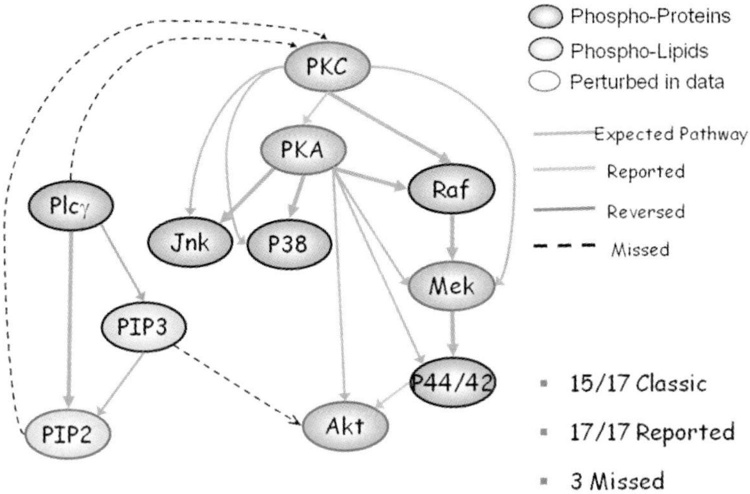

Figure 7. Model results contrasted to known influence connections. Literature reports were used to assess the retrieved network. "Classic/Expected" edges indicate edges that are well established in the literature, "Reported" edges are not well established and have never been reported in T-cells, "Reversed" indicates an edge which is directed incorrectly, and "Missed" indicates well established edges that were not found by the modeling approach.

Mek to Erk, Plc to PIP2) and one a relationship of recruitment leading to phosphorylation (Plc to PIP3). In almost all cases, the direction of causal influence was correctly inferred (an exception was Plc to PIP3, in which case the arc was inferred in the reverse direction). All the influences are contained within one global model, thus the causal direction of arcs is often compelled so that these are consistent with other components in the model. These global constraints allowed detection of causal influences from molecules that were not perturbed in our assay. For instance, although Raf was not perturbed in any of the measured conditions, the method correctly inferred a directed arc from Raf to Mek-as expected for the well characterized Raf-Mek-Erk signal transduction pathway. In some cases, the influence of one molecule on another is mediated by intermediate molecules that were not measured in the dataset. In the results, these indirect connections were detected as well (Fig. 8). For example, the influence of PKA and PKC on the MAPKs p38 and Jnk likely proceeded via their respective (unmeasured) MAPK kinase kinases. Thus, unlike some other approaches used to elucidate signaling networks (for example, protein-protein interaction maps[58,59] that provide static biochemical association maps with no causal links, the Bayesian network method can detect both direct and indirect causal connections and therefore provide a more contextual picture of the signaling network.

Another important feature demonstrated is the ability to dismiss connections that are already explained by other network arcs (Fig. 8). This is seen in the Raf-Mek-Erk cascade. Erk, also known as p44/42,

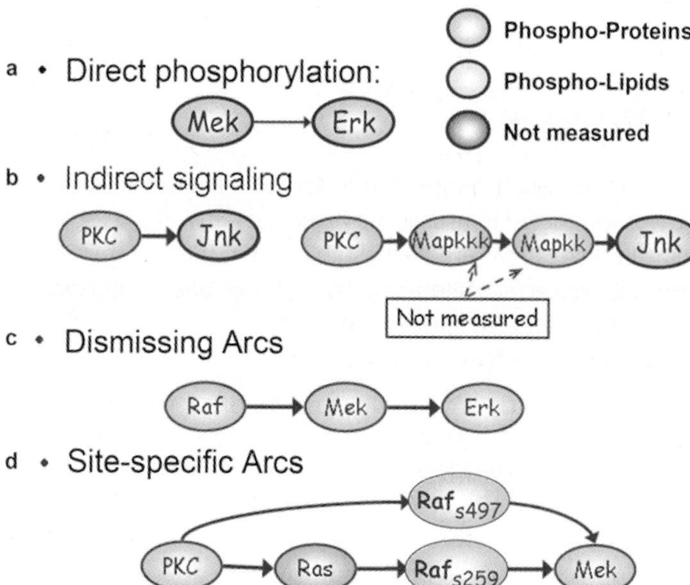

Figure 8. Inferred network demonstrates several features of Bayesian networks. (a) Arcs in the network may correspond to direct events or (b) indirect influences. (c) When intermediate molecules are measured in the data set, indirect influences rarely appear as an additional arc. No additional arc is added between Raf and Erk because the dependence between Raf and Erk is dismissed by the connection between Raf and Mek, and between Mek and Erk. (d) Connections in the model contain phosphorylation site–specificity information. Because Raf phosphorylation on S497 and S499 was not measured in our data set, the connection between PKC and the measured Raf phosphorylation site (S259) is indirect, likely proceeding via Ras. The connection between PKC and the undetected Raf phosphorylation on S497 and S499 is seen as an arc between PKC and Mek.

is downstream of Raf and therefore dependent on Raf, yet no arc appears from Raf to Erk, as the connection from Raf to Mek, and from Mek to Erk, explains the dependence of Erk on Raf. Thus, an indirect arc should appear only when one or more intermediate molecules is not present in the dataset, otherwise the connection will proceed via this molecule.

This is an example of *explaining away* (see Bayesian network tutorial, above). The intervening molecule may also be a shared parent. For example, phosphorylation status of p38 and Jnk are correlated (Fig. 8), yet they are not directly connected, as their shared parents (PKC and PKA) mediate the dependence between them. Although we cannot know if an arc in our model represents a direct or indirect influence, it is unlikely that our model contains an indirect arc that is mediated by any molecule observed in our measurements. As can occur with closely connected pathways, correlation exists between most molecule pairs in this dataset. Therefore, the relative "lack" of arcs in our model contributed greatly to the accuracy and interpretability of the inferred model.

A more complex example is the influence of PKC upon Mek, known to be mediated by Raf (Fig. 8). PKC is known to affect Mek through two paths of influence, each mediated by a different active, phosphorylated, form of the protein Raf. Although PKC phosphorylates Raf directly at S499 and S497, this event is not detected by our measurements, as we use only an antibody specific to Raf phosphorylation at S259 (Fig. 8). Therefore, our algorithm detects an indirect arc from PKC to Mek, mediated by the presumed unmeasured intermediate Raf phosphorylated at S497 and S499.[6] The PKC to Raf arc represents an indirect influence that proceeds via an unmeasured molecule, presumed to be Ras.[54,55] We discuss above the ability of our approach to dismiss redundant arcs. In this case there are two paths leading from PKC to Mek because each path corresponds to a separate means of influence from PKC to Mek — one via Raf phosphorylated at S259, and the other through Raf phosphorylated at S497 and S499. Thus, neither path is redundant. This result demonstrates the important distinction that this analysis is sensitive to specific phosphorylation sites on molecules and is capable of detecting more than one route of influence between molecules.

Four well-established influence connections do not appear in our model: PIP2 to PKC, PLC to PKC, PIP3 to Akt, and Raf to Akt. Bayesian networks are constrained to be a-cyclic, so if the underlying

Table 1. Nodes measured in pathway and specificity antibodies used. The left-hand column shows target molecules measured in this study that were assayed using monoclonal antibody to the target residues (site of phosphorylation or phosphorylated product as described).

Measured molecule	Antibody specificity
Raf	Phosphorylation at S259
Erk1 and Erk2	Phosphorylation at T202 and Y204
p38	Phosphorylation at T180 and Y182
Jnk	Phosphorylation at T183 and Y185
AKT	Phosphorylation at S473
Mek1 and Mek2	Phosphorylation at S217 and S221
PKA substrates	Detects proteins and peptides containing a phospho-Ser/Thr residue with arginine at the −3 position
PKC	Detects phosphorylated PKC-α, -βI, -βII, -δ, -ε, -η, and -0 isoforms only at C-terminal residue homologous to S660 of PKC-βII
PLC-γ	Phosphorylation at Y783
PIP_2	Detects PIP_2
PIP_3	Detects PIP_3

network contains feedback loops we cannot necessarily expect to uncover all connections. For example, in our model the path from Raf to Akt (via Mek and Erk) precludes the inclusion of an arc from Akt to Raf, due to this acyclicity constraint. Availability of suitable temporal data could possibly permit this limitation to be overcome using dynamic Bayesian networks.[60,61] Alternatively, we could employ a cyclic extension that we are currently developing (see Extensions, below).

5.4.2.2 Experimental confirmation of predicted network causality

Three influence connections in our model are not well established in the literature: PKC on PKA, Erk on Akt, and PKA on Erk. To probe the validity of these proposed causal influences, we searched for prior reports in the literature. Of these 3 connections, 2 have previously been reported, the PKC to PKA connection in rat ventricular

Table 2. Possible molecular pathways of influence represented by arcs in the model. Shown are the possible pathways of influence inferred from the data, with the connection shown in Fig. 8(a) and the unmeasured molecules (in bold) that might mediate indirect influences. E, expected; R, reported. See main text for further discussion. Specific phosphorylation sites are included as subscripts.

Connection	Influence path	Type	Category
PKC→Raf	PKC→Ras→Raf_{S259}	Indirect	E
PKC→Mek	PKC→$Raf_{S497/S499}$→Mek	Indirect	E
PKC→Jnk	PKC→→MKKs→Jnk	Indirect	E
PKC→p38	PKC→→MKKs→p38	Indirect	E
PKC→PKA	PKC→cAMP→PKA	Indirect	R
PKA→Raf	PKA→Raf_{S259}	Direct	E
PKA→Mek	PKA→Raf_{S621}→Mek	Indirect	E
PKA→Erk	PKA→HePTP→Erk	Indirect	E
PKA→Jnk	PKA→→MKKs→Jnk	Indirect	E
PKA→p38	PKA→→MKKs→p38	Indirect	E
Raf→Mek	Direct phosphorylation	Direct	E
PKA→Akt	PKA→CaMKK→Akt_{T308} →Akt_{S473}	Indirect	E
Mek→Erk	Direct phosphorylation	Direct	E
Plc-γ→PIP_2	Direct hydrolysis to IP3	Direct	E
Plc-γ→PIP_3	Recruitment leading to phosphorylation	Reversed	E
PIP_3→PIP_2	Precursor-product		E
Erk→Akt	Direct or indirect		R

myocytes, and the Erk to Akt connection in colon cancer cell lines.[56,57] An important goal of our work was to test the ability of Bayesian network analysis of flow cytometry data to correctly infer causal influences from unperturbed molecules within a network. For example, Erk was not acted upon directly by any activator or inhibitor in the sample sets, yet Erk showed an influence connection to Akt. Our model thus predicts that direct perturbation of Erk would influence Akt (Fig. 9). On the other hand, although the Erk and PKA are correlated the model predicts that perturbation of Erk should not influence PKA.

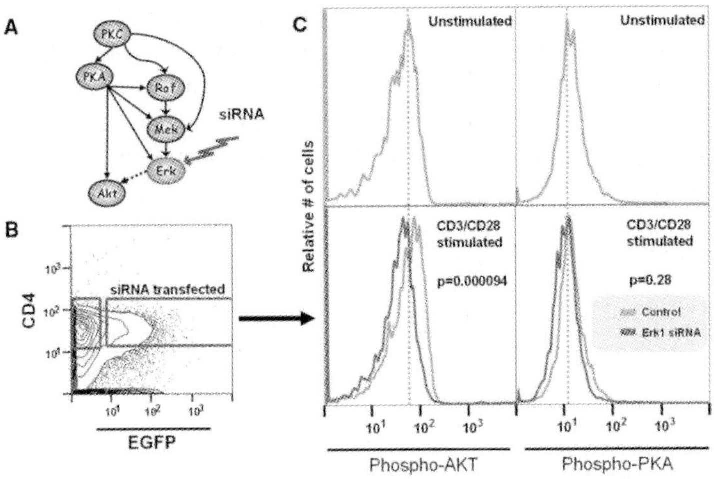

Figure 9. Validation of model prediction. (A) The model predicts that an intervention on Erk will affect Akt, but not PKA. (B) To test the predicted relationships, Erk1 and Erk2 were inhibited using siRNA in cells stimulated with antibody to CD3 (anti-CD3) and anti-CD28. (C) Amounts of Akt phosphorylation in transfected CD4 cells [enhanced green fluorescent protein (EGFP) cells] were assessed, and amounts of phosphorylated PKA are included as a negative control. When Erk1 expression is inhibited, phosphorylated Akt is reduced to amounts similar to those in unstimulated cells, confirming our prediction (P 0 0.000094). PKA is unaffected (P 0 0.28). (See page 339 for color illustration.)

As a test of these predictions (Fig. 9(A)), we used siRNA inhibition of either Erk1 or Erk2 and the amount of S473 phosphorylated Akt and phosphorylated PKA were then measured. In accord with the model predictions, Akt (p = 9.4e-5) phosphorylation was reduced after siRNA knockdown of Erk1 but activity of PKA ($p = 0.28$) was not (Fig. 9(C)). Akt phosphorylation was not affected by the knock down of Erk2. The connection between Erk 1 and Akt may be direct or indirect, involving mediatory molecules yet to be understood, but the connection is supported by both the model and the validation experiment.

5.4.2.3 Methods: Data preprocessing

Because a complete description of materials and methods for this work are given elsewhere,[52] we include here only a discussion of data preprocessing. First, data points that fell more than three standard deviations from the mean were eliminated. This was done to clean up the data by removing any suspicious datapoints (potentially debris, clumps of cells or other noise). Next, preprocessing of perturbed values was performed where needed. Under conditions of chemical intervention, more data preprocessing was often required. This is because our study uses the measured phosphorylation level of molecules as surrogates for their activity. In most cases, the inhibitors employed affected the activity of the molecules, but did not affect their phosphorylation level. For instance, when Mek is inhibited, its measured level actually increases (possibly the system responding to a lack of Mek activity), but its activity is inhibited. Therefore, in the presence of inhibitor, the measured level is no longer a legitimate surrogate for its activity level. To model these interventions, we assume that the level reflects the inhibited (or activated) level, setting inhibited molecules to level 1 ("low"), and activated molecules to level 3 ("high").

Specifically, we assume that inhibition completely removes activity (setting raw values to zero) and that activation increases activity by ten fold (we multiply the raw values by 10 — this value was chosen somewhat arbitrarily). While these appear to be strong assumptions, note that because of the smoothing affect of the discretization, the precise values by which we modify the raw values in the pre-processing phase should not have significant impact. This would be an interesting point to address specifically, by varying the degree of fold change assumed as a result of perturbation, but this sensitivity analysis has not been performed. Note that this preprocessing was applied only in cases in which, as a result of the inhibition, the measured value no longer reflects the activity level of the molecule. This was not always the case: under Psitectorigenin treatement, the level rather than the activity of Pip2 is affected. Therefore, the raw values of Pip2 were not altered. Data were then discretized to three levels (low, medium or high levels of the phosphorylated protein), using an agglomerative approach that seeks to minimize loss of pairwise mutual information among variables. This algorithm is thoroughly described, including pseudocode, in Ref. 62.

5.5 EXTENSIONS AND FUTURE WORK

In this final section, we briefly describe two algorithms which we developed to extend the functionality and usability of Bayesian network analysis of signaling pathways, with emphasis on the use of flow cytometry data. We conclude with a discussion of directions for future work.

5.5.1 Extensions

5.5.1.1 Markov neighborhood algorithm

To build a Bayesian network model of a signaling pathway with n variables, the global nature of the model necessitates measurement of all n variables in each sample, in order to enable elucidation of complex relationships (and conditional independencies) involving multiple variables. However, in some measurement technologies (and in particular, in flow cytometry), the number of variables which can feasibly be measured in a sample is limited. (Note that in the case of flow cytometry, each cell is actually a "sample", since we treat each cell as a datapoint). Considering that the extreme cutting edge of flow cytometry can handle at most ~a dozen signaling pathway molecules, but that signaling pathways can involve dozens and even hundreds of molecules, this is a serious limitation, severely curtailing our ability to effectively model signaling pathways. To address this limitation, we developed an algorithm that extends the modeling capability from m (the number of variables we can measure in each cell) to n (the number of variables we wish to model), for $n > m$. This approach, called the *Markov neighborhood algorithm*,[63] works by selecting candidate parents for each variable, and then measuring the variable with all candidate parents (called its neighborhood). See Fig. 10 for overview of the approach.

5.5.1.1.1 Sketch of algorithm

Step 0. A set of preliminary experiments are performed, in which each pairwise (or more generally, each p-wise, for $p = 2, 3, \ldots$) measurement among the variables is obtained.

Step 1. Define neighborhoods. A neighborhood is defined for each variable by selecting, for each variable X_i, those variables which have a statistical dependence with X_i that is greater than some threshold. The dependence metric can be any reasonable metric- for example, correlation may be used, in which case, the neighborhood for X_i will contain all other

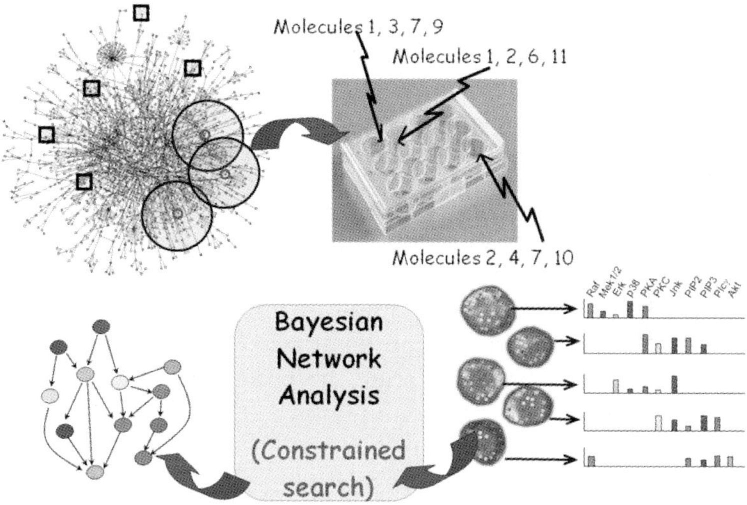

Figure 10. *Overview of Markov neighborhood algorithm.* A set of preliminary experiments are used to determine variable neighborhoods; these include variables which appear to be reasonable candidate parents based on a dependence metric (e.g. correlation). Sets of m molecules from each neighborhood are profiled under various stimulus conditions, and a constrained Bayesian network structure learning analysis is performed, in which candidate variable parents are selected from the (profiled portion of the) variable's neighborhood.

variable found to have a sufficiently large correlation with X_i (larger than a threshold).

Step 2. Select a set of size m from each neighborhood and perform measurements on each set. Note that the selection can be done cleverly to maximize the number of neighborhoods covered by each experiment (since the neighborhoods are not mutually exclusive sets), thus minimizing the number of experiments.

Step 3. Prune neighborhoods by detecting conditional independencies among neighborhood variables. For example, if in X_i's neighborhood, it is found that X_i and X_k are conditionally independent given X_j (as in $X_k \rightarrow X_j \rightarrow X_i$), then it is not necessary to measure X_k when X_j is measured in X_i's neighborhood. Note that step 3 is another form of step 1, as our goal is to define neighborhoods.

Step 4. Perform structure learning on available neighborhood measurements. Search space is constrained such that the parents of variable X_i are selected from X_i's neighborhood.

Step 5. Repeat steps 2–4 until neighborhoods are fully characterized, or until it is not possible to prune neighborhoods further.

This algorithm describes a method to scale up the number of variables modeled, to a number greater than the maximal ability of the measurement technology. As technology improves and this maximal ability grows, this approach can be applied to model increasingly large networks on a systems scale, potentially enabling modeling and prediction of complex phenomenon in the cell.

5.5.1.2 *Learning cyclic structures with bayesian network based cyclic networks*

Signaling networks make extensive use of feedback loops, employing them to tune phenotypic outcome by amplifying and dampening key pathway points. When a signaling pathway structure contains cycles, an accurate portrayal of pathway structure cannot be achieved with a Bayesian network.[c] To overcome this often serious limitation, we developed methods that augment Bayesian Networks to cyclic domains. The perturbations that we used in that work affect the activity level of the perturbed variable (and not the actual amount of that variable), though this approach can be extended for other forms of inhibition.

The first idea used is that when a variable is inhibited, its parents are separated from its children. We thus can use inhibitors to break open cyclic paths; once they are open and the pathway is acyclic, standard Bayesian network structure learning methods can be applied. Now the idea for detecting cycles is straightforward: If there is a cycle through a certain variable, then that variable is its own descendent. Therefore, we first use the inhibitors to detect the presence of cycles, by identifying those variables whose inhibition leads to a change in their *own* distribution. Since perturbation of the variable's activity affects the level of the variable itself, a feedback loop must be present. Next, we use the inhibitors to break open the

[c] Cycles *can* be represented in a dynamic model, such as a dynamic Bayesian network or a continuous time Bayesian network. However, as our approach cannot observe cells over time, it does not provide the dynamic data required for inference of such models.

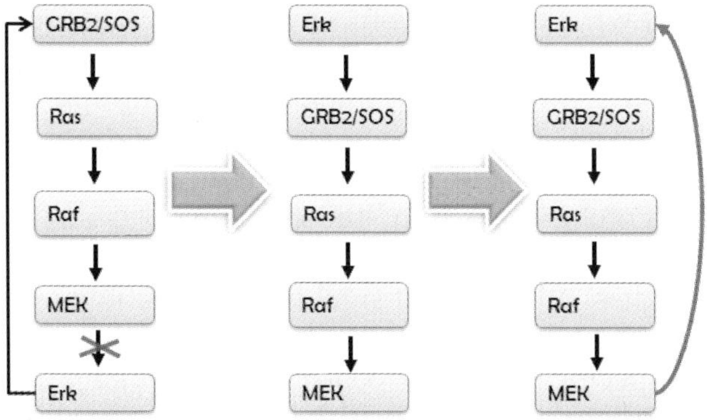

Figure 11. Overview of algorithm for learning cyclic structures. The method relies upon system controls, or perturbations, in the form of small molecule inhibitors. First the inhibitor is applied to break open cycles, next, the acyclic graph is discovered using standard Bayesian network structure learning techniques. Finally, the cycle is closed by considering the effect of inhibition conditioned on existing parents in the graph.

detected loops and perform structure learning, yielding the acyclic signaling diagram. Finally, we close the loops by determining which variables are direct descendents (children) of the perturbed variables.

These cannot be found in the structure learning stage, because the activity of the perturbed variable is blocked. We can assess this by asking which variables are affected by each perturbation, but this information does not differentiate between indirect descendants (grandchildren, etc.) and direct descendants. We *can* find these children of perturbed variables by determining if they are affected by the perturbation *conditioned on their parents* (their parents as determined by the graph learned in the structure learning step). At press time, this algorithm is in the development stage, and results so far have been theoretical, applied in the synthetic domain.[64] We are currently pursuing applications in the biological domain.

5.5.2 Discussion and Future Work

The study of signaling pathways with probabilistic models is a field still in its infancy, due largely to the shortage of appropriate data sources. In this chapter, we discussed a candidate data source, with many advantages as well as some limitations. In Sec. 5.2, we allude to a variety of ways to improve this methodology, by incorporating information from heterogeneous data sources, such as protein protein interaction predictions. The integration of multiple datasources can increase the validity of the inferred structure dramatically, particularly when no datasource can stand alone.

Protein–protein interaction studies, as well as signaling interaction datasources, attempt to find all potential interactions (for the latter studies, all potential signaling interactions), rather than specific interactions occurring in a particular cell or cell type, in response to specific stimuli or conditions. In this way, they differ markedly from our approach, which extracts interactions in a particular dataset and can be catered to a specific biological condition, cell type, disease or other cell state. The set of influence connections in a particular dataset (from our work) and the set of potential signaling interactions (from the studies described above) can be extremely complementary datasets. Not only can we take advantage of predicted signaling interactions to aid in the selection of an optimal Bayesian network (using priors on potential edges, as described above), and to select an initial set of molecules to model (based on which molecules are thought to interact), but we can also use potential signaling interaction data to help us elucidate the signaling events that underlie newly discovered connections in the Bayesian network. This is because many connections in the Bayes net structure are indirect — in other words, they do not include an enzyme and its direct substrate, but rather an upstream regulator and a molecule that is eventually affected via several other intermediaries. This will occur whenever the intermediate molecules that mediate the affect of the upstream regulator on the downstream target are not measured as part of the dataset, a common problem in this data-limited domain. (See for example the arc PKC → Jnk in Fig. 8. In this connection, the influence of the upstream regulator, PKC, on the downstream molecule, Jnk, is likely mediated by the unmeasured molecules, MAPKKK and MAPKK.)

One way to overcome this problem is to include more molecules in the dataset, a direction we are pursuing (see Extensions, above). However, in cases with insufficient prior biological knowledge, we may not have a good guess regarding which molecules may act as intermediates in an interaction (or indeed if the interaction is indirect or direct). In such cases, predicted signaling interactions that connect the upstream molecule to the downstream one can be used to complete the hypothesis of pathway structure, and to direct experiments for future Bayesian network analyses (with data which includes candidate intermediate molecules) or for direct wet lab verification. This too is an exciting direction for future extensions of these approaches.

As noted, this approach is not limited to flow cytometry or even to single cell data approaches. However, the use to single cell data confers unique advantages to our approach, permitting the analysis of rare cellular populations and the partitioning of samples into biologically relevant subsets. It also helps in a clinical setting, as a small amount of sample yields a large dataset with substantial analysis potential. One important application is the use of this approach to pinpoint disregulation in disease states, by comparing signaling pathway structure between normal and patient samples.

Causal representation is compromised even when cycles are permitted, because of hidden variables in the domain. From the algorithmic perspective, an approach that incorporates prior knowledge of the signaling structure to enable inference of hidden variables is needed. Technological advances may enable dynamic (time course) measurements in single cells; in the meantime, we are limited to "snapshots" in time. Nevertheless, it is possible to learn dynamic models from these static datasets, to some extent, this task constitutes an interesting area of research. Ultimately, appropriate computational and experimental tools that extend this approach may lead to improvements in the understanding of cellular biology and disease states, in accurate and specific diagnostics, and in more specifically tailored and perhaps personalized therapies for human disease.

REFERENCES

1. T. Ito, T. Chiba, R. Ozawa, M. Yoshida, M. Hattori, and Y. Sakaki, "A comprehensive two-hybrid analysis to explore the yeast protein interactome," *Proc. Natl. Acad. Sci. USA*, **98** (2001).
2. T. Ito, K. Tashiro, S. Muta, R. Ozawa, T. Chiba, M. Nishizawa, K. Yamamoto, S. Kuhara, and Y. Sakaki, "Toward a protein–protein interaction map of the budding yeast: A comprehensive system to examine two-hybrid interactions in all possible combinations between the yeast proteins," *Proc. Natl. Acad. Sci. USA*, **97** (2000).
3. P. Uetz, L. Giot, G. Cagney, T. A. Mansfield, R. S. Judson, J. R. Knight, D. Lockshon, V. Narayan, M. Srinivasan, P. Pochart, A. Qureshi-Emili,

Y. Li, B. Godwin, D. Conover, T. Kalbfleisch, G. Vijayadamodar, M. Yang, M. Johnston, S. Fields, and J. M. Rothberg, "A comprehensive analysis of protein–protein interactions in Saccharomyces cerevisiae," *Nature*, **403** (2000).
4. A. C. Gavin, M. Bosche, R. Krause, P. Grandi, M. Marzioch, A. Bauer, J. Schultz, J. M. Rick, A. M. Michon, C. M. Cruciat, M. Remor, C. Hofert, M. Schelder, M. Brajenovic, H. Ruffner, A. Merino, K. Klein, M. Hudak, D. Dickson, T. Rudi, V. Gnau, A. Bauch, S. Bastuck, B. Huhse, C. Leutwein, M. A. Heurtier, R. R. Copley, A. Edelmann, E. Querfurth, V. Rybin, G. Drewes, M. Raida, T. Bouwmeester, P. Bork, B. Seraphin, B. Kuster, G. Neubauer, and G. Superti-Furga, "Functional organization of the yeast proteome by systematic analysis of protein complexes," *Nature*, **415** (2002).
5. Y. Ho, A. Gruhler, A. Heilbut, G. D. Bader, L. Moore, S. L. Adams, A. Millar, P. Taylor, K. Bennett, K. Boutilier, L. Yang, C. Wolting, I. Donaldson, S. Schandorff, J. Shewnarane, M. Vo, J. Taggart, M. Goudreault, B. Muskat, C. Alfarano, D. Dewar, Z. Lin, K. Michalickova, A. R. Willems, H. Sassi, P. A. Nielsen, K. J. Rasmussen, J. R. Andersen, L. E. Johansen, L. H. Hansen, H. Jespersen, A. Podtelejnikov, E. Nielsen, J. Crawford, V. Poulsen, B. D. Sorensen, J. Matthiesen, R. C. Hendrickson, F. Gleeson, T. Pawson, M. F. Moran, D. Durocher, M. Mann, C. W. Hogue, D. Figeys, and M. Tyers, "Systematic identification of protein complexes in Saccharomyces cerevisiae by mass spectrometry," *Nature*, **415** (2002).
6. C. V. Mering, R. Krause, B. Snel, M. Cornell, S. G. Oliver, S. Fields, and P. Bork, "Comparative assessment of large-scale data sets of protein–protein interactions," *Nature*, **417** (2002).
7. M. E. Cusick, N. Klitgord, M. Vidal, and D. E. Hill, "Interactome: Gateway into systems biology," *Hum. Mol. Genet.*, **14**(2) (2005).
8. J. S. Bader, A. Chaudhuri, J. M. Rothberg, and J. Chant, "Gaining confidence in high-throughput protein interaction networks," *Nat. Biotechnol.*, **22** (2004).
9. A. Jaimovich, G. Elidan, H. Margalit, and N. Friedman, "Towards an integrated protein–protein interaction network: A relational Markov network approach," *J. Comput. Biol.*, **13** (2006).
10. R. Jansen, H. Yu, D. Greenbaum, Y. Kluger, N. J. Krogan, S. Chung, A. Emili, M. Snyder, J. F. Greenblatt, and M. Gerstein, "A Bayesian networks approach for predicting protein–protein interactions from genomic data," *Science*, **302** (2003).

11. I. Lee, S. V. Date, A. T. Adai, and E. M. Marcotte, "A probabilistic functional network of yeast genes," *Science*, **306** (2004).
12. N. Lin, B. Wu, R. Jansen, M. Gerstein, and H. Zhao, "Information assessment on predicting protein–protein interactions," *BMC Bioinformatics*, **5** (2004).
13. Y. Qi, J. Klein-Seetharaman, and Z. Bar-Joseph, "Random forest similarity for protein–protein interaction prediction from multiple sources," *Pac. Symp. Biocomput.* (2005).
14. I. Xenarios, L. Salwinski, X. J. Duan, P. Higney, S. M. Kim, and D. Eisenberg, "DIP, the Database of Interacting Proteins: A research tool for studying cellular networks of protein interactions," *Nucleic Acids Res.*, **30** (2002).
15. L. V. Zhang, S. L. Wong, O. D. King, and F. P. Roth, "Predicting co-complexed protein pairs using genomic, and proteomic data integration," *BMC Bioinformatics*, **5** (2004).
16. H. Dinkel, and H. Sticht, "A computational strategy for the prediction of functional linear peptide motifs in proteins," *Bioinformatics*, **23**, 3297–3303 (2007).
17. R. R. Joshi and V. V. Samant, "Bayesian data mining of protein domains gives an efficient predictive algorithm and new insight," *J. Mol. Model.*, **13**, 275–282 (2007).
18. S. Y. Huang and X. Zou, "An iterative knowledge-based scoring function for protein-protein recognition," *Proteins* (2008).
19. R. Linding, L. J. Jensen, G. J. Ostheimer, M. A. van Vugt, C. Jorgensen, I. M. Miron, F. Diella, K. Colwill, L. Taylor, K. Elder, P. Metalnikov, V. Nguyen, A. Pasculescu, J. Jin, J. G. Park, L. D. Samson, J. R. Woodgett, R. B. Russell, P. Bork, M. B. Yaffe, and T. Pawson, "Systematic discovery of *in vivo* phosphorylation networks," *Cell*, **129**, 1415–1426 (2007).
20. R. Singh, J. Xu, and B. Berger, "Struct2Net: Integrating structure into protein–protein interaction prediction," (2006).
21. M. Deng, S. Mehta, F. Sun, and T. Chen, "Inferring domain–domain interactions from protein–protein interactions," *Genome. Res.*, **12** (2002).
22. H. Wang, E. Segal, A. Ben-Hur, D. Koller, and D. Brutlag, "Identifying protein–protein interaction sites on a genome-wide scale," presented at *Advances in Neural Information Processing Systems (NIPS 2004)*, Vancouver, Canada (2005).

23. A. Campagna, L. Serrano, and C. Kiel, "Shaping dots and lines: Adding modularity into protein interaction networks using structural information," *FEBS Lett.* (2008).
24. Y. Shi, and J. Wu, "Structural basis of protein–protein interaction studied by NMR," *J. Struct. Funct. Genomics*, **8**, 67–72 (2007).
25. H. Lu, L. Lu, and J. Skolnick, "Development of unified statistical potentials describing protein–protein interactions," *Biophys. J.*, **84** (2003).
26. M. Narayanan, and R. M. Karp, "Comparing protein interaction networks via a graph match-and-split algorithm," *J. Comput. Biol.*, **14**, 892–907 (2007).
27. R. Singh, J. Xu, and B. Berger, "Global alignment of multiple protein interaction networks," *Pac. Symp. Biocomput.*, 303–314 (2008).
28. M. B. Yaffe, G. G. Leparc, J. Lai, T. Obata, S. Volinia, and L. C. Cantley, "A motif-based profile scanning approach for genome-wide prediction of signaling pathways," *Nat. Biotechnol.*, **19** (2001).
29. J. C. Obenauer, L. C. Cantley, and M. B. Yaffe, "Scansite 2.0: Proteome-wide prediction of cell signaling interactions using short sequence motifs," *Nucleic. Acids. Res.*, **31** (2003).
30. Z. Songyang, S. E. Shoelson, M. Chaudhuri, G. Gish, T. Pawson, W. G. Haser, F. King, T. Roberts, S. Ratnofsky, R. J. Lechleider *et al.*, "SH2 domains recognize specific phosphopeptide sequences," *Cell*, **72** (1993).
31. M. B. Yaffe, K. Rittinger, S. Volinia, P. R. Caron, A. Aitken, H. Leffers, S. J. Gamblin, S. J. Smerdon, and L. C. Cantley, "The structural basis for 14-3-3: Phosphopeptide binding specificity," *Cell*, **91** (1997).
32. M. B. Yaffe and L. C. Cantley, "Mapping specificity determinants for protein-protein association using protein fusions and random peptide libraries," *Methods Enzymol*, **328** (2000).
33. H. R. Hoogenboom and P. Chames, "Natural and designer binding sites made by phage display technology," *Immunol. Today*, **21** (2000).
34. V. Neduva, R. Linding, I. Su-Angrand, A. Stark, F. D. Masi, T. J. Gibson, J. Lewis, L. Serrano, and R. B. Russell, "Systematic discovery of new recognition peptides mediating protein interaction networks," *PLoS. Biol.*, **3** (2005).
35. I. Shmulevich, E. R. Dougherty, S. Kim, and W. Zhang, "Probabilistic Boolean Networks: A rule-based uncertainty model for gene regulatory networks," *Bioinformatics*, **18**, 261–274 (2002).

36. K. Amonlirdviman, N. A. Khare, D. R. Tree, W. S. Chen, J. D. Axelrod, and C. J. Tomlin, "Mathematical modeling of planar cell polarity to understand domineering nonautonomy," *Science*, **307**, 423–426 (2005).
37. I. T. Luna, Y. Huang, Y. Yin, D. P. Padillo, and M. C. Perez, "Uncovering gene regulatory networks from time-series microarray data with variational bayesian structural expectation maximization," *EURASIP J. Bioinform. Syst. Biol.*, **71312** (2007).
38. J. Pearl, "Probabilistic reasoning in intelligent systems: Networks of plausible inference," San Mateo, Calif.: Morgan Kaufmann Publishers (1988).
39. D. Heckerman, "A Tutorial on Learning With Bayesian Networks," (1995).
40. D. Pe'er, "Bayesian network analysis of signaling networks: A primer," *Sci. STKE*, **2005**, pl4 (2005).
41. C. Yoo and G. F. Cooper, "Causal discovery from a mixture of experimental and observational data," presented at *Uncertainty in Artificial Intelligence* (1999).
42. J. Pearl, *Causality: Models, Reasoning and Inference*, Cambridge, UK: Cambridge University Press (2000).
43. C. M. D. Heckerman and G. Cooper, "A Bayesian Approach to Causal Discovery in Computation, Causation, and Discovery," G. F. C. C. Glymour, Ed.: MIT Press, Cambridge, MA., pp. 141–166 (1999).
44. K. Sachs, D. Gifford, T. Jaakkola, P. Sorger, and D. A. Lauffenburger, "Bayesian network approach to cell signaling pathway modeling," *Sci. STKE*, **2002**, PE38 (2002).
45. L. A. Herzenberg, R. G. Sweet, and L. A. Herzenberg, "Fluorescence-activated cell sorting," *Sci. Am.*, **234**, 108–117 (1976).
46. D. F. Far, J. F. Peyron, V. Imbert, and B. Rossi, "Immunofluorescent quantification of tyrosine phosphorylation of cellular proteins in whole cells by flow cytometry," *Cytometry*, **15** (1994).
47. C. Muller, J. Kremerskothen, M. Zuhlsdorf, U. Cassens, T. Buchner, A. Barnekow, and O. M. Koch, "Rapid quantitative analysis of protein tyrosine residue phosphorylation in defined cell populations in whole blood and bone marrow aspirates," *Br. J. Haematol.*, **94** (1996).
48. P. O. Krutzik, J. M. Irish, G. P. Nolan, and O. D. Perez, "Analysis of protein phosphorylation and cellular signaling events by flow cytometry: Techniques and clinical applications," *Clin. Immunol.*, **110** (2004).
49. L. P. Kane, J. Lin, and A. Weiss, "Signal transduction by the TCR for antigen," *Curr. Opin. Immunol.*, **12** (2000).

50. J. M. Mullins, "Overview of fluorophores," *Methods Mol. Biol.*, **34** (1994).
51. M. Roederer, "http://www.drmr.com/compensation/," May 24 (2000).
52. K. Sachs, O. Perez, D. Pe'er, D. A. Lauffenburger, and G. P. Nolan, "Causal protein-signaling networks derived from multiparameter single-cell data," *Science*, **308**, 523–529 (2005).
53. M. P. Carroll, and W. S. May, "Protein kinase C-mediated serine phosphorylation directly activates Raf-1 in murine hematopoietic cells," *J. Biol. Chem.*, **269**, 1249–1256 (1994).
54. R. Marais, Y. Light, C. Mason, H. Paterson, M. F. Olson, and C. J. Marshall, "Requirement of Ras-GTP-Raf complexes for activation of Raf-1 by protein kinase C," *Science*, **280**, 109–112 (1998).
55. R. Marais, Y. Light, H. F. Paterson, and C. J. Marshall, "Ras recruits Raf-1 to the plasma membrane for activation by tyrosine phosphorylation," *Embo J.*, **14**, 3136–3145 (1995).
56. W. M. Zhang and T. M. Wong, "Suppression of cAMP by phosphoinositol/$Ca2^{+}$ pathway in the cardiac kappa-opioid receptor," *Am. J. Physiol.*, **274**, C82–C87 (1998).
57. R. Fukuda, B. Kelly, and G. L. Semenza, "Vascular endothelial growth factor gene expression in colon cancer cells exposed to prostaglandin E2 is mediated by hypoxia-inducible factor 1," *Cancer Res.*, **63**, 2330–2334 (2003).
58. Y. B. Kelley, B. P., F. Lewitter, R. Sharan, B. R. Stockwell and T. Ideker, "PathBLAST: A tool for alignment of protein interaction networks," *Nucleic Acids Res.*, **32**, W83–W88 (2004).
59. P. A. Steffen, M., J. Aach, P. D'haeseleer, and G. Church, "Automated modelling of signal transduction networks," *BMC Bioinformatics.*, **1**, 34 (2002).
60. K. M. Nir Friedman and S. Russell, "Learning the structure of dynamic probabilistic networks," presented at *Uncertainty in Artificial Intelligence*, Madison, Wisconsin (1998).
61. I. M. Ong, J. D. Glasner, and D. Page, "Modelling regulatory pathways in *E. coli* from time series expression profiles," *Bioinformatics*, **18**, S241–S248 (2002).
62. A. J. Hartemink and Massachusetts Institute of Technology, Department of Electrical Engineering and Computer Science, "Principled computational methods for the validation discovery of genetic regulatory networks," **206** (2001).

63. K. Sachs, D. Pe'er, D. A. Lauffenburger, and G. P. Nolan, "Markov neighborhood algorithm: Learning larger networks with measurements of overlapping subsets," Manuscript in Preparation.
64. S. Itani, M. I. Ohannessian, K. Sachs, G. Nolan, and M. Dahleh, "Formalism and structure learning for cyclic structures," Uncertainty in Artificial Intelligence (Submitted) (2008).

Chapter Six

Computational Modeling of Tumor Biobarriers: Implications for Delivery of Nano-Based Therapeutics

Hermann B. Frieboes, Paolo Decuzzi, John P. Sinek, Mauro Ferrari and Vittorio Cristini

6.1 INTRODUCTION

Cancer drug resistance due to both nano- and tumor-scale effects needs to be considered when modeling nanovector transport and drug release. While resistance in solid tumors is often a consequence of genetic factors, such as upregulation of anti-apoptotic proteins or overexpression of efflux mechanisms, elements at coarser physiological scales may also profoundly influence tumor therapeutic response.[1,2] These two scales are exquisitely linked, with effects in one influencing the other. A tumor could be a three-dimensional composite of fibrous and connective tissues, stromal components, vasculature, and even multiple cancer cell genetic clones. Atop this intrinsic heterogeneity is layered the anatomical and functional irregularity of tumoral vasculature, characterized by intermittent flow, collapsed vessels, diminished oxygen tension, and a large mean tissue-to-vessel distance.[3–7] Consequently, the tumor microenvironment can be highly variable, marked by gradients of cell substrates (e.g. nutrient and oxygen), with regions of hypoxia, acidity, and

Nanomedicine: *A Systems Engineering Approach* edited by Mingjun Zhang & Ning Xi

www.panstanford.com
978-981-4241-36-6

necrosis, and heterogeneous cell proliferation, all of which have effects at the physical scale of nanovectors.

Nanovectors must first flow through the vasculature, either attach to the vascular endothelium or extravasate at the right location and diffuse through the lesion tissue, and then release sufficient drug to be transported into tumor cells in lethal doses. The drug must bind to its target, e.g. DNA, and induce cell apoptosis or mitotic inhibition. The vessel bed's blood flow and spatial distribution can severely hinder uniform extravasation, calling into question the capability of nanovectors and drug molecules to adequately distribute throughout tissue. Experiments *in vitro*[8–10] and *in vivo*[11,12] demonstrate limited free drug penetration through tumors, especially highly protein-bound molecules for common drugs such as doxorubicin and paclitaxel. Once a drug molecule has traversed lesion tissue from its point of release and is presented to a cancer cell, the path from extracellular space to intracellular target is fraught with difficulties ranging from protonation due to the acidic environment, which, for example, renders anthracyclines such as doxorubicin incapable of traversing membranes, to intracellular removal by drug efflux pumps, to cellular processes that effect DNA repair and drug clearance.[13–18] In addition to pharmacokinetics, drug pharmacodynamics can also be impaired. Significant hypoxia and hypoglycemia throughout may induce cell quiescence, reducing the efficacy of cell-cycle chemotherapeutic agents like doxorubicin and cisplatin.[19,20] Hypoglycemia further causes the glucose-regulated stress response, detrimental to the action of topoisomerase II-directed drugs like doxorubicin.[21–24]

The heterogeneity and physical three-dimensionality of the tumoral microenvironment and its influence at the nano-scale thus present serious challenges to drug assessment for both traditional therapies and nano-based therapeutics. This is a major reason why a particular drug showing marked activity against a particular specimen in the laboratory *in vitro* may have disappointing potency *in vivo*, as evidenced by the differential between positive predictive accuracy of *in vitro*-assisted therapy selection (around 70%) and negative predictive accuracy (around 90%) — a situation not remarkably changed over the years.[25,26] Supraoptimal delivery of drug to cultured cells may ameliorate some of the biobarriers, but such delivery *in vivo* is oftentimes impractical due to patient toxicity. A drug that consistently works *in vitro* can therefore be expected to only sometimes work *in vivo*.

Unraveling the myriad interactions of therapeutic determinants within the complex three-dimensional tumoral environment is difficult, resulting in high costs of drug development and patient suffering. A significant capability of computational (*in silico*) modeling is the ability to integrate components into a virtual system capable of reproducing complex behavior, incorporating circuits of information flow difficult to analyze explicitly, thus providing better control over and monitoring of the simulated *in vivo* tumor environment. The power of *in vitro* experimentation lies in its ease of implementation while remaining in the biological realm. By its very nature, *in vitro* experimentation attempts to refine and isolate. Yet much of what happens *in vivo* is the result of a nonlinear system whose behavior is more than the sum of its parts. Moreover, computer modeling can create hypothetical environments and conditions impossible to achieve otherwise, the study of which is nonetheless instrumental in unraveling disease and drug mechanisms. This expansive control, founded upon an adequately mathematical basis, could facilitate the discovery of hypotheses to ascertain efficacy of drugs and nano-based therapies, potentially on a patient-by-patient basis. The relative ease and cost-efficiency of performing simulations could furthermore allow an exhaustive investigation of strategies, revealing the optimal among them. Accurately calibrated and rigorously validated, models of nanovectored drug delivery to tumors could provide a "dry-lab" to be used as a powerful complement to the traditional wet-lab in fundamental research, drug discovery, and the clinic.[27] Results could then suggest supportive *in vitro* and *in vivo* experimentation, the end result being new therapeutic targets or strategies. Simultaneously, weaknesses and strengths of the computational models could be uncovered and addressed. Such an approach has the potential to facilitate an era of discovery and progress in understanding and treating cancer via nanotherapeutics, and provide new hope to its sufferers.

6.2 BACKGROUND

The computational modeling of tumor biobarriers is dependent on adequate mathematical and *in silico* modeling of vascular and avascular tumors. The past two decades have witnessed impressive work in modeling tumor growth.[28–39] As the basic elements of these models have matured, specialized models have been developed of angiogenesis and flow,[40–43] drug delivery and response,[11,44–47] and effects of the three-dimensional tumor microenvironment.[48–50] Simulation and analysis using one-dimensional modeling (employing cylindrical or spherical symmetry) have also provided insight (*sans* discrete vasculature). More recently, powerful numerical methods have been developed to simulate multi-dimensional complex morphological progression and its relation to cell phenotype and the microenvironment involving, for example, nutrient and biomechanical tissue response.[38,49,50] We briefly review the implications of some of this work on nano-based therapeutics.

A spatio-temporal model of tumor response to sequestered, intracellular drug (doxorubicin) predicted that the long-term response to repeated therapy rounds is very sensitive to changes in the threshold level of drug required to initiate apoptosis at the maximum rate.[44] Perturbations of this parameter mediated the difference between effective tumor regression and minimal growth delays. Sensitivity analysis showed that decreasing cellular permeability, as opposed to decreasing sequestration rate or increasing cellular efflux, may be the most effective way for tumor cells to overcome the growth control afforded by successive rounds of treatment. These results imply that cell permeability to drug is a key parameter in the design of nanovectors.

The action of a single chemotherapeutic drug and how different drug kinetics and treatment regimes may affect the final treatment outcome were investigated through a mathematical model that showed that a single infusion of drug could be more effective than repeated short applications.[45] The model enabled predictions of the strength of drug required to achieve tumor regression. This result presents a lower bound on the amount of drug to be delivered via nanovectors. In another study, reduction in volume of a vascular tumor in response to specific chemotherapeutic administration strategies was modeled via partial differential equations governing intratumoral drug concentration and cancer cell density.[51] In the model the tumor was treated as a continuum of two types of cells that differ in proliferation rates and responses to the chemotherapeutic agent. Insight into the tumor's response to therapy was gained by applying a combination of analytical and numerical techniques to the model equations, e.g. bolus injection and continuous drug infusion resulted in similar times to cure a tumor containing only one kind of drug sensitive cells. However, when the tumor contained a drug resistant population, continuous infusion could significantly increase the time to cure. Since drug release kinetics of nanovectors more resembles continuous infusion, this result may directly apply to nano-based therapeutics.

In order to study drug penetration in the laboratory through three-dimensional tumor tissue for commonly used drugs, the multicellular layer (MCL) method can be employed.[9] Results have suggested a limited ability of anticancer drugs to reach tumor cells that are distant from blood vessels. Confirming these findings, an avascular tumor growth model was adapted to compare the effects of drug application on multi-cell spheroids and monolayer cultures, showing an enhanced survival rate in spheroids, consistent with experimental observations, indicating that the key factor determining this effect is drug penetration.[47] This fundamental biobarrier may be alleviated through nanovectors designed to extravasate and actively transport drug to their target.

A theoretical model for the cellular pharmacodynamics of cisplatin that takes into account the kinetics of drug uptake by cells and intracellular binding was used to predict the dependence of survival on the time course of extracellular exposure.[52] Cellular pharmacokinetic parameters were derived from uptake data for human ovarian and head and neck cancer cell lines, and survival was assumed to depend on the peak concentration of DNA-bound intracellular platinum. The model provided a better fit to experimental data sets including long exposure times, such as would occur with nano-based therapeutics, as well as a possible explanation for the fact that cell kill correlates well with area under the extracellular concentration-time curve in some data sets, but not in others. This model may help to optimize delivery schedules and dosing of cisplatin in cancer therapy.

A mathematical model for cellular uptake and cytotoxicity of doxorubicin assumed sigmoidal, Hill-type dependence of cell survival on drug-induced damage.[53] Since experimental evidence indicates distinct intracellular and extracellular mechanisms of doxorubicin cytotoxicity, drug-induced damage was expressed as the sum of two terms, representing the peak values over time of concentrations of intra- and extracellular drugs. Dependence of cell kill on peak values of concentration rather than on an integral over time is consistent with observations that dose-response curves for doxorubicin converge to a single curve as exposure time is increased. The model provided good fits to *in vitro* cytotoxicity data, in agreement with experiments showing how saturation of cellular uptake or binding with concentration can result in plateaus in the dose-response curve at high concentrations and short exposure. These findings also affect drug delivered via nanovectors, which would implement lower concentrations and longer exposure times.

Modeling of fluid flow through tumor vascular networks highlighted issues that may have major implications for delivery of chemotherapeutic drugs to tumors, such as, under certain conditions, simulations predicted that an injected chemotherapy drug may bypass the tumor altogether,[42] which could also severely diminish nanovector delivery to the tumor site. In further studies, a mathematical model that simultaneously couples vessel growth with blood flow through the vessels was created to examine the effects of various changing physical and biological model parameters on the developing vascular architecture and the delivery of chemotherapeutic drugs to the tumor.[41] Simulations of treatments under different parameter regimes indicated that networks characterized by increased capillary branching would be poorly suited to supplying cell substrates to a developing tumor. Conversely, it was shown that treatment targeting tumor cells by injection into such a capillary network would yield poor treatment efficacy, since the highly dilated shunt removes the drug from the network before it can reach the tumor. These same issues would affect not only drug delivered by nanovectors, but the nanovectors themselves.

6.3 TUMOR BIOBARRIERS

The first biobarrier that nanovectors encounter is the flow of blood in the vasculature. Vascular diffusivity is therefore a key parameter in nano-based therapeutics. Nanovectors can be designed to adhere preferentially to the endothelium of tumor vessels, and then release drug at a constant rate. Compared with boundary conditions for free drug, mainly involving the convection in the blood and followed by diffusion through tumor tissue, the boundary conditions for nanotherapeutics must consider different convection dynamics, plus interaction with the vessel endothelium. From a computational modeling standpoint, once nanovectors have adhered to the desired site, the diffusion of drug into tumor tissue follows the same dynamics as a bolus application except that the amount per time may be smaller and the release time longer. Nanovectors can also be designed to extravasate from the vessels and diffuse into the tumor tissue first before releasing the drug. In the next subsections, we describe computational modeling of these biobarriers in further detail.

6.3.1 Vascular Flow

The complex interaction between vascular blood flow and tumor growth was recently examined[54] through an improved continuum model of solid tumor invasion[55] with a model of tumor-induced angiogenesis.[41] This multi-scale model of vascular solid tumor growth coupled invasion and angiogenesis through both tumor angiogenic factors (TAF) released by the tumor cells and cell substrates from the neo-vascular network. As blood flows through this network, oxygen and nutrient extravasate and diffuse through the extra-cellular matrix (ECM), furthering tumor growth, which in turn influences TAF expression. Extravasation is mediated by the hydrostatic stress (solid pressure) generated by the growing tumor. This pressure also affects vascular remodeling by restricting the radii of the vessels and hence the flow pattern and wall shear stresses. Tumor progression and vascular network were further coupled via the ECM as both tumor and endothelial cells upregulate matrix degrading enzymes that degrade the ECM, in turn affecting haptotactic migration. Simulations demonstrated the importance of the nonlinear coupling between growth and remodeling of the vascular network, the blood flow through the network, and the tumor progression, all of which would directly affect delivery, extravasation, and drug release by nanovectors. Significantly, the solid pressure created by tumor cell proliferation shuts down large portions of the vascular network, thereby dramatically affecting flow, network remodeling, delivery of cell substrates to the tumor, and tumor progression. ECM degradation by tumor cells had a dramatic effect on both development of the vascular network and tumor growth response. When the degradation was significant, the neovasculature tended to encapsulate, rather than penetrate the tumor, and hence would severely impair nanovector delivery through the vascular flow.

6.3.2 Vascular Diffusivity

The effective longitudinal diffusion of nanovectors along non-permeable and permeable capillaries considering the contribution of molecular and convective diffusion is an important issue in the transport of nanovectors. Based on Taylor's theory of shear dispersion for non-permeable and permeable capillaries as a function of hemodynamic conditions and nanovector size, it has been shown by Decuzzi and coworkers[56] that for a given capillary size there exists a critical radius a_{cr} for which the effective longitudinal diffusion along the capillary has a minimum: nanovectors with $a < a_{cr}$ diffuse mainly by Brownian diffusion whereas nanovectors with $a > a_{cr}$ diffuse mainly by convection and the effective diffusion coefficient D_{eff} grows with a. In normal capillaries, the critical radius ranges between few nanometers and 150 nm, whereas in tumor vessels a_{cr} can be much larger depending on the wall permeability. In permeable conduits, the effective diffusion reduces significantly compared to normal non-leaky vessels and a_{cr} grows almost linearly with the hydraulic permeability L_p of the blood vessels, which depends on the organ and location in the body.[56] This implies that the longitudinal diffusion of nanovectors within tumor microvasculature is significantly reduced, which can be interpreted as a new physiological barrier to the delivery of therapeutics and diagnostic agents into tumors. Based on this, possible strategies can be devised[56] to increase the number of vectors reaching into tumor vessels: (i) use vectors with a critical radius for normal vessels, so that the difference between the effective diffusivity of normal and tumoral vessels is reduced, (ii) inject solutions of vectors with different radii, which can be targeted simultaneously to different capillaries, given the large variability of vessel radii and flow conditions; (iii) normalize the tumor vasculature, i.e. decrease vessel permeability.

6.3.3 Adhesion to Vascular Endothelium

Three main governing parameters have been identified which influence adhesion probability P_a of a nanovector to the endothelium: the *shear stress*, proportional to the propulsive force kV_0 exerted by the fluid (k is effective viscosity of blood and V_0 is the free velocity of the particle), the *loading rate*, inversely proportional to the time constant $\tau = (nk_2 + k)/nk_1$ (where n is number of active bonds and k_1 and k_2 are elastic and viscous parameters of the particle,

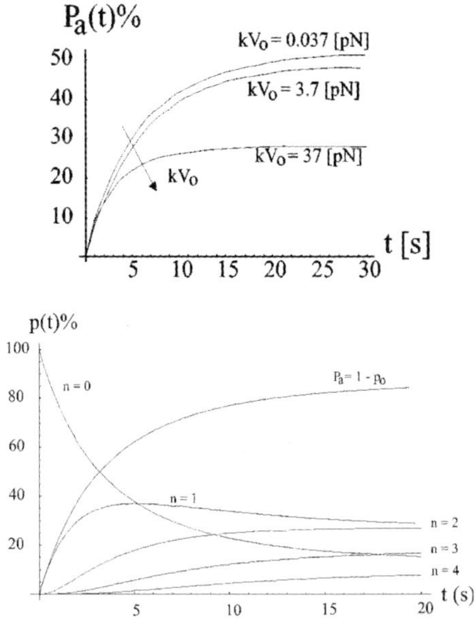

Figure 1. Top: Effect of wall shear stresses on the adhesion probability as a function of time ($kV_0 = 0.037$ [pN], 3.7 [pN], and 37 [pN]). Bottom: Variation of adhesion probability $P_a(t)$ as a function of time for different values of number of simultaneously active bonds n. Reprinted with permission from Decuzzi *et al.*, *Annals of Biomed. Eng.*, 32, 797–798 (2004). Copyright © Kluwer Academic Publishers. With kind permission of Springer Science and Business Media.

respectively), and the *ligand/receptor density* ratio m.[57] The effect of each of these parameters on the adhesion probability can be estimated (Figs. 1 and 2). As the hemodynamic force kV_0 acting over the particle increases, the probability of adhesion decreases, and a linear relation of the form $P_a \approx A_1 - A_2(kV_0)$ exists between the long-term adhesion probability P_a and kV_0, where A_1 and A_2 depend on the system properties.[57]

6.3.4 *Endocytosis*

Decuzzi and Ferrari[58] developed a mathematical model to predict both the adhesive and endocytotic performances of a nanoparticle,

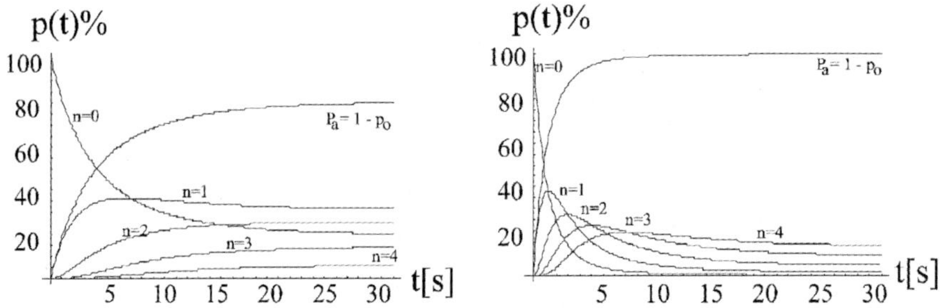

Figure 2. Variation of adhesion probability with time for two different values of ligand density. Left: $m_l = 700$ [μm^{-2}]; Right: $m_l = 2500$ [μm^{-2}]. Reprinted with permission from Decuzzi *et al.*, *Annals of Biomed. Eng.*, 32, 800 (2004). Copyright © Kluwer Academic Publishers. With kind permission of Springer Science and Business Media.

identifying three different categories of governing parameters: geometrical (radius of the particle a); biophysical (ligand-to-receptor surface density ratio β; non-specific interaction parameter F; force parameter H_1) and biological (ligand–receptor binding affinity K_A^0). A sample model result is shown in Fig. 3, linking ligand-to-receptor surface density ratio to the value of the radius. By tailoring the surface properties of the particle to generate an attractive non-specific force, the number of ligands required to achieve firm adhesion can be significantly reduced under fixed hydrodynamic and biological conditions (fixed H_1 and K_A^0). Also, the model suggests engineering ligands with smaller reactive compliances rather than larger binding affinities to enhance fast firm adhesion.

Combining the nanoparticle adhesion and endocytic performances, Decuzzi and Ferrari[58] also developed *Design Maps* (Fig. 4) relating the ligand-to-receptor surface density ratio β with the non-specific attractive parameter F for given values of the particle size, force parameter H_1 and binding affinity. Based on the nanoparticle properties that can be controlled during their fabrication (radius a, ligand-to-receptor surface density ratio β, and ligand–receptor binding affinity K_A^0), these Design Maps enable *a priori* determination of whether a proposed particle design can adhere to the targeted vasculature and undergo internalization by endothelial cells.

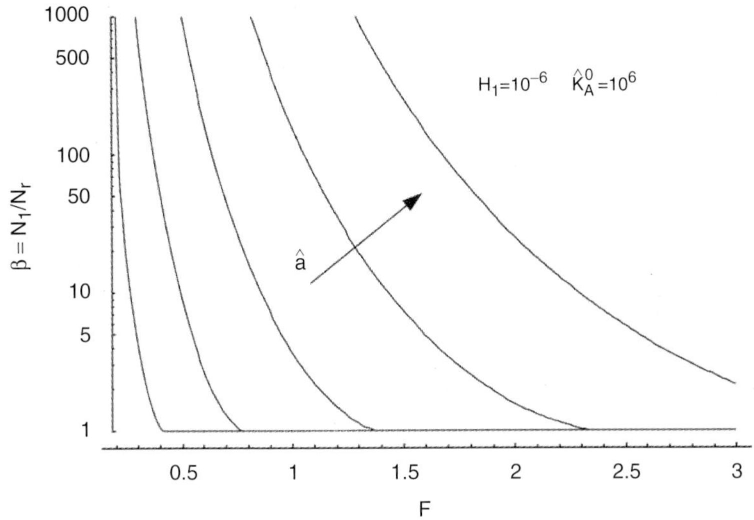

Figure 3. Ligand-to-receptor surface density ratio β as a function of the non-specific parameter F for different values of the dimensionless particle size $\bar{a}$, for a 50% probability of adhesion. Reprinted with permission from Decuzzi and Ferrari, *Biomaterials*, 28, 382 (2008). Copyright © Elsevier.

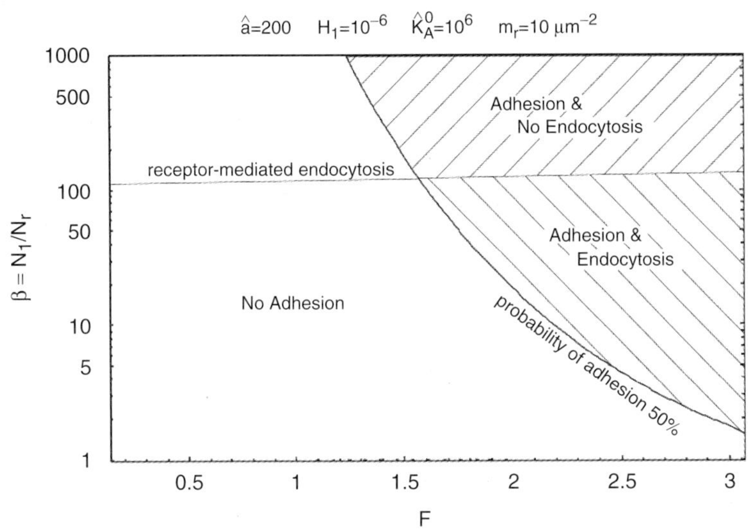

Figure 4. A typical design map showing areas for no adhesion; adhesion and no endocytosis; adhesion and endocytosis. Reprinted with permission from Decuzzi and Ferrari, *Biomaterials*, 28, 382 (2008). Copyright © Elsevier.

6.3.5 Drug Release

Once nanovectors have successfully adhered or extravasated at their intended site, cellular level drug kinetics and pharamacodynamics become the main concerns of computational modeling. Most important is the sustained release of a sufficient concentration of drug, which is also be a concern for ligand-conjugated vectors since some of their drug will be released into the tumor interstitium. The physics of nanoparticle drug release is well researched, with the Higuchi, power law, and Weibull models sometimes used as phenomenological approximations. Nanovector release profiles frequently evince a simple bi-exponential release pattern described by $C_t = C_\infty - (Ae^{-\alpha t} + Be^{-\beta t})$, where C_∞ is the total drug, C_t is the amount of drug released by time t, A is the rapidly released portion of drug with rate constant α, and $B = C_\infty - A$ is the slowly released portion of drug with rate constant β.[46,59,60] If the release can be sustained long enough, then the bi-exponential becomes approximately linear with release rate $B\beta$. Even with this simplification, cellular level drug kinetics and transport are highly non-uniform not only because of the inhomogeneous transport of vectors through and extravasation from tumoral vasculature, but because of drug gradients due to cellular uptake and metabolism.

In order to better understand the interaction between tumor-scale effects and drug release, Sinek *et al.*[61] examined in detail the therapeutic efficacy of two common drugs, cisplatin and doxorubicin, in relation to phenotypic and microenvironmental conditions. Model parameters governed extracellular drug/tissue diffusivity; cellular uptake, efflux, and metabolism; cellular density; and the effect of cell substrate heterogeneity on drug action. Simulations were performed in a two-dimensional (non-symmetric) setting employing discrete vasculature, which enables the incorporation of morphological and topological influence on drug and cell substrate distributions. The effect of these distributions on therapeutic efficacy is of special interest.

For example, extravasation, diffusion and cellular uptake of both drug and cell substrate is simulated according to the quasi-steady state reaction-diffusion equations

$$\begin{aligned} 0 &= \nu_s \delta_V + D_s \nabla^2 s - \eta_s s, \\ 0 &= \nu_n \delta_V + D_n \nabla^2 n - \eta_n s, \end{aligned} \tag{6.1}$$

where s and n are the local concentrations of drug and substrate, respectively, the ν's are (spatially and temporally variable) production rates related to release of drug and supply of substrates, the η's are uptake rates by cancer cells, and the D's are diffusion coefficients. δ_V is the Dirac delta function indicating the location of the vasculature. Drug action is then modeled as cell kill being proportional to normalized drug concentration $\bar{s}$ acting on the fraction of cycling cells, given by the normalized cell substrate $\bar{n}$. When combined with the growth of cells, modeled as the product of a mitosis constant and normalized substrate $\lambda_M \bar{n}$, the net local growth or regression of tumor cells (the velocity field divergence) becomes

$$\nabla \bullet \mathbf{u} = \lambda_M \bar{n} - \lambda_D \overline{sn}, \tag{6.2}$$

where λ_D is the killing power of the drug.

6.3.6 Tumor Response

Sinek *et al.*[46] performed simulations to study extravasational difficulties due not only to irregular vascular topology generated using Anderson and Chaplain's model, but also due to pressure variations within tumor interstitium. The latter was accomplished by using $\nu_n = \nu_n'(p_V - p)(n_V - n)$ in Eq. (6.9), where ν_n' is constant, p_V and p are the pressures in the vasculature and tumor, respectively, and n_V and n are the cel substrate concentration in the vasculature and tumor, respectively. A similar function for ν_s was used for the initial extravasation of particles. This model qualitatively demonstrated that inhomogeneities in drug delivery and action, even using nanoparticles releasing drug at a constant rate, had the potential to diminish chemotherapeutic efficacy, leaving substantial regions of tumor unharmed after weeks of simulated therapy (Fig. 5).

The work by Sinek *et al.*,[46] by assuming one homogenous lesion compartment and not based upon experimentally acquired parameter values, employed a rudimentary pharmacokinetics and pharmacodynamics (PKPD) component. In contrast, Sinek *et al.*[61] recently implemented an extensive multi-compartment PKPD component whose parameter values were calibrated via published experimental data. This enabled a comparison of the tissue- and cell-level drug dynamics of the two drugs, and facilitated the generation of hypotheses to explain their *in vivo* characteristics. Indeed, the methodology presented could, with additional development, be

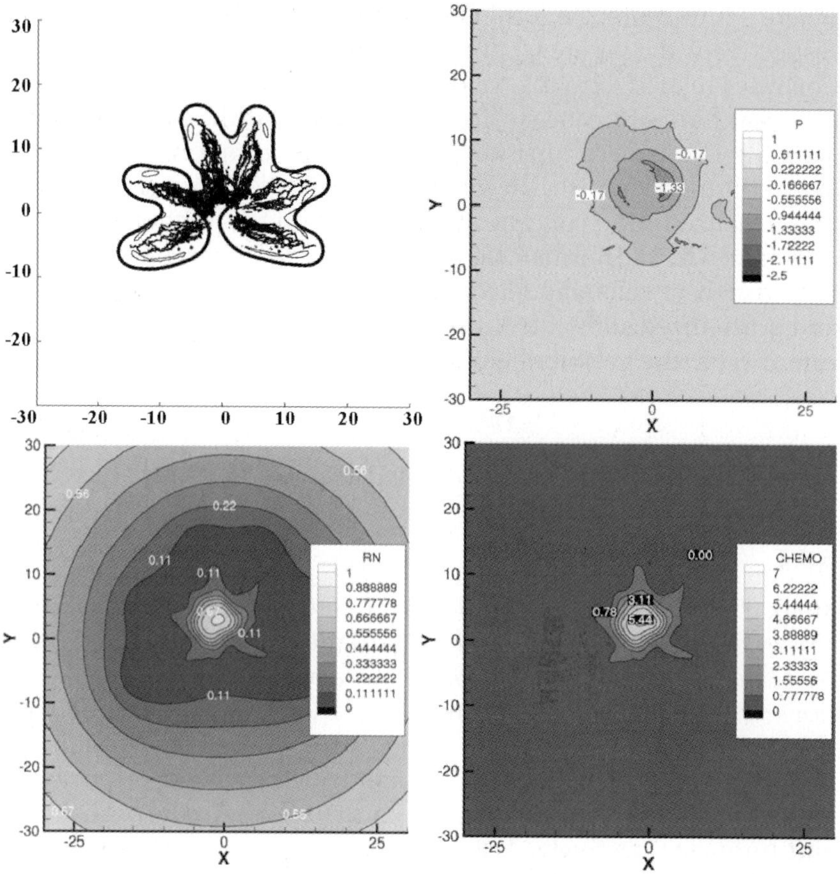

Figure 5. Simulations of nanovector chemotherapy. Clockwise, from upper left corner: tumor at equilibrium after weeks of simulated continuous therapy, pressure contours, drug concentration contours, and cell substrate contours. Inhomogeneities in substrate delivery and initial nanoparticle extravasation and subsequent diffusion and uptake of drug result in stable equilibrium at significant tumor mass. Reprinted with permission from Sinek *et al.*, *Biomed. Microdev.*, 6, 306 (2004). Copyright © Kluwer Academic Publishers. With kind permission of Springer Science and Business Media.

applied to both established and nascent drugs to the end of refining clinical trials and assisting in clinical therapeutic strategy. It is this recent work that will be the focus for the remainder of this chapter, as we evaluate implications of the results for nano-based therapeutics delivery.

6.4 COMPUTATIONAL MODEL

6.4.1 Model Description

The multiscale tumor growth and angiogenesis simulator developed by Zheng *et al.*[38] was used to grow the lesions upon which chemotherapy was simulated. This is a nonlinear, continuum scale, two-dimensional growth engine whose accuracy is made possible by an adaptive finite element mesh due to Cristini *et al.*[62] The mesh enables multi-scale computation for finely resolving tumor morphology, especially around important areas such as the necrotic/tumor and tumor/host interfaces and around capillary sprouts. Without true multi-dimensionality and discrete vasculature, it would be difficult to simulate the heterogeneities of cell substrate and drug that profoundly affect therapeutic efficacy. The mathematical model is derived from first principles describing physical conservation laws (e.g. diffusion equation), with conserved variables representing the known characteristics of tumor behavior. Variables include rates of proliferation, apoptosis and in consequence cell substrate consumption and cell inhibition; cell substrate and drug concentrations and diffusion rates; mobility function and velocity of tumor cells; and tumor mass effects. Realistic and heterogenous vasculature through which substrates and drug are provided is generated via the angiogenesis model of Anderson and Chaplain.[40] The simulation field incorporates three phases: viable cancerous tissue, normal host tissue, and necrotic debris. By solving the model equations numerically tumor progression and therapy outcome are predicted by quantifying their link to combinations of relevant variables. At any given time during tumor growth and treatment, the computational model outputs the computed values of all relevant variables at every location within the three-dimensional tumor tissue, e.g. the spatial distributions of oxygen, cell substrates, drug and tumor cells.

Briefly, cell substrates are provided through the discrete microvasculature, which is generated in response to angiogenic regulators produced from perinecrotic cells. This results in proliferation and tumor growth. The simple steady-state diffusion equation

$$0 = k_v(1 - n)\delta + D_n \nabla^2 n - k_n n \tag{6.3}$$

is used to model cell substrate delivery and uptake, where n is the local substrate normalized by the intravascular level, k_v is a measure of vascular porosity (0 is impermeable, ∞ is completely porous), δ is the Dirac delta function located along the vasculature, D_n is cell substrate diffusivity, and k_n is the local rate of consumption by cells.[38] The characteristically high porosity of tumor vasculature implies a very high setting of k_v so that, essentially, vasculature provides a constant boundary condition of 1. Experiments[63] demonstrate that oxygen penetrates approximately 150 μm into tumor tissue before falling to about 10% of serum level. Combining this with a diffusivity D_n of around 60,000 $\mu m^2 \min^{-1}$,[64,65] the substrate uptake rate is calculated to be $k_n = 24 \min^{-1}$. Waste resulting from necrotic cell degradation is assumed to be removed via convection towards and through the tumor-host interface as well as via scavenger cell phagocytosis. In regions where cell substrates are sufficient to maintain viability, mitosis is assumed to be directly proportional to their concentration, with the proportionality constant dependent upon the average cell cycle time of the malignant population.

Once the tumors are grown, drug administration released through the vasculature is simulated by a multi-compartment pharmacokinetics model, based upon earlier work.[14,32,53,66] For cisplatin, there are three compartments corresponding to (6.1) extracellular, (6.2) cytosolic, and (6.3) DNA-bound drug. For doxorubicin, there is a fourth compartment corresponding to intracellular organelles, e.g. lysosomes. The system of equations governing transport for both drugs (with different parameter values) is[61]:

$$\begin{aligned}
\dot{s}_1 &= k_v(s_v - s_1)\delta + D_s\nabla^2 s_1 - k'_{12}s_1 + k'_{21}(s_2/10^6 V_C)\\
\dot{s}_2 &= k_{12}V_C s_1 - k_{21}s_2 + k_{32}s_3 - k_{23}s_2(1 - s_3/s_M)\\
&\quad + k_{42}s_4 - k_{24}s_2 \\
\dot{s}_3 &= k_{23}s_2(1 - s_3/s_M) - k_{32}s_3 - k_3 s_3\\
\dot{s}_4 &= k_{24}s_2 - k_{42}s_4
\end{aligned} \tag{6.4}$$

where s_i represents drug concentration in compartment i, k_{ij} represents a transfer rate from compartment i to j, and k_i represents a rate of permanent removal from compartment i and the system. s_v is intravascular drug concentration during bolus, and s_m is a DNA saturation parameter relevant to doxorubicin. V_C is the volume of a cell (assumed spherical with diameter 10 μm, yielding $V_C = 520$ fL-cell^{-1}) and appears in the first two equations to reconcile the dimensions of s_v and s_1 (μM) with the dimensions of all other compartments (fmoles/cell). k_v and δ are the same as in Eq. (6.3).

The primed rates appearing in the first equation are related to their unprimed counterparts via $k'_{ij} = k_{ij} = F$ where F is the extracellular fraction of whole tissue. Taking a baseline tumor density of $\rho = 1.0\text{E}9$ cells-mL^{-1}, a well-known representative value, in combination with the cell volume previously quoted results in

$$F = 1 - \rho V_C(10^{-12}\,\text{mL-fL}^{-1}) = 0.48, \qquad (6.5)$$

also a reasonable value. Finally, D_S is the diffusivity of the drug through extracellular space.

Both cisplatin and doxorubicin pass through cell membrane according to k_{12} (which includes possible pump and transporter activity, as do all other rates). From there, the drugs may efflux according to k_{21} or may bind to DNA according to k_{23}. The kinetics differ from here for the two drugs. Cisplatin may be removed according to the rate k_3, which destroys the functioning of the drug and repairs the DNA.[18] Doxorubicin, however, has an off rate given by k_{32}, and moreover may be sequestered and released by lysosomes according to k_{24} and k_{42}.[13,15,67] Although lysosomal flow to membrane and exocytosis of sequestered drug plays a role in some drug resistant cell lines, drug resistance is not necessarily modeled via this function, and this process is assumed to be negligible in accordance with Ref. 68. On the other hand, the quantity of drug lysosomes can sequester is important, as this contributes to the cellular uptake of drug, and hence, its penetration characteristics.

The pharmacodynamics model consists of the Hill-type equation along the lines of those employed in Refs. 52 and 53

$$E = \frac{N(n)}{1 + A^{-1}x^{-m}} \qquad (6.6)$$

where E is cell inhibition (1 minus surviving fraction), x is DNA-bound drug-time product (area under the curve, or AUC), and A and m are phenomenologically fit parameters. $N(n)$ is a function of cell substrate n ranging from 0 to 1 used to mimic the effect of hypoxia and hypoglycemia. Results with doxorubicin show that cells in deeper layers of tumors do not respond as well to drug as do cells on the surface, even when intracellular drug levels are taken into account.[19,20] Other experiments demonstrate reduced response in monolayer when cells are forced into quiescence due to reduced oxygen.[24] Still others show that hypoglycemia can deplete topoisomerase II, thus reducing the effect of some anthracyclines.[23] These results imply that the response of cells to doxorubicin *in vivo* might correlate to the local level of cell substrates, a phenomenon that we herein call the "substrate effect." For our purposes, the exact form of N is not important. For simplicity, we choose $N = n^p$, where p is a phenomenological parameter derived from the data of Ref. 20, and equals 0.4. Since in the model n is normalized with respect to the intravascular level, it runs from 0 to 1, and thus so does N. Furthermore, at full substrate levels, $N = 1$, and so cell inhibition is maximal. In the simulations, drug pharmacokinetics [Eqs. (6.4)] is allowed to proceed from the time of drug release to washout 20 hours later. During this time the locally varying DNA-bound AUC is calculated and used to find cell inhibition [Eq. (6.6)].

6.4.2 Pharmacokinetics Model Parameters

A generally acceptable theoretical setup for performing experiments to measure compartmental concentrations (and therefore to derive the rate constants) is either a suspension or monolayer in an inexhaustible drug-laden medium corresponding to s_1. Under these conditions, the relevant model consists of the last three equations in Eqs. (6.4), with s_1 held constant. We will refer to this model as the modified version of Eqs. (6.4). All model parameters and values are summarized in Table 1. These will be referred to as the baseline values, some of which will be adjusted later to simulate different tumor characteristics and therapeutic treatments. We emphasize that parameter values, having been derived from a variety of published experimental data spanning many years and cell types, correspond to a prototypical tumor and cancer cells suitable for the simulations herein, but not necessarily representative of any particular clinical specimen.

6.4.2.1 Cisplatin parameters

We begin with cisplatin, setting k_{24} and k_{42} to 0 since we assume only three compartments, and k_{32} to 0 since we assume the repair rate k_3 is the dominant removal rate of DNA-bound drug. k_3 is next obtained as follows. In experiments performed by Sadowitz *et al.*,[69] adducts per million nucleotides on isolated peripheral blood mononuclear cell DNA fall from 75 to 5 and 185 to 40 in two hours in two different experiments. Thus, assuming the exponential repair model $\dot{s}_3 = k_3 s_3$, we calculate the repair rate to be about 0.015 min^{-1}. An initial estimate of k_{23} is then made as follows. Sadowitz shows that for 7 μM cisplatin, in two hours peripheral blood mononuclear cells accumulate from about 25 (non-thiol-blocked cells) to 175 (thiol-blocked cells) adducts per million nucleotides. Assuming that DNA consists of about 1.25E6 kbp, this converts to from 1.04E-4 to 7.27E-4 fmoles of Pt docked on the DNA (1 atom/adduct). Neglecting the cell membrane and supposing DNA to be exposed directly to the drug, we have the ODE $\dot{s}_3 = 7\lambda_{23} - k_3 s_3$, where λ_{23} is a clearance parameter (fL-min^{-1}). The solution is $s_3 = 7(\lambda_{23}/k_3)(1 - \exp(-k_3 t))$. Substituting values of $k_3 = 0.015\,\text{min}^{-1}$, $t = 120\,\text{min}$, and $1.04\text{E-}4 \leq s_3 \leq 7.27\text{ E-}4$ fmole yields $0.27 \leq \lambda_{23} \leq 1.9$ fL-min^{-1}. To convert this to a rate we use the relation $k_{23} = \lambda_{23}/V_C$, arriving at $5.19\text{E-}4 \leq k_{23} \leq 3.65\text{E-}3\,\text{min}^{-1}$. The assumption that DNA was exposed directly to the cisplatin solution means that this rate is only a bootstrap approximation and must be refined. We note that the extremely low ratio of adducts per kbp implies that the saturation capacity of DNA with respect to cisplatin is never approached, and so set s_m to ∞.

Next, we estimate k_{12} and k_{21}. While doing this we will refine the initial estimate of k_{23}. The whole procedure involves fitting the best curves to data from Troger *et al.*[70] (Fig. 1(a)). Troger exposed human tongue carcinoma CAL-27 cells in monolayer to four different concentrations of cisplatin and then measured total intracellular amount of Pt at selected times. This corresponds to $s2 + s3$ in the model. Beginning with the previous estimate of k_{23} and setting s_1 to concentrations used by Troger, we adjust k_{12} and k_{21} in the modified version of Eqs. (6.4) until a good fit of Troger's data is obtained. Simultaneously, we adjust k_{23} to keep the DNA-bound drug true to results of Sadowitz previously discussed. We remark that the disparity between the inward and outward rates derived for cisplatin may be due in part to carrier-mediated transport, e.g. the CTR1 influx transporter.

6.4.2.2 Doxorubicin parameters

Proceeding to doxorubicin we first obtain an acceptable range for k_{12} and k_{21} from the literature. For a variety of anthracyclines, including doxorubicin, initial estimates of cell membrane permeability P are taken from experiments with SU-4 and SU-4R wildtype and resistant human lymphoma cells,[66] from experiments with EHR2 and

EHR2/DNR+ wildtype and resistant Ehrlich ascites tumor cells,[14,68] and from experiments with MDA-468 breast cancer cells.[11] The range reported is $2.4 \leq P \leq 1000$ mm-min^{-1}. The relation $k_{12} = PA_C/V_C$, where A_C represents the cell membrane area, can then be used to arrive at an initial range of $1.4 \leq k_{12} \leq 600\,\text{min}^{-1}$, which will be refined later. In the case of passive diffusion, $k_{21} = k_{12}$. Note that these values are far larger than those obtained for cisplatin previously. More generally, it has been remarked that cell membrane permeability for cisplatin is much lower than for doxorubicin, etoposide, and vinblastine, although all four drugs are thought to enter cells by passive diffusion.[71]

We next turn our attention to DNA-binding affinity. Given the great DNA affinity of the anthracyclines, saturability of the DNA must be taken into account, requiring an estimate of s_m. There is evidence a typical anthracycline molecule intercalation occludes from 3 to 10 binding sites in a manner that cannot be corrected exactly by a factor;[67,72,73] however, to a first approximation we assume that such a correction can be applied.

Demant and Friche[14] report a DNA binding site concentration of about 5 mM within a cell volume of 1000 fL, yielding 5 fmoles of sites. A low value of 0.7 fmoles is obtained by using the assumed value of 1.25E6 kbp and the reported site exclusion parameter of about 3 from Rizzo *et al.*[67] Tarasiuk *et al.*[73] find that the DNA of human lymphocytes is comprised of about 6.0E6 kbp and that one intercalating molecule of doxorubicin requires 10 base pairs. Thus, Tarasiuk's data implies a factor-corrected quantity of 1 fmole of binding sites, which can be taken as a representative value of s_m.

DNA binding kinetics of the anthracyclines is nontrivial, perhaps requiring multiple steps and demonstrating sequence specificity.[67,74] Bearing this in mind, as an approximation it will suffice to assume non-specific, one-step binding and unbinding according to the chemical reaction

$$\text{Drug Molecule} + \text{DNA bp} \rightleftharpoons \text{Intercalated bp}$$

where the forward rate is k_{on} and the reverse rate is k_{off}. A representative value for the binding coefficient in the above equation for doxorubicin is reported as $k_{on} = 4.2E8\,\text{M}^{-1}\,\text{min}^{-1}$ and a value of the unbinding coefficient (identical with k_{32}) as $k_{off} = 1800\,\text{min}^{-1}$.[67] From k_{on} we calculate a clearance parameter (as with cisplatin) given as $\lambda_{23} = k_{on}s_m = 4.2\text{E}8\,\text{fL-min}^{-1}$ (being cautious with the scales of

Table 1. A complete summary of baseline pharmacokinetics and pharmacodynamics parameters. Tumor growth and angiogenesis parameters can be found in (Zheng *et al*., 2005). Reprinted with permission from Sinek *et al.*, *J. Theor. Biol.* (in press). Copyright © Elsevier.

Parameter	Description	Baseline Value	
		Dox	Cis
V_C	Cell Volume (fL)	520	520
ρ	Cell Density (cells-ml^{-1})	1.0E9	1.0E9
F	Interstital Fraction	0.48	0.48
D_n	Nutrient/ECM Diffusivity (μm^2 min^{-1})	60E3	60E3
D_s	Drug/ECM Diffusivity (μm^2 min^{-1})	1.0E3	30E3
k_n	Nutrient Metabolism (min^{-1})	24	24
k_{12}	Drug Uptake (min^{-1})	5.40	0.054
k_{21}	Drug Efflux (min^{-1})	5.40	1.56E-3
k_{23}	Drug-DNA Binding (min^{-1})	8.02E5	3.82E-4
k_{32}	Drug-DNA Release (min^{-1})	1.80E3	0.0
k_3	Drug-DNA Repair (min^{-1})	0.0	0.015
k_{24}	Lysosomal Sequestration (min^{-1})	10.0	0.0
k_{42}	Lysosomal Release (min^{-1})	0.07	0.0
s_M	Drug-DNA Capacity (fmole)	1.00	∞
A	Phenomenological Parameter for PD Model	0.188	7.75
m	Phenomenological Parameter for PD Model	1.14	1.58
p	Phenomenological Parameter for PD Model (Nutrient Effect)	0.4	0.0

the dimensions). k_{23} can then be calculated as $\lambda_{23} = V_C$, given in Table 1.

We next turn our attention to the rates k_{24} and k_{42} governing lysosomal sequestration. Experiments by Hurwitz *et al.*[15] using U-937 myeloid leukemia cells and their doxresistant variant U-A10 show that the ratio of DNA-bound to lysosomally-sequestered drug is about 3. (Hurwitz uses daunorubicin, an anthracycline related to doxorubicin.) In the modified model equations with all other parameters set as described above, the amount of sequestered drug at equilibrium is dependent only upon the ratio k_{24}/k_{42}. This ratio furthermore does not affect the equilibrium quantity of DNA-bound drug. Arbitrarily selecting $k_{24} = 1\,\text{min}^{-1}$, we find that the appropriate DNA-bound to lysosomally-sequestered ratio is obtained by setting k_{42} to 0.007. Considering that lysosomal membrane permeability is quite high,[68] the lysosomally-bound drug must achieve equilibrium quickly, which can be modified by changing k_{24} while keeping the ratio $k_{24} = k_{42}$ constant. We find that increasing k_{24} by a factor of 10 reduces the time required for the system [Eqs. (6.4)] to achieve 95% of equilibrium value (max_{95}) to about 300 minutes, below which further increases in k_{24} only reduce this time negligibly. Thus, we set $k_{24} = 10$ and $k_{42} = 0{:}07$.

To refine the initial range of k_{12} and k_{21}, we use the modified version of Eqs. (6.4) to compare the simulated monolayer uptake profiles of total intracellular drug with those of DeGregorio *et al.*[75] using human Ewing's sarcoma and rhabdomyosarcoma cells. At $5.40\,\text{min}^{-1}$ both uptake profiles and equilibrium values compare favorably at three test concentrations.

6.4.2.3 Drug-tissue diffusivity

The last pharmacokinetics parameter values needed are the diffusivities D_S of cisplatin and doxorubicin through tumor interstitium. For molecules of their size (dox M.W. = 544, cis M.W. = 300), diffusivity should be about 30,000 $\mu\text{m}^2\text{-min}^{-1}$.[64,65] However, doxorubicin faces particularly severe barriers due to binding to extracellular constituents such as hyaluronic acid,[8,76] and its diffusivity in some tissues has been estimated to be as low as 1000 $\mu\text{m}^2\text{-min}^{-1}$,[11] which we take as the baseline value.

6.4.3 Pharmacodynamics Model Parameters

In order to calibrate the pharmacodynamics model [Eq. (6.6)], we use *in vitro* data of Levasseur *et al.*[77] with A2780 ovarian cancer cells exposed in monolayer to both doxorubicin and cisplatin over a range of times and concentrations. We assume the previously discussed modified pharmacokinetics model along with the values derived, and simulate Levassuer's exposures followed by approximately 24 hours of drug washout in drug-free medium (s_1 is set to 0). During this time, DNA-bound AUC is calculated. These data are then used in conjunction with Levasseur's surviving fraction data to fit the parameters A and m in Eq. (6). During this process, cell substrates are assumed plentiful ($n = 1$), thus bypassing the substrate effect.

6.5 COMPUTATIONAL SIMULATIONS

6.5.1 Non-Dimensionalization and Numerical Methods

Non-dimensionalization of Eqs. (4) is via the length- and time-scales

$$L = \sqrt{D_S/k_{12}}, \quad T = k_{12}^{-1}$$

resulting in

$$\begin{aligned}
\bar{s}_1 &= \bar{k}_v(1 - \bar{s}_1)\delta + \bar{\nabla}^2\bar{s}_1 - \bar{s}_1/F + \bar{k}_{21}s_2/F \\
\bar{s}_2 &= \bar{s}_1 - \bar{k}_{21}\bar{s}_2 + \bar{k}_{32}\bar{s}_3 - \bar{k}_{23}\bar{s}_2(1 - \bar{s}_3/\bar{s}_M) + \bar{k}_{42}\bar{s}_4 - \bar{k}_{24}\bar{s}_2 \\
\bar{s}_3 &= \bar{k}_{23}\bar{s}_2(1 - \bar{s}_3/\bar{s}_M) - \bar{k}_{32}\bar{s}_3 - \bar{k}_3\bar{s}_3 \\
\bar{s}_4 &= \bar{k}_{24}\bar{s}_2 - \bar{k}_{42}\bar{s}_4.
\end{aligned} \tag{6.7}$$

The numerical methods for tumor growth and angiogenesis have been described in detail in Zheng *et al.*[38] For the reaction-diffusion equations [Eqs. (6.4)] we first use Strang splitting. Then the trapezoidal rule is applied to the reaction part, and the Crank-Nicolson scheme, to the diffusion part. For a description of these methods, see Tyson *et al.*[78] and references therein.

6.5.2 In Silico Experiments

In silico experiments were performed as follows:

(1) Three simulated tumors are grown using the model of Zheng *et al.*[38] Each lesion represents one replication of each experiment.
(2) The pharmacokinetics model [Eqs. (6.7)] is used to deliver drug to the lesions. In each case the intravascular concentration of drug s_V is held constant for two hours, then set it to zero for eighteen more hours to allow washout. This simulates drug release from nanovectors bound to the tumor vascular endothelium.
(3) DNA-bound AUC is computed by re-dimensionalizing s_3 and time and integrating using Matlab. The result is then used in the pharmacodynamics model [Eq. (6.6)] to compute cell inhibition. Cell substrates [Eq. (6.3)] are relevant when the substrate effect is employed.

Each lesion is produced based upon the same set of growth and vasculature parameters (see Zheng *et al.*[10,38] for a complete description), but randomness in the angiogenesis algorithm and slightly different initial shapes produce different vasculatures and morphologies. It is assumed that a tumor *in vivo* does not grow or regress appreciably during the 20 h course of the therapy, hence tumor and vascular growth are stopped during the *in silico* therapies. Intravascular concentrations are calibrated in each case to produce a total cellular growth inhibition of 50%. This concentration is referred to as the IC50. The sharp "square wave" delivery of drug is perhaps a caricature of drug administration, but it allows for consistent analysis and comparison of results.

The first set of experiments compares DNA-bound drug AUC distributions of doxorubicin and cisplatin under the baseline conditions in Table 1. We furthermore show the homogenizing effect of doxorubicin retention on final DNA-bound AUC.[20] We next investigate the impact of inhibition heterogeneity on dosing requirements, paying particular attention to the substrate effect for doxorubicin under baseline conditions and improved penetration by, for example, removing hyaluronic acid.[8,76] The third set of simulations more deeply investigates the effect of doxorubicin penetration therapies under three circumstances: baseline tumor density, high tumor density, and baseline tumor density with Pgp efflux activity. These are chosen because they demonstrate a spectrum of possibilities due to their effect on cellular drug uptake. High tumor density increases uptake, while Pgp efflux decreases it. In order to simulate increased penetration, we increase D_S for doxorubicin from its baseline value to 5000 μm^2-min^{-1} for a moderate increase, and 30,000 for the maximum increase, thus matching the performance of cisplatin. To simulate high tumor density we increase ρ by 50 percent to 1.5E9 cells-ml^{-1}. This has the effect of lowering the interstitial fraction F to 0.22, which in turn increases k'_{12} and k'_{21} while leaving all other rates unchanged. Pgp efflux is simulated by increasing k_{21} by a factor of 10, which has the effect of reducing all intracellular compartment concentrations by approximately the same factor. This is consistent with previous results[79] that show Pgp activity can reduce intracellular concentrations of daunorubicin (an anthracycline related to doxorubicin) by up to a factor of 100. In the fourth and final set of experiments we investigate permeabilization therapy with respect to cisplatin, whereby a detergent, such as

digitonin, or electropermeabilization is used to increase the permeability of cell membrane.[71,80] We take an extreme case, increasing the rate constants k_{12} and k_{21} from baseline both by a factor of 100. Note that this does not increase the limiting intracellular or DNA-bound levels of drug attained in simulated monolayer, only the rate at which these come to equilibrium. Thus highly permeabilized, DNA-bound max95 is attained at 3.4 h of exposure; further permeabilization reduces this negligibly. For comparison, max95 is greater than 27 h for unpermeabilized cells. This therapy is simulated under both *in vivo* baseline and very high cell densities achieved by increasing the baseline density 75% to 1.75E9 cells-ml^{-1}. At this density, the interstitial fraction F drops to a mere 0.08. Both of these are further compared to monolayer results to probe the conditions under which *in vitro* assays can be used to predict clinical efficacy.

Although all treatments described are duplicated in each of the three *in silico* tumors, only representative plots with appropriate summaries of all data are displayed. The substrate effect is only used where noted.

6.5.3 Results

6.5.3.1 First experiment

Examine DNA-bound AUC distributions at various times in the baseline simulated lesions (each lesion corresponding to a column, I, II, or III), shown in Fig. 6. The lesion/host interface is demarked by thick black contours, while the microvasculature appears as a web of thin red curves. Dark interior regions are necrotic debris. From top to bottom, the times correspond to two hours, eight hours, fourteen hours, and twenty hours post drug release. Levels are normalized relative to the average AUC within viable lesion for comparison of heterogeneity. Although surrounding host tissue cells uptake and bind with drug differently than cancer cells, we make no distinction in these color plots; however, quantitative analytical results only consider DNA-bound drug within viable lesion. The two left column sequences (Lesions I and II) show doxorubicin AUC, while the rightmost column shows cisplatin.

For both Lesions I and II, at 2 h doxorubicin AUC is seen to be about 3 times the average (dark red) in the vicinity of the vasculature, and almost 0 (blue) elsewhere. The distribution is only slightly more homogeneous by 8 h. By 14 h the heterogeneity has lessened, with

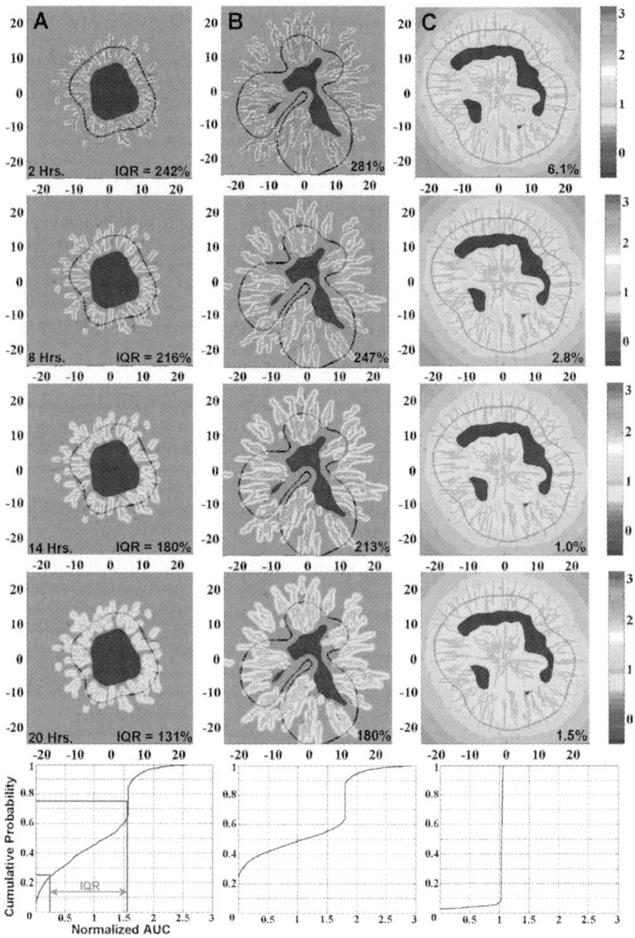

Figure 6. DNA-bound AUC at four times (rows: 2, 8, 14, and 20 h) post bolus initiation for three two dimensional simulated baseline tumor lesions (columns). I and II are doxorubicin, while III is cisplatin. Results are normalized to average lesion AUC at the time taken to enable comparison of distribution heterogeneities. Thick black contours are tumor boundaries. Thin red curves are vasculature. Dark regions are necrotic areas. Each unit represents 200 μm. Bottom probability distributions show final AUC distribution at 20 h. A concise measure of heterogeneity is given by the inter-quartile range (IQR), depicted in the lower left graph and explained in the text. Although AUC in host tissue is also shown in plots, the analysis considers only DNA-bound drug in viable lesion. Reprinted with permission from Sinek *et al.*, *J. Theor. Biol.* (in press). Copyright © Elsevier. (See page 340 for color illustration.)

the peaks close to the vasculature reaching only about 2.2. Finally, at the conclusion of washout 20 h after drug release, the distribution has become much more homogeneous, with peaks only reaching about 1.7 times the average. In contrast, cisplatin distribution within Lesion III remains extremely homogeneous, right at the average, throughout the entire treatment.

The probability distributions at the bottom, corresponding to AUC at 20 h post drug release, allow for a more quantitative comparison. The two corresponding to doxorubicin show much heterogeneity relative to cisplatin on the right. Using the leftmost distribution as an example, the average DNA-bound AUC is found to be 6.04 fmole-min. Twenty five percent of tumor cells receive less than 1.66 fmole-min each, while 25% of tumor cells receive more than 9.54 fmole-min. The remaining 50% of the tumor cells receive between these two values, a range of 7.88 fmole-min. When normalized with respect to the average and expressed as a percent, this yields 131% (the interquartile range, or IQR), and gives a concise measure of distribution heterogeneity (the closer to 0, the more homogeneous). IQR's are given at each of the other time points as well. All three tumors, despite varied lesion and vasculature morphologies, demonstrate similar results (not all shown). Doxorubicin AUC IQR's typically lessen from about 250% at 2 h to 150% at 20 h; cisplatin AUC IQR's drop from about 10 to 2%. Interestingly, in the run shown, the heterogeneity for cisplatin increases slightly in the last frame. This happens in one of the other two tumors as well.

6.5.3.2 Second experiment

Investigate the impact of drug and cell substrate heterogeneity on cell inhibition distributions and IC50's. Drug administration is simulated for cisplatin using baseline lesions exactly as in Fig. 6. The PD model [Eq. (6.6)] is then used to calculate cell inhibition. For doxorubicin we use baseline lesions as well as lesions in which drug penetration therapy is applied. Experiments for doxorubicin are run both with and without the substrate effect.

A table of average IC50's and log(IC50/IC50;mono)'s for these experiments is given in Table 2.

"IC50;mono" refers to baseline cells exposed in monolayer and serves as a reference. Note that, as these are simulated monolayer exposures, IC50;mono is deterministic. Figure 7 shows a typical cell

Table 2. Means ± SD's of the IC50's and the logs of their ratios with respect to monolayer treatments for experiments to investigate the impact of drug and cell substrate heterogeneity. IC50, mono is the IC50 of baseline cells in monolayer. At the 5% significance level using a one-tailed *t*-test, the average log ratio for cisplatin does not exceed 0. On the other hand, in three of the four experiments with doxorubicin, they do. Paired one-tailed *t*-tests show that the average log IC50 ratios for doxorubicin with the substrate effect are greater than that without regardless of penetration therapy. Reprinted with permission from Sinek *et al.*, *J. Theor. Biol.* (in press). Copyright © Elsevier.

Drug	$IC_{50,mono}$ (μM)	IC_{50} (μM)	$\log(IC_{50}/IC_{50,mono})$
Dox Baseline Nut. Eff. Off	0.175	0.482 ± 0.163	0.424 ± 0.138 $^{*}p < 0.05$
Dox Baseline Nut. Eff. Off	0.175	1.34 ± 0.874	0.830 ± 0.261 $^{*}p < 0.05$
Dox w/Penetration Nut. Eff. Off	0.175	0.197 ± 0.0172	0.0511 ± 0.0371 $^{*}p < 0.05$
Dox w/Penetration Nut. Eff. Off	0.175	0.371 ± 0.0356	0.325 ± 0.0407 $^{*}p < 0.05$
Cis Baseline	7.05	7.14 ± 0.0757	0.00529 ± 0.00462 $^{*}p < 0.05$

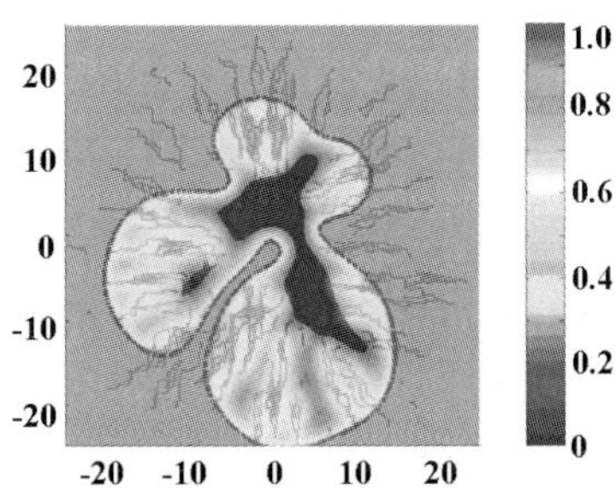

Figure 7. Contour plot shows cell substrate distribution in Lesion B demonstrating significant heterogeneity. Other lesions are similar. Reprinted with permission from Sinek *et al.*, *J. Theor. Biol.* (in press). Copyright © Elsevier. (See page 341 for color illustration.)

substrate profile, using Lesion II as an example with an IQR of 36%. This measurement is completely analogous to that used in Fig. 6 except that here it is applied to substrate distribution and there is no normalization since substrate levels are bounded absolutely from 0 to 100 percent, the level within the vasculature, itself. Cell substrate IQR's for the other two lesions are within 2% of this value.

At the 5% significance level, one-tailed *t*-tests show that the average logIC50 ratio is not greater than 0 for cisplatin, underscoring the homogeneity of its distribution. In contrast, out of the four experiments performed for doxorubicin from the combinations of substrate effect and penetration therapy, three indicate that the average log ratios are greater than 0 at the 5% significance level. Within this group of four we can analyze the strength of the substrate effect. For the baseline lesion, the substrate effect increases the logIC50 ratio by 0.406 units (a factor of about 2.5). For the lesion with penetration therapy, the increase is 0.274 units (a factor of about 1.9). Paired *t*-tests show that these differences are significant at the 5% level. Cell inhibition distributions closely mirror their AUC distributions, with that of cisplatin being virtually uniform at 50% inhibition throughout. Conversely, doxorubicin displays heterogeneity, increased with the addition of the substrate effect.

Using Lesion II as a representative example for doxorubicin, the upper block of frames in Fig. 8 demonstrates the inhibition distributions for the baseline lesion with and without the substrate effect. While the broadening of the cumulative probability plot as well as a comparison of the color distribution plots indicate that the substrate effect increases heterogeneity, inhibition IQR is reduced from 81 to 77% (again, not normalized). The effect of penetration therapy in the lower block of frames is readily apparent. IQR's, color plots, and probability graphs all indicate more uniform inhibition, ranging moderately from 35 to 65%.

Again we see increased heterogeneity in the plots with the addition of the substrate effect. This time the IQR also reflects the increase. Lesions I and III yield similar results.

6.5.3.3 Third experiment

Investigate the effect of therapies designed to improve doxorubicin penetration under several combinations of drug/interstitium diffusivities, cell densities, and drug efflux activities (e.g. Pgp). Figure 9

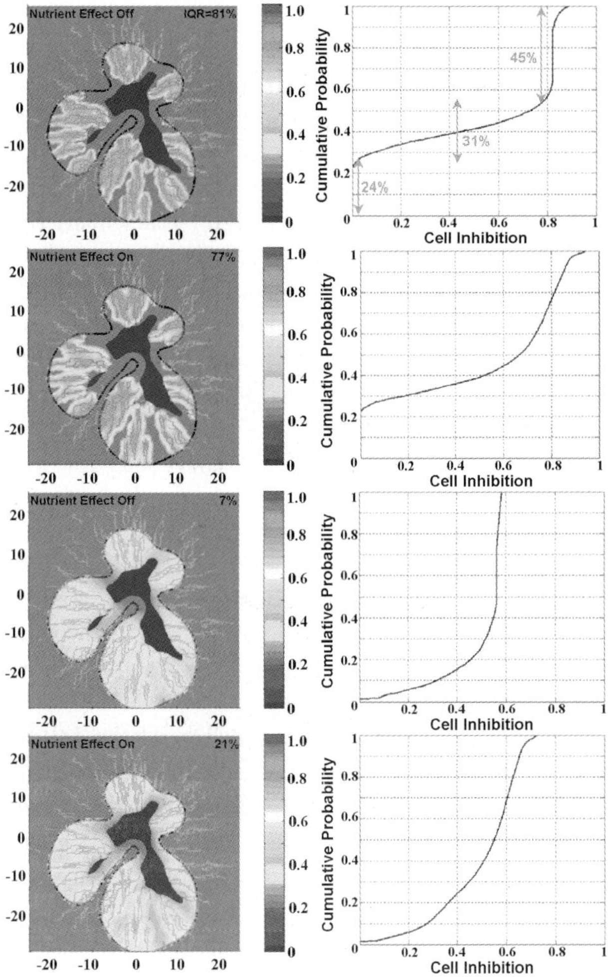

Figure 8. (Upper block) Cell growth inhibition profile of Lesion II at baseline settings with and without the substrate effect after bolus administration depicted in Fig. 2. Probability plot and IQR are now of inhibition distribution and are not normalized with respect to any average. Although the IQR indicates decreased heterogeneity with the substrate effect, both the color distribution plot and the probability plot indicate increased heterogeneity as is evidenced by the broadening of the curve. (Lower block) The same experiment, except with doxorubicin penetration increased. Now both the plots and IQR show increased heterogeneity. The appropriate IC50 is used in each experiment. Reprinted with permission from Sinek *et al.*, *J. Theor. Biol.* (in press). Copyright © Elsevier. (See page 341 for color illustration.)

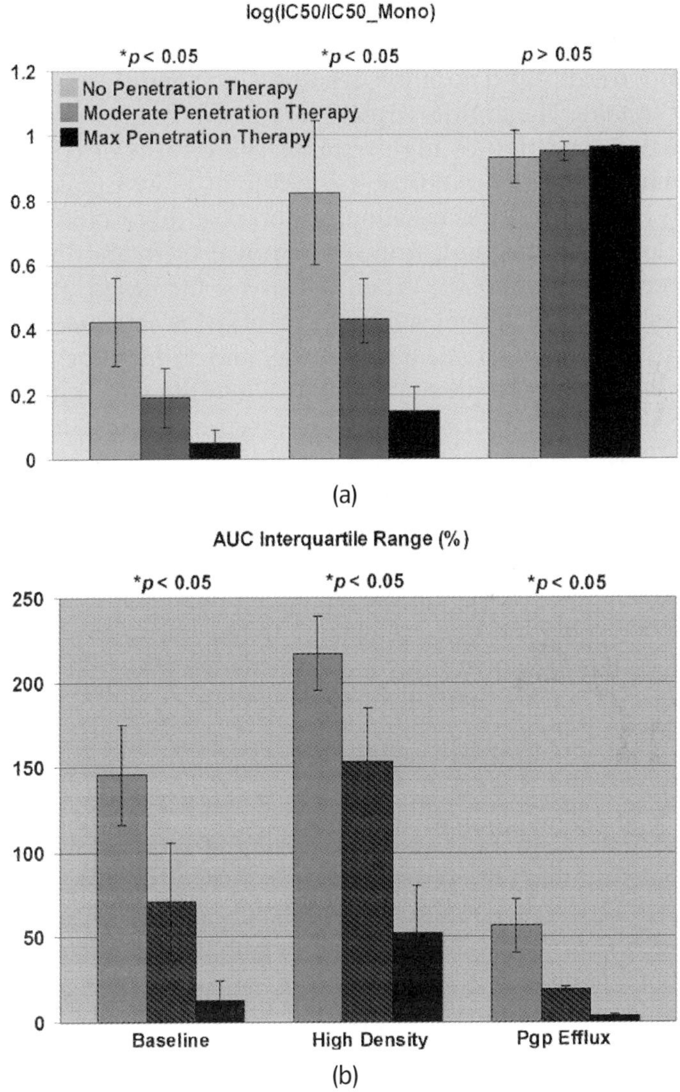

Figure 9. The effect of increasing doxorubicin penetration on (a) log(IC50 = IC50;mono) and (b) interquartile range shown in three cases: baseline tumor (excepting penetration therapy), high-density tumor, and normal density tumor with Pgp efflux. High density has the effect of increasing drug uptake, while Pgp efflux has the opposite effect. Three replications per bar are displayed with standard deviations and results of ANOVA. Reprinted with permission from Sinek *et al.*, *J. Theor. Biol.*, (in press). Copyright © Elsevier.

gives bar graphs of (a) logIC50 ratios and (b) AUC interquartile ranges for three scenarios. The leftmost triplet corresponds to baseline tumor density and no efflux, resulting in a condition of "normal" cellular uptake. The middle triplet corresponds to high density with no efflux, a condition of high uptake. The rightmost corresponds to baseline density with efflux, a condition of low uptake. For the logIC50 ratio (a), in the baseline tumor case there is a change of −0.388 log units in going from no removal of hyaluronic acid to almost complete removal. When density is increased, the change increases to −0.709; however, when Pgp efflux is activated, ANOVA reveals there is no statistical difference, and in fact, the measured change is positive. Results are similar when the substrate effect is included, with all bars essentially increased by a constant, approximately 0.37. For the AUC interquartile range (b), it is seen that heterogeneity is greatest in the high density case and least in the Pgp efflux case. Within each triplet the heterogeneity decreases with increasing penetration therapy, as expected. The magnitudes of change mirror those for the logIC50 ratios, with the baseline case experiencing a moderate change (from 146 to 13%), the high density case experiencing a dramatic change (from 217 to 52%), and the Pgp efflux case experiencing the least change (from 57 to 4%).

6.5.3.4 Fourth experiment

Investigate the effect of permeabilization therapy vis-a-vis cisplatin. Figure 10 shows log(IC50;perm = IC50;unperm) for three cases: monolayer, *in vivo* with baseline cell density, and *in vivo* with high cell density. Here, the subscripts "perm" and "unperm" denote the application or withholding of permeabilization therapy. Permeabilization results in a decrease of 0.154 logIC50 units for simulated monolayers, i.e. a reduction of IC50 by a factor of 0.7, and is thus effective *in vitro*. An interesting question is whether this carries over *in vivo*, i.e. whether a monolayer assay can be used to predict clinical efficacy. Improvements for the two *in vivo* simulations are comparable to monolayer results, with all three log-differences about −0.14, and no statistical difference between improvement for the baseline case and for monolayer at the 5% significance level using a two-tailed *t*-test.

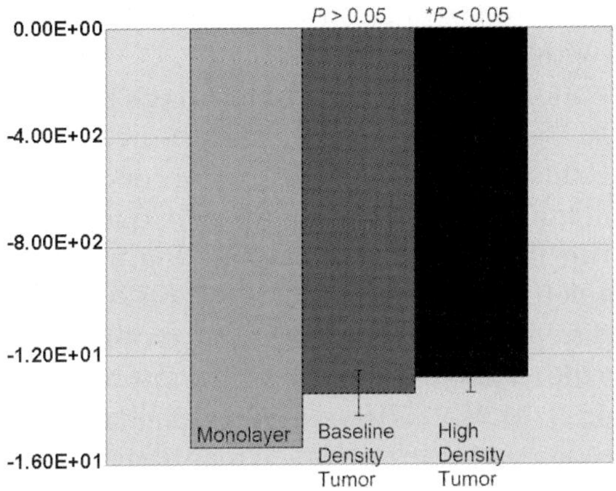

Figure 10. Effect of permeabilization therapy with respect to cisplatin is shown in three cases. Bars are of log(IC50, perm/IC50, unperm) where IC50, perm and IC50, unperm correspond to permeabilized and unpermeabilized conditions. Three replications per bar with results of two-tailed *t*-tests relative to monolayer displayed. While there is a statistical difference at the 0.05 significance level for the high-density tumor, this disappears at the 0.01 significance level. Reprinted with permission from Sinek *et al.*, *J. Theor. Biol.* (in press). Copyright © Elsevier.

6.6 DISCUSSION

In agreement with experimental observation, the simulations reviewed herein show that heterogeneities of drug, nutrient, and oxygen, caused in part by irregular vasculature and lesion morphology, exist and may significantly impact therapy. Moreover, the sense and magnitude of their influence is not intuitively obvious. A good example of this is that, despite its well-noted penetration difficulties, doxorubicin performs well clinically. This may be somewhat explained by its retention in tissue removed from vasculature, causing homogeneity of exposure to increase long after the drug exposure has been terminated (Fig. 6). This phenomenon has been experimentally verified in Ref. 20 with spheroids. Because of this, the resulting cell inhibition distribution is more homogenous than would otherwise be expected. On the other hand, cisplatin maintains a homogenous DNA-bound distribution at all times from drug release to 20 hours later, resulting in an extremely uniform cell inhibition distribution. This result, as well as the near equality of its IC50 and IC50;mono demonstrated in Table 2, has also been experimentally verified with spheroids.[19,71,81]

While retention in tissue contributes to the performance of doxorubicin *in vivo*, Table 2 demonstrates that its heterogeneity of distribution contributes to increased serum drug concentrations to match the same cell inhibition in monolayer. In one case, the average amount of drug increases by nearly one log unit. It is reasonable to expect that heterogeneity of cell substrates, resulting in hypoxia and hypoglycemia, should compound this problem for doxorubicin. Indeed, this is the case as can be seen by the approximate doubling of the IC50's (0.482 μM vs. 1.34 and 0.197 vs. 0.371) when the substrate effect is applied. By graphically and quantitatively showing corresponding cell inhibition distributions Fig. 8 offers further insight into these phenomena. It is easily seen that cell inhibition distributions are as heterogeneous as their corresponding DNA-bound AUC distributions, with areas of lesion removed from vasculature experiencing reduced cell inhibition. An examination of the probability plots in the upper block, corresponding to baseline lesions, shows that a full 24% of viable lesion undergoes no inhibition at all. It is clear from these graphs that penetration therapy greatly decreases heterogeneity of cell inhibition (and commensurately, IC50) as does removal of the substrate effect.

One puzzling behavior is that while both the color and probability plots demonstrate consistently increased heterogeneity brought about by the substrate effect (as is evidenced by the broadening of the probability curves), the IQR actually decreases in the baseline case from 81 to 77%. This occurs with Lesions I and III as well. A solution to the mystery is obtained by noticing that, absent the substrate effect, a large portion of the tumor either experiences no inhibition (about 24%, which is distant from the vasculature) or an already heterogeneous inhibition (about 31%). The remaining 45% receives a near homogeneous level of inhibition (the vertical portion of the curve), and this fraction corresponds to tissue close to the vasculature. Thus, the only significant heterogeneity that can be induced by the substrate effect is within this fraction. Indeed, it is just this part of the curve that broadens in the second probability distribution, indicating greater heterogeneity, as expected. As the IQR is designed to measure heterogeneity somewhat more globally, it misses-in fact, misdiagnoses–the change occurring within this fraction.

In addition to dosing requirements, there is a second and subtler reason to consider heterogeneity of the inhibition distribution. Heterogeneities in microenvironmental conditions have been linked to increased lesion fragmentation and invasiveness.[50,82–85] While the mechanisms underlying this phenomenon are complex, involving myriad protein signaling events and activities at the cellular level, they may at least partly rely on gross lesion effects, including those caused by drug administration.

While the simulations herein fail to account for many critical aspects of tumor growth and drug response (such as clonal heterogeneity, cell phase sensitivity, and signaling pathways) and parameter settings may in some cases be inexact, it should not be concluded that these shortcomings invalidate characteristics the simulations have revealed. The computational modeling of biobarriers has provided evidence that not only do macroscopic environmental conditions, namely, drug and cell substrate distribution heterogeneity, potentially greatly impact therapeutic efficacy, but also that the outcome of therapeutic strategies can depend upon them in nonlinear and a priori unpredictable ways. The results of the third set of experiments (doxorubicin penetration therapy) provide an example. In light of this, it would be prudent to devote attention to factors residing at coarser and more global scales than solely the genetic when evaluating nano-based therapeutics.

6.7 CONCLUSION

One of the broader goals of computational modeling of nano-based therapies in cancer treatment is to demonstrate how increasingly sophisticated *in silico* technology, driven by mathematical modeling and calibrated with experimental data, can be developed to provide an alternate investigative and clinical tool complementary to traditional methods.[86–90] In particular, the results reviewed here provide important information regarding cellular-scale pharmacokinetics and pharmacodynamics. In clinical application, the results could be used to guide therapeutic strategy. For example, risks associated with doxorubicin penetration therapy could be minimized if it were known that a patient's tumor was expressing Pgp or otherwise had lowered cellular uptake. Nanovectors could be designed to circumvent or minimize these barriers, and simulations could verify and optimize their design. With further development, we anticipate that such *in silico* models and methods will become increasingly accurate and useful. Towards that end a more veridical model of vasculature along the lines of McDougall and Stephanou, in which blood flow plays a key role in the formation of vasculature[41,43] would be needed. Further development of 3D models, which are proving successful at accurately simulating morphological evolution,[49] is also foreseen. Liposomal and nanovectored drug delivery, with the potential of unprecedented accuracy and specificity of delivery,[27] will require PKPD modeling as shown herein, especially paying particular attention to transport in and extravasation from blood vessels.[91] The computational modeling of biobarriers will thus be an essential component of eventual successful implementation of nano-based therapeutics.

REFERENCES

1. A. Minchinton, and I. Tannock, "Drug penetration in solid tumors," *Nature Rev. Cancer*, **6**, 583–592 (2006).
2. O. Trédan, C. M. Galmarini, K. Patel, and I. F. Tannock, "Drug resistance and the solid tumor microenvironment," *J. Natl. Cancer Inst.*, **99**, 1441–1454 (2007).
3. J. Baish, Y. Gazit, D. Berk, M. Nozue, L. Baxter, and R. Jain, "Role of tumor vascular architecture in nutrient and drug delivery: An invasion percolation-based network model," *Microvasc. Res.*, **51**, 327–346 (1996).
4. C. Gulledge, and M. Dewhirst, "Tumor oxygenation: A matter of supply and demand," *Anticancer Res.*, **16**, 741–749 (1996).
5. Z. Haroon, K. G. Peters, C. S. Greenberg, and M. W. Dewhirst, "Angiogenesis and blood flow in the solid tumors," in B. A. Teicher (ed.), *Antiangiogenic Agents in Cancer Therapy*, pp. 3–21, Humana Press Inc. Totowa, NJ (1999).
6. R. Jain, "Delivery of molecular medicine to solid tumors: Lessons from *in vivo* imaging of gene expression and function," *J. Control. Release*, **74**, 7–25 (2001).
7. T. Padera, B. Stoll, J. Tooredman, D. Capen, E. di Tomaso, and R. Jain, "Pathology: Cancer cells compress intratumour vessels," *Nature*, **427**, 695 (2004).
8. N. Kohno, T. Ohnuma, and P. Truog, "Effects of hyaluronidase on doxorubicin penetration into squamous carcinoma multicellular tumor spheroids and its cell lethality," *J. Cancer Res. Clin. Oncol.* **120**, 293–297 (1994).

9. I. Tannock, C. Lee, J. Tunggal, D. Cowan, and M. Egorin, "Limited penetration of anticancer drugs through tumor tissue: A potential cause of resistance of solid tumors to chemotherapy," *Clin. Cancer Res.*, **8**, 878–884 (2002).
10. J. Zheng, C. Chen, J. Au, and M. Wientjes, "Time- and concentration-dependent penetration of doxorubicin in prostate tumors," *AAPS Pharm. Sci.*, **3**, E15 (2001).
11. J. Lankelma, H. Dekker, F. Luque, S. Luykx, K. Hoekman, P. van der Valk, P. van Diest, and H. Pinedo, "Doxorubicin gradients in human breast cancer," *Clin. Cancer Res.*, **5**, 1703–1707 (1999).
12. A. Primeau, A. Rendon, D. Hedley, L. Lilge, and I. Tannock, "The distribution of the anticancer drug doxorubicin in relation to blood vessels in solid tumors," *Clin. Cancer Res.*, **11**, 8782–8788 (2005).
13. G. Arancia, A. Calcabrini, S. Meschini, and A. Molinari, "Intracellular distribution of anthracyclines in drug resistant cells," *Cytotechnology*, **27**, 95–111 (1998).
14. E. Demant, and E. Friche, "Kinetics of anthracycline accumulation in multidrug-resistant tumor cells: Relationship to drug lipophilicity and serum albumin binding," *Biochem. Pharmacol.*, **56**, 1209–1217 (1998).
15. S. Hurwitz, M. Terashima, N. Mizunuma, and C. Slapak, "Vesicular anthracycline accumulation in doxorubicin-selected U-937 cells: Participation of lysosomes," *Blood*, **89**, 3745–3754 (1997).
16. S. Simon, and M. Schindler, "Cell biological mechanisms of multidrug resistance in tumors," *Proc. Natl. Acad. Sci. USA*, **91**, 3497–3504 (1994).
17. Y. Takemura, H. Kobayashi, H. Miyachi, K. Hayashi, S. Sekiguchi and T. Ohnuma, "The influence of tumor cell density on cellular accumulation of doxorubicin or cisplatin *in vitro*," *Cancer Chemother. Pharmacol.*, **27**, 417–422 (1991).
18. D. Wang, and S. Lippard, "Cellular processing of platinum anticancer drugs," *Nat. Rev. Drug Discov.*, **4**, 307–320 (2005).
19. R. Durand, "Chemosensitivity testing in V79 spheroids: Drug delivery and cellular microenvironment," *J. Natl. Cancer Inst.*, **77**, 247–252 (1986).
20. R. Durand, "Slow penetration of anthracyclines into spheroids and tumors: A therapeutic advantage?" *Cancer Chemother. Pharmacol.*, **26**, 198–204 (1990).
21. A. Lee (1987). Coordinated regulation of a set of genes by glucose and calcium ionophores in mammalian cells. *Trends Biochem. Sci.*, **12**, 20-23.

22. H. Mellor, D. Ferguson, and R. Callaghan, "A model of quiescent tumour microregions for evaluating multicellular resistance to chemotherapeutic drugs," *Br. J. Cancer*, **93**, 302–309 (2005).
23. J. Shen, J. Subjeck, R. Lock, and W. Ross, "Depletion of topoisomerase II in isolated nuclei during a glucose-regulated stress response," *Mol. Cell. Biol.*, **9**, 3284–3291 (1989).
24. W. Siu, T. Arooz, and R. Poon, "Differential responses of proliferating versus quiescent cells to adriamycin," *Exp. Cell Res.*, **250**, 131–141 (1999).
25. J. P. Fruehauf, "In vitro assay-assisted treatment selection for women with breast or ovarian cancer," *Endocr. Relat. Cancer*, **9**, 171–182 (2002).
26. J. P. Fruehauf, and A. Bosanquet, "In vitro determination of drug response: A discussion of clinical applications," *Princ. Pract. Oncol. Updates*, **7**, 1–16 (1993).
27. M. Ferrari, "Cancer nanotechnology: Opportunities and challenges," *Nature Rev. Cancer*, **5**, 161–171 (2005).
28. T. Alarcón, H. Byrne, and P. Maini, "A cellular automaton model for tumour growth in inhomogeneous environment," *J. Theor. Biol.*, **225**, 257–274 (2003).
29. R. Araujo, and D. McElwain, "A history of the study of solid tumour growth: The contribution of mathematical modelling," *Bull. Math. Biol.*, **66**, 1039–1091 (2004).
30. N. Bellomo, and L. Preziosi, "Modelling and mathematical problems related to tumor evolution and its interaction with the immune system," *Math. Comput. Model.*, **32**, 413–452 (2000).
31. C. Breward, H. Byrne, and C. Lewis, "A multiphase model describing vascular tumour growth," *Bull. Math. Biol.*, **65**, 609–640 (2003).
32. M. Chaplain, "Avascular growth, angiogenesis and vascular growth in solid tumors: The mathematical modelling of the stages of tumour development," *Math. Comput. Model.*, **23**, 47–87 (1996).
33. V. Cristini, J. S. Lowengrub, and Q. Nie, "Nonlinear Simulation of Tumor Growth," *J. Math. Biol.*, **46**, 191–224 (2003).
34. Y. Jiang, J. Pjesivac-Grbovic, C. Cantrell, and J. Freyer, "A multiscale model for avascular tumor growth," *Biophys. J.*, **89**, 3884–3894 (2005).
35. P. Macklin, and J. S. Lowengrub, "Evolving interfaces via gradients of geometry-dependent interior Poisson problems: Application to tumor growth," *J. Comput. Phys.*, **203**, 191–220 (2005).
36. P. Macklin, and J. S. Lowengrub, "An improved geometry-aware curvature discretization for level set methods: Application to tumor growth," *J. Comput. Phys.*, **215**, 392–401 (2006).

37. C. P. Please, P. G., and D. McElwain, "A new approach to modelling the formation of necrotic regions in tumours," *Appl. Math. Lett.*, **411**, 89–94 (2005).
38. X. Zheng, S. Wise, and V. Cristini, "Nonlinear simulation of tumor necrosis, neovascularization and tissue invasion via an adaptive finite-element/level-set method," *Bull. Math. Biol.*, **67**, 211–259 (2005).
39. V. Cristini, H. B. Frieboes, X. Li, J. S. Lowengrub, P. Macklin, S. Sanga, S. M. Wise, and X. Zheng, "Nonlinear modeling and simulation of tumor growth," in *Modelling and Simulation in Science, Engineering and Technology* (Birkhauser, Boston), in press.
40. A. Anderson, and M. Chaplain, "Continuous and discrete mathematical models of tumor induced angiogenesis," *Bull. Math. Biol.*, **60**, 857–899 (1998).
41. S. McDougall, A. Anderson, and M. Chaplain, "Mathematical modelling of dynamic adaptive tumour-induced angiogenesis: Clinical implications and therapeutic targeting strategies," *J. Theor. Biol.*, **241**, 564–589 (2006).
42. S. McDougall, A. Anderson, M. Chaplain, and J. Sherratt, "Mathematical modelling of flow through vascular networks: Implications for tumour-induced angiogenesis and chemotherapy strategies," *Bull. Math. Biol.*, **64**, 673–702 (2002).
43. A. Stéphanou, S. McDougall, A. Anderson, and M. Chaplain, "Mathematical modelling of the influence of blood rheological properties upon adaptive tumour-induced angiogenesis," *Math. Comput. Model*, **44**, 96–123 (2006).
44. T. Jackson, "Intracellular accumulation and mechanism of action of doxorubicin in a spatio-temporal tumor model," *J. Theor. Biol.*, **220**, 201–213 (2003).
45. E. S. Norris, K. J., and H. Byrne, "Modelling the response of spatially structured tumours to chemotherapy: Drug kinetics," *Math. Comput. Model*, **43**, 820–837 (2006).
46. J. P. Sinek, H. B. Frieboes, X. Zheng, and V. Cristini, "Two-dimensional chemotherapy simulations demonstrate fundamental transport and tumor response limitations involving nanoparticles," *Biomed. Microdevices*, **6**, 297–309 (2004).
47. J. Ward, and J. King, "Mathematical modelling of drug transport in tumour multicell spheroids and monolayer cultures," *Math Biosci.*, **181**, 177–207 (2003).
48. S. Sanga, H. B. Frieboes, J. P. Sinek, and V. Cristini, "A multi-scale approach for computational modeling of biobarriers to cancer

chemotherapy via nanotechnology," in *Cancer Nanotechnology*, Ch. 10, pp. 1–21 American Scientific (2006).
49. H. B. Frieboes, J. S. Lowengrub, S. M. Wise, X. Zheng, P. Macklin, E. L. Bearer, and V. Cristini, "Computer simulation of glioma growth and morphology," *NeuroImage*, **37**, S59–S70 (2007).
50. P. Macklin, and J. S. Lowengrub, "Nonlinear simulation of the effect of microenvironment on tumor growth," *J. Theor. Biol.*, **245**, 677–704 (2007).
51. T. L. Jackon, and H. M. Byrne, "A mathematical model to study the effects of drug resistance and vasculature on the response of solid tumors to chemotherapy," *Math. Biosci.*, **164**, 17–38 (2000).
52. A. El-Kareh, and T. Secomb, "A mathematical model for cisplatin cellular pharmacodynamics," *Neoplasia*, **5**, 161–169 (2003).
53. A. El-Kareh, and T. Secomb, "Two-mechanism peak concentration model for cellular pharmacodynamics of Doxorubicin," *Neoplasia*, **7**, 705–713 (2005).
54. P. Macklin, S. R. McDougall, A. R. A. Anderson, M. A. J. Chaplain, V. Cristini, and J. Lowengrub, "Multiscale modelling and nonlinear simulation of vascular tumour growth," *J. Math. Biol.*, in press.
55. P. Macklin, and J. S. Lowengrub, "A new ghost cell/level set method for moving boundary problems: Application to tumor growth," *J. Sci.Comput.*, **35**, 266–299.
56. P. Decuzzi, F. Causa, M. Ferrari, and P. A. Netti, "The effective dispersion of nanovectors within the tumor microvasculature," *Annals Biomed. Eng.*, **34**, 633–641 (2006).
57. P. Decuzzi, S. Lee, M. Decuzzi, and M. Ferrari, "Adhesion of microfabricated particles on vascular endothelium: A parametric analysis," *Annals Biomed. Eng.*, **32**, 793–802 (2004).
58. P. Decuzzi, and M. Ferrari, "Design maps for nanoparticles targeting the diseased microvasculature," *Biomaterials*, **29**, 377–384 (2008).
59. J. Kreuter, "Nanoparticles," in J. Kreuter (ed.), *Colloidal Drug Delivery Systems*, Marcel Dekker, Inc., New York, Basel, Hong Kong (1994).
60. S. S. Feng, and S. Chien, "Chemotherapeutic engineering: Application and further development of chemical engineering principles for chemotherapy of cancer and other diseases," *Chem. Eng. Sci.*, **58**, 4087–4114 (2003).
61. J. P. Sinek, S. Sanga, X. Zheng, H. B. Frieboes, M. Ferrari, and V. Cristini, "Predicting drug pharmacokinetics and effect in vascularized tumors using computer simulation," *J. Theor. Biol.*, in press.

Available via open access at http://dx.doi.org/10.1007/s00285-008-0214-y.

62. V. Cristini, J. Blawzdziewicz, and M. Loewenberg, "An adaptive mesh algorithm for evolving surfaces: Simulations of drop breakup and coalescence," *J. Comput. Phys.*, **168**, 445–463 (2001).
63. W. Mueller-Klieser, "Microelectrode measurement of oxygen tension distributions in multicellular spheroids cultured in spinner flasks," *Recent Results Cancer Res.*, **95**, 134–149 (1984).
64. L. Nugent, and R. Jain, "Extravascular diffusion in normal and neoplastic tissues," *Cancer Res.*, **44**, 238–244 (1984).
65. E. Swabb, J. Wei, and P. Gullino, "Diffusion and convection in normal and neoplastic tissues," *Cancer Res.*, **34**, 2814–2822 (1974).
66. M. Dordal, A. Ho, M. Jackson-Stone, Y. Fu, C. Goolsby, and J. Winter, "Flow cytometric assessment of the cellular pharmacokinetics of fluorescent drugs," *Cytometry*, **20**, 307–314 (1995).
67. V. Rizzo, N. Sacchi, and M. Menozzi, "Kinetic studies of anthracycline-DNA interaction by fluorescence stopped flow confirm a complex association mechanism," *Biochemistry*, **28**, 274–282 (1989).
68. E. Demant, M. Sehested, and P. Jensen, "A model for computer simulation of P-glycoprotein and transmembrane delta pH-mediated anthracycline transport in multidrug-resistant tumor cells," *Biochim. Biophys. Acta*, **1055**, 117–125 (1990).
69. P. Sadowitz, B. Hubbard, J. Dabrowiak, J. Goodisman, K. Tacka, M. Aktas, M. Cunningham, R. Dubowy, and A. Souid, "Kinetics of cisplatin binding to cellular DNA and modulations by thiol-blocking agents and thiol drugs," *Drug Metab. Dispos.*, **30**, 183–190 (2002).
70. V. Troger, J. Fischel, P. Formento, J. Gioanni, and G. Milano, "Effects of prolonged exposure to cisplatin on cytotoxicity and intracellular drug concentration," *Eur. J. Cancer*, **28**, 82–86 (1992).
71. A. Jekunen, D. Shalinsky, D. Hom, K. Albright, D. Heath, and S. Howell, "Modulation of cisplatin cytotoxicity by permeabilization of the plasma membrane by digitonin *in vitro*," *Biochem. Pharmacol.*, **45**, 2079–2085 (1993).
72. J. McGhee, and P. von Hippel, "Theoretical aspects of DNA-protein interactions: Cooperative and non-co-operative binding of large ligands to a one-dimensional homogeneous lattice," *J. Mol. Biol.*, **86**, 469–489 (1974).
73. J. Tarasiuk, F. Frzard, A. Garnier-Suillerot, and L. Gattegno, "Anthracycline incorporation in human lymphocytes. Kinetics of uptake and nuclear concentration," *Biochim. Biophys. Acta*, **1013**, 109–117 (1989).

74. X. Qu, C. Wan, H. Becker, D. Zhong, and A. Zewail, "The anti-cancer drug-DNA complex: Femtosecond primary dynamics for anthracycline antibiotics function," *Proc. Natl. Acad. Sci. USA.*, **98**(14), 212–14,217 (2001).
75. M. DeGregorio, G. Lui, B. Macher, and J. Wilbur, "Uptake, metabolism, and cytotoxicity of doxorubicin in human Ewing's sarcoma and rhabdomyosarcoma cells," *Cancer Chemother. Pharmacol.*, **12**, 59–63 (1984).
76. N. Kohno, T. Ohnuma, M. Kaneko, and J. Holland, "Interactions of doxorubicin and cisplatin in squamous carcinoma cells in culture," *Br. J. Cancer*, **58**, 330–334 (1988).
77. L. Levasseur, H. Faessel, H. Slocum, and W. Greco, "Implications for clinical pharmacodynamic studies of the statistical characterization of an *in vitro* antiproliferation assay," *J. Pharmacokinet. Biopharm.*, **26**, 717–733 (1998).
78. R. Tyson, L. Stern, and R. LeVeque, "Fractional step methods applied to a chemotaxis model," *J. Math. Biol.*, **41**, 455–475 (2000).
79. Y. Chen, and S. Simon, "In situ biochemical demonstration that P-glycoprotein is a drug efflux pump with broad specificity," *J. Cell Biol.*, **148**, 863–870 (2000).
80. T. Tanaka, Y. Kaneda, T. Li, T. Matsuoka, N. Zempo, and K. Esato, "Digitonin enhances the antitumor effect of cisplatin during isolated lung perfusion," *Ann. Thorac. Surg.*, **72**, 1173–1178 (2001).
81. S. Inoue, O. T. H. J., and L. Wasserman, "Susceptibility of multicellular tumor spheroids (MTS) to doxorubicin (DXR) and cisplatin," *Proc. Am. Assoc. Cancer Res.*, **26**, 341 (1985).
82. V. Cristini, H. B. Frieboes, R. Gatenby, S. Caserta, M. Ferrari and J. P. Sinek, "Morphologic instability and cancer invasion," *Clin. Cancer Res.*, **11**, 6772–6779 (2005).
83. H. B. Frieboes, X. Zheng, C. Sun, B. Tromberg, R. Gatenby, and V. Cristini, "An integrated computational/experimental model of tumor invasion," *Cancer Res.*, **66**, 1597–1604 (2006).
84. P. Kunkel, U. Ulbricht, P. Bohlen, M. Brockmann, R. Fillbrandt, D. Stavrou, M. Westphal, and K. Lamszus, "Inhibition of glioma angiogenesis and growth *in vivo* by systemic treatment with a monoclonal antibody against vascular endothelial growth factor receptor-2," *Cancer Res.*, **61**, 6624–6628 (2001).
85. S. Pennacchietti, P. Michieli, M. Galluzzo, M. Mazzone, S. Giordano, and P. Comoglio, "Hypoxia promotes invasive growth by

transcriptional activation of the met protooncogene," *Cancer Cell*, **3**, 347–361 (2003).
86. A. Bangs, and T. Paterson, "Finding value in in silico biology," *Biosilico*, **1**, 18–22 (2003).
87. H. B. Frieboes, J. P. Sinek, O. Nalcioglu, J. P. Fruehauf, and V. Cristini, "Nanotechnology in cancer drug therapy: A biocomputational approach," in M. Ferrari, A. P. Lee, and L. J. Lee (eds.), *BioMEMS and Biomedical Nanotechnology*, Vol. I, chap. 15, pp. 435–460, Springer, New York, NY (2006).
88. S. Sanga, J. P. Sinek, H. B. Frieboes, M. Ferrari, J. P. Fruehauf and V. Cristini, "Mathematical modeling of cancer progression and response to chemotherapy," *Expert Rev. Anticancer Ther.*, **6**, 1361–1376 (2006).
89. S. Sanga, H. Frieboes, X. Zheng, R. Gatenby, E. Bearer, and V. Cristini, "Predictive oncology: A review of multidisciplinary, multiscale in silico modeling linking phenotype, morphology and growth," *Neuroimage*, **37**, S120–S134 (2007).
90. J. P. Sinek, H. B. Frieboes, B. Sivaraman, S. Sanga, and V. Cristini, "Mathematical and computational modeling: Towards the development of nanodevices for drug delivery," in C. S. S. R. Kumar (ed.), *Nanotechnologies for the Life Sciences Volume 4: Nanodevices for the Life Sciences*, Ch. 2, pp. 29–66, Wiley-VCH, Weinheim, Germany (2006).
91. F. Gentile, M. Ferrari, and P. Decuzzi, "Transport of nanoparticles in blood vessels: The effect of vessel permeability and blood rheology," *Ann. Biomed. Eng.*, **36**, 254–261 (2007).
92. H. Byrne, and M. Chaplain, "Growth of nonnecrotic tumors in the presence and absence of inhibitors," *Math. Biosci.*, **130**, 151–181 (1995).
93. H. Byrne, and M. Chaplain, "Growth of necrotic tumors in the presence and absence of inhibitors," *Math. Biosci.*, **135**, 187–216 (1996).
94. H. J. Kuh, S. H. Jang, G. Wientjes, and J. L. S. Au, "Computational model of intracellular pharmacokinetics of paclitaxel," *J. Pharmacol. Exp. Ther.*, **293**, 761–770 (2000).
95. S. M. Wise, J. S. Lowengrub, H. B. Frieboes, and V. Cristini, "Nonlinear simulations of three-dimensional multispecies tumor growth-I: Model and numerical Method," *J. Theor. Biol.*, **253**, 524–543.

Chapter Seven

Multiscale-Multiparadigm Modeling and Simulation of Nanometer Scale Systems and Processes for Nanomedical Applications

Andres Jaramillo-Botero, Ravinder Abrol, Adri van Duin and William A. Goddard III

7.1 INTRODUCTION

The US National Institutes of Health define the word Nanomedicine as "an offshoot of nanotechnology" in the context of "specific medical intervention at the molecular scale for curing diseases or repairing tissues".[1] Other broader definitions promote the term as the application of nanotechnology into the realm of medicine[2] for the "preservation and improvement of human health using molecular tools and molecular knowledge of the human body".[3] It is no surprise then, that the field of medicine has seen the most natural applications of nanotechnology as most biological features and structures are found in the range of 1–500 nanometer.[4] DNA and proteins, which can be considered as the nanoscale building blocks of life, function within this length scale and their aggregation into larger supramolecular systems with the help of other biomolecules allows for complex higher-level biological functions to be performed in coordinated fashion. During disease processes, biological structure and function are disrupted and any medical intervention aimed at sensing, diagnosing and controlling that disease without side-effects

Nanomedicine: *A Systems Engineering Approach* edited by Mingjun Zhang & Ning Xi

www.panstanford.com
978-981-4241-36-6

needs to be functional on the same nanometer length scale. The appeal of this length scale is that the physical and chemical properties of matter can change dramatically by comparison with those at the macroscopic continuum, and that by having control over these will enable our ability to impact significantly the way we live healthy lives.

7.1.1 Conceptual Challenges for Novel Nanomedicine Applications

In spite of the appealing nature of nanoscale biomedical development, the field represents a daunting interdisciplinary scientific and engineering challenge which requires formidable advances in many areas in order to deepen the understanding of the underlying complexity of biological systems, from the many networks of molecules that comprise cells and tissues as well as the processes that regulate them and their interactions all the way down to the molecular level events. It also requires the development of bio-amenable and controllable nanometer scale devices and systems capable of affecting the biological systems.

A few of the hurdles that lay ahead and that need to be surmounted in order to fully take advantage of the incredible opportunities for development bestowed by nanoscale phenomena are: (a) The lack of non-intrusive experimental tools to probe biological systems *in-vivo* with atomic precision; (b) our inability to "see" in real-time the consequences of nanometer scale experiments; (c) our precarious ability to build and control nanoscale devices and systems using either a bottom-up self-assembly approach or a top-down synthesizes approach, added to the massive amount of concurrent information flow that takes place during any given biological function; and (d) the absence of scalable, forward and inverse, predictive capabilities to determine and control the molecular structure-property relationships that govern functional biology at the scales of interest (from the scale of atoms up to the continuum, from femtoseconds to milliseconds and beyond).

Notwithstanding the current limitations, the potential range and reach of nanomedical applications is huge and it is only going to increase further in the years to come[4,5] thanks to its transcendence and to the steady and increasing support for the development of Nanotechnology from research agencies and private resources worldwide.[6] While the current major thrust in the field has been on the development of targeted drug-delivery systems (DDS) based on nanoparticles (< 200 nm) and on focused molecular diagnosis,[7,8] it is expected that the "nanotools" being derived from these efforts will open the field to a wide range of novel applications, including but not limited to probing protein misfolding diseases like Parkinson's, Alzheimer's, Down's syndrome, Huntington disease, and others which result from misfolded and aggregated protein nanostructures that disrupt a range of biological functions,[9] controlling protein activation and protein over-expression (e.g. bcl-2 protein over-expression is believed to affect the apoptotic pathway in cancer cells),[10] to name a few.

7.1.2 The Role and Importance of First-Principles-Based Multiscale-Multiparadigm Modeling and Simulation Approaches

Progress towards an understanding of complex phenomena in biological sciences and processes from engineering is now regarded as an equal and indispensable trilogy between theory, experiment, and computation. Computer simulations have become an imperative partner given the high costs, danger, and, in some cases, impossibility to study biological structures and phenomena by direct experimentation. Relying on fast computational methods to accurately predict the chemistry and physics, as a function of operating conditions and time, will enable experimentalists to judiciously orient and gauge such progress. Computational science has filled some of the open gaps which theory alone cannot resolve via closed form analytical models, primarily due to the size and complexity of the variable space and to the inherently multiscale space-time interactions.

Unfortunately, the nanoscale systems of interest to nanomedicine are often too large for standard atomistic simulation approaches — for example, if we assume one cell has approximately one meter of DNA, then the complete set of genes in each cells could be an estimated 3-billion base pairs or around 150-billion atoms. This requires a level of description coarser than atomic that must still contain the atomistic information responsible for the chemical properties (e.g. bonds and reaction processes, among others). For this to happen we must extend existing atomistic simulations to

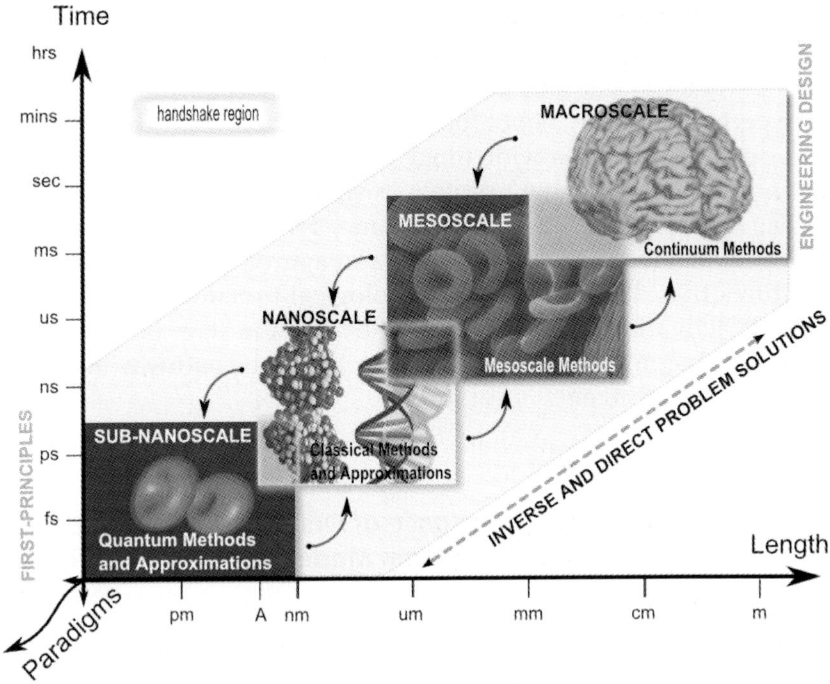

Figure 1. First-principles-based multiscale, multiparadigm modeling and simulation.

handle engineering problems using a hierarchical *multiscale, multiparadigm* approach (see Fig. 1) that seamlessly traverses length and time scales without critical information or accuracy loss from finer scales. The disparate physicochemical methods used to describe the properties and phenomena of matter need to converse fluently and seamlessly between inter-scales and intra-scale methods using the appropriate boundary and handshake conditions to avoid any fundamental discontinuities.

The conjunction of theory and computational modeling is not only useful as a prediction tool, but it also serves to perform model based control of structures, dynamics, and properties of the biological mesoscale systems of interest to nanomedicine. These systems include both the biological processes targeted for recovery/improvement and the nanometer scale devices and systems required for modifying or complementing these to achieve a prolonged state of well-being.

7.1.3 Monograph Organization and Overall Objectives

The sections that follow include a description of the principal nanoscale properties and phenomena of interest in organic and synthetic materials suitable for Nanomedicine, a general overview of the theoretical and multiscale modeling and simulation advances that are providing a complementary and enabling tool for the growth and development of Nanoscale science, and a sample description of nanomedical applications being tackled by the research community and by industry with the use of multiscale modeling methods and tools.

The content of this document is organized to touch progressively upon the following general questions regarding modeling and simulation for nanomedical applications:

- what needs to be modeled (systems, processes, and phenomena)?,
- at what level of accuracy (resolution per scale)?,
- how (appropriate methods for solving)?, and
- what has been the recent progress (example applications)?

It is by no means an exhaustive survey of all current efforts in the field or a presentation of all possible future applications, but an overview intended to provide a modeling and simulation perspective on the nanomedical problems and to stimulate a multidisciplinary interest in the solution of those problems.

7.2 FUNDAMENTAL PROPERTIES OF ORGANIC MATTER AT THE NANOSCALE

It is at the sub-nanometer length scale ($\sim$0.1–0.2 nm) where the building blocks of matter are established, i.e. atoms. These provide material unity and the potential for application convergence between anything and everything, natural or man-made. The spatial and temporal relationships that exist between the atomic constituents of matter define the properties that regulate their collective behavior at larger scales. As noted previously, the nanometer scale brings about important phenomena that are not present in bulk material at the continuum level, phenomena that can be exploited in favor of novel solutions applicable to medicine; some of which are still not fully understood and hence are not yet exploitable. As the size of an entity approaches that of its building blocks, several key factors influence the existence of such phenomena which in turn leads to significant differences in mechanical, thermodynamic, optical, and magnetic and transport properties for nanometer scaled systems. Among these factors are:

- *A larger surface area to volume ratio as a function of entity size.* As an object increases in size its volume increases as the cube of its linear dimensions while surface area increases as linear dimension squared. This leads to an increase in the number of interfaces at which such an entity can interact with its surroundings (i.e. a larger percentage of atoms are at the interacting surfaces relative to the number of atoms that compose the entity) — thus affecting chemical, mechanical, thermal and transport properties. In biological cells for example, the surface must allow sufficient exchange to support the contents of the cell, and hence this ratio limits its size (e.g. for an animal Eukaryotic cell $\sim$5–100 μm). This ratio is also particularly relevant to reaction processes in biological systems and the rate at which they can occur, because there is more surface available to react (e.g. in enzymes), transport ions across highly heterogenous medium (e.g., cell membranes), and heat transfer properties, among others.
- *Effects resulting from quantized energy states for matter at the atomic and subatomic levels.* Thus the discrete quantum levels of a nanocluster might be tuned to modulate the electron transport normally modulated by the pH, ions, and redox centers. Important Quantum effects include electron tunneling for Scanning Tunneling Microscopy,[11,12] quantum Hall effect for resistance calibration instruments,[13] spin polarization in Magnetic Resonance Imaging.[13] Furthermore, radiation induced processes such as photoisomerization in vision and photosynthesis in plants[14] depend on the quantum yield which depends in turn on the molecular structure. These quantized energies lead to novel optical and magnetic properties for nanometer scale devices potentially, useful for medical diagnosis or intervention.
- *Thermal fluctuations can be commensurate with the size of a nanometer scaled system.* For example, the magnetic anisotropy energy can become comparable to thermal energy so that thermal noise can alter the electric or magnetic dipoles of macromolecules. These thermal fluctuations can affect the positional variance and conformations of biological structures or bio-amenable actuators and sensors, requiring precise characterization and control of the

stiffness properties.[15] Furthermore, at the nanometer scale, thermal transport properties depend on the number of interfaces (the change in surface area to volume ratio) which can dampen phonon vibrations without altering electronic conduction. Such effects could be used for thermal regulation through novel thermoelectric devices.[16] Here one needs to assess the effect of thermal carrier dimensional confinement/localization to determine how thermoelectric properties depend on size and how these effects can be utilized for nanomedical applications.

- *The discreteness of atomic structures leads to irregular surfaces and interfaces*. This obscures bulk materials concepts as "surface tension", dielectric constant, and pH. In designing active nanodevices and nanosystems with coupled mechanical degrees of freedom, such ill-defined contact surfaces may act to disable normal operation due to excessive short range Pauli repulsive forces between atoms at short distances and loss of coupling a large separations.[17] For example long alkane molecules are used as lubricants in macrosystems, but for nanometer scale system such molecules may act as dirt to render the nanosystem inoperable (see Fig. 2).

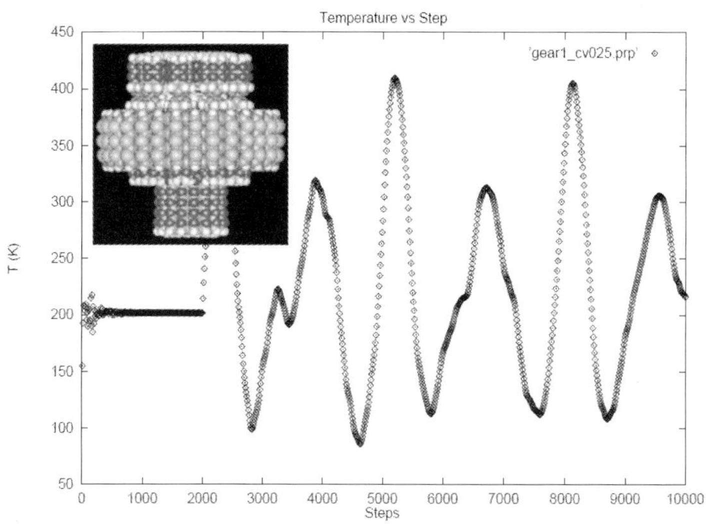

Figure 2. Tooth slippage in a nanometerscale replica of a planetary gear due to ill-conditioned atomic surfaces (Kinetic T profile during constant angular velocity dynamics). From Ref. 17.

Our limited ability to perceive in real-time the dynamics of these phenomena in nanometer scale systems makes multiscale computational modeling and simulation particularly valuable. However, to establish the appropriate computational methods for calculating the properties at the scales of interest, one must understand at which scale a particular property becomes relevant (refer to Fig. 1). Table 1 lists properties that appear at the nanoscale and which subsequently condition the length scale.

Since all of the properties of interest are fundamentally defined from the atomic composition and arrangement in a system, the following section reviews the essential atomic constituents in biological organisms.

Table 1. Properties of interest in materials for nanomedicine associated to length scale.

Property class	Property
Structural	Inter-atomic structure (bond topology, distances, angles)
Mechanical	Vibrational modes, Elastic moduli, Yield limits, Strength, Toughness, Creep
Surface	Oxidation, Adhesion, Friction, Wear, Hydrophobicity, Defects
Transport (electrical, thermal, etc.)	Chemical potential (electrons and protons), Band gaps, Fermi energy, Electron transfer (tunneling), Proton transfer, Phonon modes, Thermal conductivity, Conductance, Intensity, Permeability, Permittivity, Temperature, Reflectivity
Optical	Density of electronic states, Polarizability, Dielectric constant (frequency dependent), Optical absorption, Reflectivity, Photoizomerization, Refraction
Magnetic	Coupling (inductive or capacitive), Electron spin, Surface dependencies
Rheological	Viscosities and anisotropic flow, Flow rates (Newtonian and Non-Newtonian fluids), Shear stress and extensional stress, Plasticity

7.2.1 Building Blocks of Organic Matter

The atomic chemical composition of a human body is mainly comprised of four elements: Carbon (C), Oxygen (O), Nitrogen (N), and Hydrogen (H). These constitute ~99% of our physical structure, with O and H alone accounting for ~87%.[18] But they lead to ~10^5 molecular constituents in the form of water, lipids, proteins, RNA, DNA and other inorganic and organic molecules.

Understanding the relationship between structure and function of these constituents is critical to developing the ability to control human health conditions. Since their typical feature size (~ 10's nm) is far below our capability to observe directly and since most tools for probing these scales in vivo cannot reach such resolution, it is necessary to develop theoretical predictive models and computation to obtain a fundamental understanding. The following section summarizes some methods used to perform such calculations using a first-principles-based multiscale approach.

7.2.2 First Principles-Based Multiscale, Multiparadigm Simulations: From the Sub-Nanoscale to the Mesoscale

The computations required for accurate modeling and simulation of nanometer scale systems for nanomedicine involve a hierarchy of levels of theory: starting with quantum mechanics (QM) to determine the electronic states; force fields to average the electronics states to obtain atom based forces (a force field, FF), molecular dynamics (MD) based on the FF; mesoscale or coarse grain descriptions that average over many atoms; and finally continuum mechanics using distributed properties for a membrane or whole cell.

It is essential that the theory and computation be based on first principles QM because there is insufficient experimental data about the atomistic details of the surfaces and interfaces essential for determining biological function to use in empirically based developments, and because it allows predictions to be carried out without such prior information. Furthermore, QM provides the fundamental principles to provide the reaction paths and barrier heights to describe chemical reactions processes. However, the practical scale for accurate QM today is ~ 100 atoms per molecule or periodic cell (a length scale of 2 nanometer) whereas the length scale may be ~ 10 nanometer for an average protein, ~ 100 nanometer for a small bacteria or $\sim$5–100 μm for a typical Eukaryotic animal cell. Thus, minimal simulations of simple organisms would involve millions of atoms.

Moreover, describing the structures is just the first step in predicting the physicochemical processes responsible for higher level biological function (signaling, energy conversion, chemical synthesis). This requires coupling seamlessly and without discontinuities from QM to the corresponding constitutive equations, requiring intermediate models and methods suitable to handle a succession of length and time scales — a strategy referred to here as *First-Principles-Based Multiscale-Multiparadigm Simulation*.

Such approaches are making it possible to routinely solve problems once thought to be intractable (e.g. enzyme catalysis and protein activation).[19–21] Such coupling of paradigms requires that one propagate information about the uncertainties as well as a complete characterization of handshake regions. The goal is a bottom-up approach, based on first principles QM, to characterize properties of materials and processes at a hierarchy of length and time scales. This will improve our ability to design, analyze and interpret experimental results, perform model-based prediction of phenomena, and to precisely control the multi-scale nature of material systems for nanomedical applications.

7.2.3 Foundations of Quantum Mechanics and its Importance to Multiscale Modeling

The fundamental theory to describe the behavior of matter at the atomic scale is Quantum Mechanics (QM), which need not require any empirical data about a particular system. Some elements of QM and the approximations and methods used to solve the QM equations follow.

In 1924, Louis de Broglie argued that since light could be seen to behave under some conditions as particles.[22] (e.g. Einstein's explanation of the photoelectric effect) and at other times as waves (e.g. diffraction), one can also consider that matter has the same ambiguity of possessing both particle and wave properties. Starting with de Broglie's idea that particles behave as waves and the fundamental (Hamilton's) equations of motion from classical mechanics, Erwin Schrödinger[23] developed the wave equation that describes the space- and time-dependence of quantum mechanical systems,[24]

$$-\frac{\hbar}{2m}\nabla^2\psi(R,r,t)+U(R,r,t)\,\psi(R,r,t)=i\hbar\frac{\partial\psi(R,r,t)}{\partial t} \quad (7.1)$$

where the first term is related to the kinetic energy and U is the potential energy. The wavefunction Ψ depends on the coordinates of all electrons (r) nucleius (R) and describes the dynamics of how they change with time t.

7.2.3.1 *Approximations to Schrödinger's equation*

A number of simplifications to Schrödinger's equation are commonly made to ease the computational costs. We will review some of the basic methods.

7.2.3.1.1 Adiabatic approximation (treat electrons separately from the nuclei)

An excellent approximation is to factor the total wavefunction in terms of an electronic wavefunction that depends parametrically on the nuclear positions and a nuclear wavefunction

$$\psi(R,r)=\psi^{el}(R,r)\times\psi^{n}(R) \quad (7.2)$$

This is known as the Born-Oppenheimer[25] approximation and suitable for essentially all problems relevant to nanomedicine. For each set of nuclear positions, R, the electronic Schrödinger equation is solved leading to an $E(R)$ that depends only on R. This is referred to as the Potential Energy Surface (PES). Modern codes also lead directly to the interatomic forces, required for the dynamics.

The methods for solving the electronic Schrödinger equation have evolved into sophisticated codes that incorporate a hierarchy of approximations that can be used as "black boxes" to achieve accurate descriptions for the PES for ground states of molecular systems. Popular codes include Gaussian,[26] GAMESS,[27] and Jaguar[28] for finite molecules and VASP,[29] CRYSTAL,[30] CASTEP,[31] and Sequest[32] for periodic systems.

The simplest electronic wavefunction involves a simple product on one particle functions, antisymmetrized to form a Slater determinant, in order to satisfy the Pauli Principle. Optimizing these orbitals leads to the Hartree-Fock (HF) method, with the optimum orbitals described as molecular orbitals (MO). HF is excellent for ground state geometries and good for vibrational frequencies, but its neglect of electron correlation[33] leads to problems in describing bond breaking and chemical reactions. In addition it cannot account for the London dispersion forces responsible for van der Waals attraction of molecular complexes. A hierarchy of methods has been developed to improve the accuracy of HF. Some of the popular methods include second-order Moller-Plesset perturbation theory (MP2),[34] coupled cluster with multiple perturbative excitations (CC), multireference selfconsistent field (MC-SCF), and multireference configuration interaction, MR-CI[35] methods (See Ref. 36 for a recent review). A form of MC-SCF useful for interpreting electron correlation and bonding is the generalized valence bond (GVB) method,[37–39] which leads to the best description in which every orbital is optimized for a single electron. These are referred to as *ab initio* methods as they are based directly on solving the Schrödinger Equation (7.1), without any empirical data. In addition, many methods have proved useful in which empirical data is used to obtain approximate descriptions for systems too large for *ab initio* methods.[40]

A non-empirical alternative to *ab initio* methods that now provides the best compromise between accuracy and cost for solving Schrödinger's equation of large molecules is Density Functional Theory (DFT). The original concept was the demonstration[41] that the it's the energy and density ρ are a function of nuclear coordinates and hence all the properties of a (molecular) system be deduced from a functional of $\rho(r)$

$$E = \varepsilon\left[\rho\left(r\right)\right] \tag{7.3}$$

DFT has evolved dramatically over the years, with key innovations including the formulation of Kohn-Sham equations[42] to develop a practical one-particle approach while imposing the Pauli principle, the Local Density Approximation (LDA) based on the exact solution of the correlation energy of the uniform electron gas, the generalized gradient approximation (GGA) to correct for the gradients in the density for real molecules, and incorporating exact exchange into the DFT. This has led to methods such as B3LYP and X3LYP that provide accurate energies ($\sim$3 kcal/mol) and geometries[43] for solids, liquids and large molecules.[44,45] Although generally providing high accuracy, there is no prescription for improving DFT when it occasionally leads to large errors. Even so it remains the method of choice for problems in nanomedicine.

An alternative method, Quantum Monte Carlo (QMC) can in principle provide the exact answer by merely running the calculations for a sufficiently long time. However, these methods have not yet proved to be fast enough for problems relevant to nanomedicine.

7.2.3.1.2 Treat the nuclei as classical particles moving on a PES

The PES is used to describe the dynamics of the molecule as the nuclei move to describe vibrations or reactions, which is referred to as Molecular Dynamics (MD), thus Schrödinger's equation is replaced by Newton's equations of motion, an ordinary differential equation,

$$F = -\frac{\partial U}{\partial R} = m\frac{d^2 R}{dt^2} \tag{7.4}$$

where, F represents the forces (the negative gradient of the PES with respect to R) and m is the corresponding atomic mass. Integrating (7.4) leads to trajectories. Of course, all information about the electrons is gone (electron tunneling, electronic exited states). Such calculations in which the Forces come directly from QM are often referred to as Car-Parrinello calculations.[46] Unfortunately, the costs of QM limit such calculations to ~ 100 atoms.

7.2.3.1.3 Force fields

The practical MD solution for systems in nanomedicine is to describe the PES, U, in terms of a force field (FF), a superposed set of analytic functions describing the potential energy (inter-atomic forces) as a function of atomic coordinates leading to the equations of motion,

$$F = m_i \ddot{x}_i = -\nabla_i U\left(x_1, x_2, \ldots, x_n\right) \tag{7.5}$$

where, U is portioned in terms of valence or bond functions and nonbond functions,

$$U = \left[U_r + U_\theta + U_\phi + U_\psi\right]_{\text{bond}} + \left[U_{\text{vdW}} + U_{Coulomb}\right]_{\text{non-bond}} \tag{7.6}$$

For non-reactive processes the bonded components are conventionally treated harmonically, e.g.

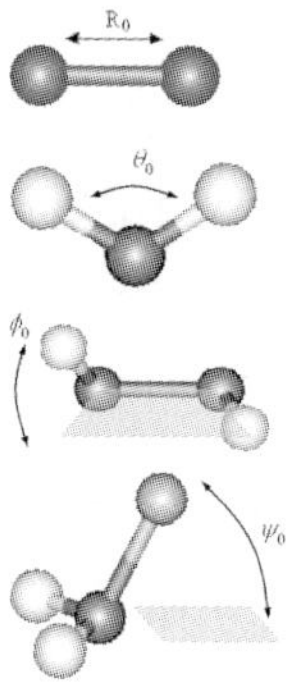

$$U_r = \sum_{bonds} k_r (r - r_0)^2$$

$$U_\theta = \sum_{angles} k_\theta (\cos(\theta) - \cos(\theta_0))^2$$

$$U_\phi = \sum_{torsions} k_\phi \sum c_n [\cos(n\phi) + 1]$$

$$U_\psi = \sum_{inversions} k_\psi (\cos(\psi) - \cos(\psi_0))^2$$

where the sub indices 0 indicate equilibrium values. The k constants are related to force constants for vibrational frequencies the c constants are related to an energy barriers and *n refers* to periodicity.

There are generally two non-valence or non-bonded terms. Here, the van der Waals (vdW) term accounts for short range repulsion arising from the Pauli Principle and for the long range attraction arising from the London Dispersion terms,

$$U_{vdW} = \sum_{\substack{R_{ij} > Rcut \\ [excl\,(1-2, 1-3)]}} \hat{U}_{vdW}(R) S\left(R_{ij}, R_{on}, R_{off}\right). \tag{7.7}$$

The electrostatic or Coulomb interactions, depend on the charges of each pair of atoms (Q_{ij}) and the dielectric constant ($\varepsilon = 1$ in a vacuum but larger values are used for various media)

$$U_{\text{Coulomb}} = C_0 \sum_{i>j} \frac{Q_i Q_j}{\varepsilon R_{ij}} S\left(R_{ij}, R_{on}, R_{off}\right). \tag{7.8}$$

Here C_o converts units (e.g. for energy in kcal/mol, distances in Å, and charge in electron units, $C_0 = 332.0637$). The most time-consuming aspect of MD simulations for large systems are the long-range nonbond interactions, Eqs. (7.7) and (7.8), which decrease slowly with R. This scales as O (N^2) for an N particle system (e.g. a protein with 600 residues would have $\sim$ 6,000 atoms requiring $\sim$ 18 million terms to be evaluated every time step). The S term in both (7.7) and (7.8) allows the long range terms to be cut off smoothly to decrease the costs. An alternative is the Cell Multipole Method[47]

(CMM) [and the Reduced CMM[48]] which scales linearly with size, reducing the cost while retaining accuracy for the million atom systems important for nanomedical applications.

Many useful FF have been developed (see Ref. 49 for a recent review) over the last 30 years specifically aimed at biological systems. Commonly used FF include: AMBER,[50] CHARMM,[51] Dreiding[52] and OPLS.[53] The parameters in these FF were adjusted to fit a combination of results from theory and experiments, but for nanomedicine it will important to develop new generations of FF in which all parameters are derived from QM calculations on small representative systems,[49] adjusting the FF descriptions to reproduce the structures, energetics, and dynamics from QM on nanoscale systems. The reason is that introduction of external nanostructures into biological systems leads to interfaces between materials for which there is little or no empirical data. Thus one needs a framework where this new data, as well as the biological components, are determined from QM.

Force Fields make it practical to apply MD simulations to the atomic-level dynamics of proteins[54] interacting with nanoscale components providing realistic descriptions of the atomic motions under physiological conditions. They allow one to carry MD simulations on systems up to a million times larger than for QM.

7.2.3.2 Solvent interactions in biological systems

Solvents have a major effect on the structure and properties of biomolecules. Including the solvent explicitly for a typical protein with 4,000 atoms may require $\sim$40,000 atoms for the solvent. This is practical today, but becomes impractical for many systems of interest to nanomedicine. The alternative of ignoring the solvent by using an effective dielectric constant, ε, in Eq. (7.8) is common, but it may not accurately represent the PES for the system. A useful compromise is the use of continuum solvent using the Poisson-Boltzmann approximation,[55] which includes solvent reorganization using explicit first solvation layer, and the Generalized Born method,[56] in large-scale QM-DFT[57,58] and MD simulations.

7.2.4 Bridging QM and MD with Full Chemistry

A major difficulty with MD using common FF is that they do not describe chemical reaction processes. The alternative is to use QM which is not practical for systems larger than ~100 atoms. Moreover even for small systems, to obtain accurate reaction barriers and mechanisms requires careful consideration of the details of how the QM is done.

7.2.4.1 Simulating reactive processes with FF

A recent breakthrough in simulation technology provides a generally valid and accurate way to capture the barriers for various chemical reaction processes (allowed and forbidden reactions) into the force fields needed for large-scale molecular dynamics (MD) simulation. This is the ReaxFF[59] reactive force field; which is derived only from QM calculations. ReaxFF is capable of reproducing the energy surfaces, structures, and reaction barriers for reactive systems at nearly the accuracy of QM but at costs nearly as low as simple FF. Thus ReaxFF providing the link between first principles QM and the atomistic simulations of the heterogeneous materials structures and interfaces required for the systems of nanomedicine.

Applications of ReaxFF have been reported for a wide range of materials, including hydrocarbons,[59] nitramines,[60] ceramics,[61] metals and metal oxides,[62,63] metal/hydrocarbon interactions[64] and metal hydrides.[65] ReaxFF has been used to simulate chemical events in many systems, including nanotube deformation and buckyball polymerization,[66,67] thermal decomposition of polymers,[68] high-energy materials initiation[69,70] crack propagation,[71] reactive oxygen — and hydrogen migration in fuel cells[72] and metal/metal oxides surface catalysis.[73] ReaxFF includes the following features:

- **Environmentally dependent charge distributions on atoms.** The charges on the atoms adjust in response to the local environment allowing them to change as bonds are broken and formed and to shield the Coulomb interaction between atoms at small distances.
- **Bond order dependent valence terms.** A general relation is provided between bond distance and bond order and between bond order and bond energy (and hence forces). The bond orders gradually go to zero as the bond lengths increase and they gradually increase for shorter distances finally saturating for the smallest distances (e.g. BO = 3 for CC bonds). This provides a smooth description of the valence terms during chemical reactions.
- **Non-bond or van der Waals interactions.** ReaxFF uses a simple Morse function to account for the short-range repulsion and steric

interactions arising from the Pauli principle (between *every* atom pair). The long range attraction accounts for vdW attraction.

- **No cutoffs.** All interactions change smoothly during reactions (which are allowed to occur at any time and place) so that ReaxFF can be used with general conditions of temperature and pressure.
- **Transferable potential.** Simple FF provide different parameters for differ atom environments (e.g. single versus double bonds, sp3 versus sp2 geometries). ReaxFF eschews such description using only a single atom type for each element which is necessary since bond orders and geometries change during reactions. This leads to good transferability of the FF.
- **QM based.** All parameters are optimized/derived directly from QM studies on a large number of reactions. This allows extensions to new materials in nanomedicine, where there may be no experimental data.

7.2.4.2 Recovering electrons in MD

Numerous biological processes involve excited electronic states. Examples include the low lying electronic states involve in the photoisomerization of Rhodopsin and the very highly excited states created under conditions of high-energy radiation fluxes;. Generally FF, including ReaxFF, describe only the ground electronic state. Indeed even with QM it is often quite difficult to obtain accurate descriptions of even the first excited state and these methods are limited to only 100s of atoms for time scales of picoseconds. In general, there is no effective way to do QM on the highly excited states involved in radiation exposure. A new methodology that has the potential to enable simulations for such systems is the *electron force field* (eFF).[74] eFF was developed to enable simulation of electron dynamics of large highly excited systems. Current work has been limited to systems involving hydrocarbons, but the results are most promising, explaining for example the Auger processes following ionizing the 1s electron from C (~ 270 eV). eFF includes the electrons explicitly in the dynamical equations, describing their dynamics simultaneous with the nuclei (no Born-Oppenheimer approximation). Thus eFF lies in between QM and MD for excited electrons, including *non-equilibrium* dynamics of electron motions.

In eFF, the electrons are described as spherical Gaussian wave packets whose size and position change dynamically. The total wavefunction is a product of these wavepackets, so that there is not explicit antisymmetrization. Instead, the Pauli principle is replaced with pairwise Pauli contributions in which same spin electrons have a repulsive interaction at short distances. eFF has just three universal parameters (for the spin dependent electron-electron interactions), but it leads to reasonable geometries and energetics for molecules involving covalent, ionic, multicenter, or metallic type bonding (for hydrogen through carbon atoms). eFF has been shown to provide a the electron and nuclear dynamics following Auger ionization of C1s electrons in hydrocarbons[74] and hydrogen at high pressures and temperatures.

7.2.5 Bridging Classical Particle Mechanics with the Mesoscale: Constrained MD and Coarse-Grain Force Fields

Successful applications of first principles methods to supramolecular design requires a scale lying in between the molecular or atomistic scale (where it is convenient to describe molecules in terms of a collection of bonded atoms) and the continuum or macroscale (where it is convenient to describe systems as continuous with a finite element mesh basis for a digital description).[75] This coarse grain or Mesoscale level is most important for determining the properties and performance of a wide range of different materials, including "soft condensed matter", amphiphilic (surfactant containing) fluids, colloids and polymers, gels, liquid crystals; proteins, and DNA. Here we need to describe such biological processes as protein activation, enzymatic transformations, ribosome activity, and general diffusive motions of biomolecules on timescales of microseconds and longer — see Sec. 7.3.6).

7.2.5.1 Constrained MD

An atomistic approach to overcome this problem involves simplifying the description of the system by introducing Cartesian constraints to reduce the number of effective degrees of freedom (DOF) of a system. An example is the SHAKE[76,77] procedure, by which the Lagrange multipliers regarding Cartesian constraints imposed on a system (typically bond stretch or angles) are explicitly added to the potential energy function governing the equations of motion (EOM) for the system. SHAKE uses iterative (i.e. approximate) corrections at each integration time-step to correct violations to the constrained space. Variants of SHAKE have been proposed to increase numerical precision and extend support for non-equilibrium dynamics, including RATTLE,[78] M-SHAKE,[79] LINCS,[80] and SETTLE[81] methods. At high temperatures these approximate solutions can lead to high fluctuations,[82] which limits the time step size and the number of explicit constraints required for scaling to the large macromolecular systems of nanomedicine.

Alternative methods that rely on computational parallelism and additional physical approximations offer new opportunities to address these problems, including accelerated molecular dynamics methods, such as hyperdynamics[83,84] for infrequent event systems and the internal coordinate constrained MD methods based on rigid body approximations[85,86] (see Fig. 3). These lead to important improvements in simulation times and system scales applicable to the study of conformations and dynamics of large biomolecular

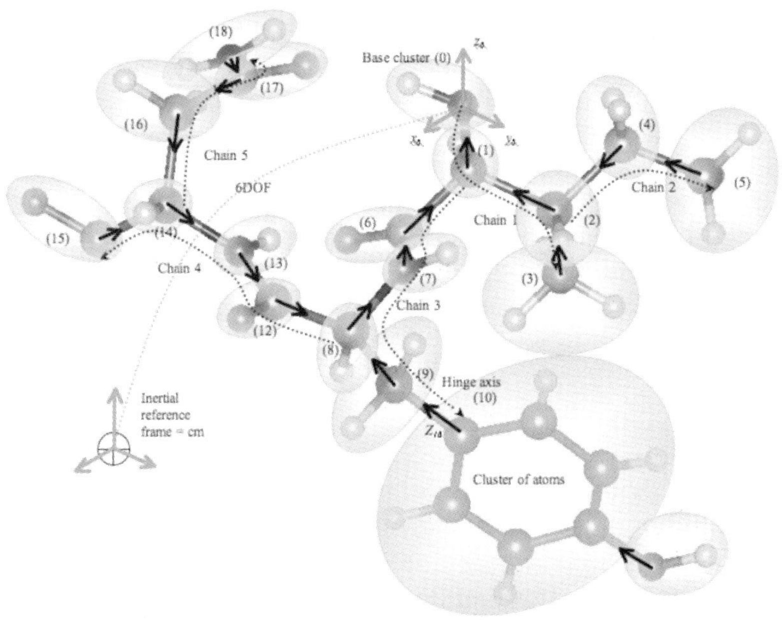

Figure 3. Constrained molecular multibody (Tripeptide with 54 atoms and 162DOF is reduced to 19 clusters and [19+6]DOF); implicit constraints result in molecular rigid bodies shown in grey. (See page 342 for color illustration.)

systems including protein–protein interactions, lipid–protein interactions, and membrane–membrane interactions.

Expressing the molecular EOM in terms of internal coordinates can avoid high frequency vibrational modes allowing longer time steps. This leads to a semi-classical dynamics formulation of the constrained equations of motion for a set of articulated rigid molecular bodies (see Fig. 3) in which a molecule is partitioned in terms of clusters of atoms (e.g. a phenyl ring, an alpha helix, or even an entire protein domain). Here one can constrain bond lengths and bond angles to focus on the torsions that distinguish conformations, dramatically reducing the DOF. This simplifies the description and analysis of the molecular system (e.g. reduces the inter-atomic force calculations) at the expense of increased computational complexity of the equations of motion.

Mazur *et al.*[87,88] used the internal coordinates representation to investigate the conformations and dynamics of bio-macromolecules. However solving the EOM with this method scaled exponentially with size and relied on a costly expression of the inter-atomic potentials in internal coordinates. Subsequently, our group pioneered the development of practical internal coordinate constrained MD methods, based on ideas initially developed by the robotics community,[82,89–91] reaching $O(n)$ serial implementations, using the Newton-Euler Inverse Mass Operator or NEIMO[92–94] and Comodin[95] based on a variant of the Articulated Body Inertia algorithm,[96] as well as a parallel implementation of $O(\log n)$ in $O(n)$ processors using the Modified Constraint Force Algorithm or MCFA.[91,97] The general state space equations of motion (EOM) for an internal coordinate constrained MD representation of a system can be written as,

$$\tau = M(Q)\ddot{Q} + C\left(Q,\dot{Q}\right)\dot{Q} \tag{7.9}$$

where, τ corresponds to the vector of generalized forces (e.g. torques), M denotes the articulated body inertia matrix, C denotes the nonlinear velocity dependent terms of force (e.g. Coriolis, centrifugal and gyroscopic forces), Q, $\dot{Q}$, $\ddot{Q}$ correspond to the generalized coordinates that define the state of the system. It then follows that the dynamics of motion for a microcanonical ensemble is obtained by solving for the hinge accelerations,

$$\ddot{Q} = M^{-1}(Q)\left[\tau - C\left(Q,\dot{Q}\right)\dot{Q}\right].$$

In general, the advantages of using internal coordinate constrained MD for biomolecular systems include:

- *A reduction in the total number of degrees of freedom (DOF).*
- *An increased integration time-step of the equations of motion.* Constraints lead to a reduction in the sampling frequency (via the Nyquist-Shannon sampling theorem).[98,99]
- *Faster energy exchange between low- and high-frequency modes.* Narrower frequency gap that reduces the differences in effective temperature calculations. In practice, this effect combined with an increased time-step, allows simulations at higher temperatures.
- *Implicit expression of relative constraints.*
- *Decoupled contributions to* $SO(3)$. Filtering DOF that are not of interest to particular simulations can lead to faster sampling of the conformational space.
- *An effective flattening (averaging) of the potential energy surface (PES).* Reduced energy fluctuations (see Fig. 4) and a reduced number of local minima in the energy landscape allowing a smoother exploration of the conformational space.

The efficiency of these methods lies in not only solving the EOM in time lower-bound computational efficiency using Cartesian-based

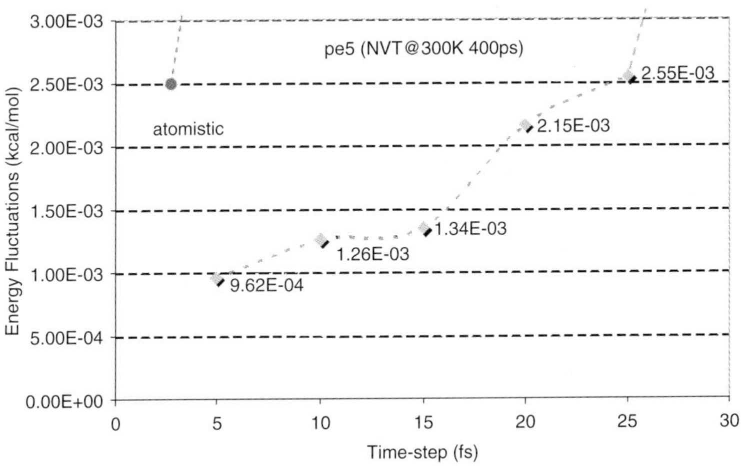

Figure 4. Reduced energy fluctuations in internal coordinate constrained MD [diamonds] for pe5 polymer 400ps NVT-CMD @300K versus atomistic Dreiding MD [dot][40] lead to increased integration time-steps.

projected FF, but in finding a systematic coarse-grain representation and a new set also optimized coarse-grain FFs of the system being modeled.

7.2.5.2 Coarse-Grain force fields

Coarse grain (CG) models are natural step toward expanding the power of MD simulations to study collective phenomena in biophysical systems. Nevertheless, these must carry enough information about the atomistic behavior while at the same time be efficient to scale in both time (> 1 ms) and length (> 100 s nm). For example, biological simulations require accurate models to represent solute-solvent interactions, in particular with water (and lipids), hence its coarse-grain representation should provide an effective medium for other molecules to exist in and interact with it, i.e. account for its momentum so that its behavior can be consistent with hydrodynamics, have the correct density at the desired temperature, and be able to maintain a liquid/vapor interface over such a temperature.[100]

To this end, several approaches have been used and demonstrated for phospholipids,[101–109] oligosaccharides and their water mixtures,[110] proteins,[111] dendrimers and polymers.[112] Coarse-grain force fields have been developed from heuristically simplified models of biomolecules (e.g. water, alkanes, lipids, etc.) and by systematic optimization procedures of a set of interaction potentials between collections of atoms treated as rigid bodies. In the former case, the fitting process relies on results from the finer atomistic MD using Monte Carlo schemes,[100,113] random search algorithms including Genetic Algorithms (GAs),[114,115] and hybrid algorithms to accelerate convergence using artificial Neural Networks (ANNs)[114] and gradient based algorithms near local minima. Our group focuses on a first-principles-based strategy in order to provide not only the accuracy from finer grain calculations, but improved scalability and a seamless coupling that does not rely on heuristics.

The major challenge in automated fitting of FF for large systems comes from solving a global optimization problem which is non-deterministic in nature (i.e. there is no analytical solution for finding a single global minima), hence systematic procedures tend to require heuristic input to minimize the effect of "bad" initial conditions or are limited in transferability. One approach to circumvent this problem involves the use of inverse solutions, based on stochastic methods (e.g. Reverse Monte Carlo [RMC] schemes) to probe the configuration space though a random walk in search of a set of parameters that are consistent with experimental data, nevertheless, this requires the availability of good experimental data. We have demonstrated preliminary successful use of hybrid stochastic GA-gradient based algorithms to systematically fit coarse-grain torsion-only force fields directly from atomistic MD energy gradients using a 1:1 mapping between Cartesian space force interactions with internal differential space torques (see Fig. 5) which leads to improved integration time-steps and overall simulation time (30–40 fs) for large biomolecules using the Constrained MD methodology described. Current applications include fast conformation search and structure prediction.

7.2.6 *Alternatives for Bridging into the Continuum*

An important method that bridges from the atomistic to the continuum is the Quasi-continuum (QC) method developed by Tadmor, Ortiz and Phillips.[116] With the goal of modeling an atomistic

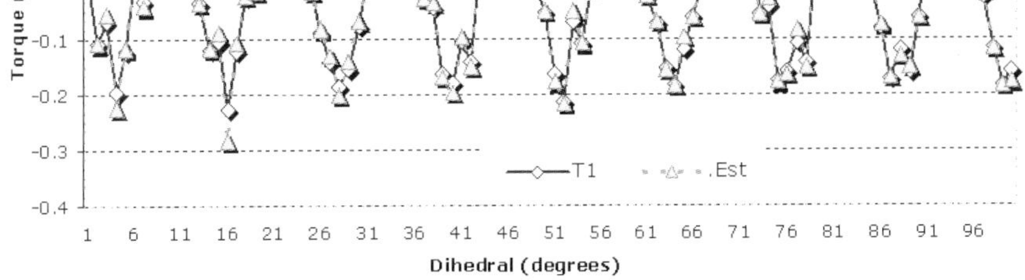

Figure 5. NVT @300 constrained MD using predicted (energies) torques in internal coordinates from the hybrid GA-gradient module in CMDF for a polymer chain (single polymer base torsion shown: atomistic Dreiding force field, blue/yellow: GA-gradient estimated). Fitness function based on RMSD measure calculated from structure derived from Dreiding force field calculations. (See page 342 for color illustration.)

system without explicitly treating every atom in the problem, the QC provides a framework whereby degrees of freedom are judiciously eliminated and force/energy calculations are expedited. This is combined with adaptive model refinement to ensure that full atomistic detail is retained in regions of the problem where it is required while continuum assumptions reduce the computational demand elsewhere.

Finite Element methods (FEM) have been used to determine the continuum level mechanical properties of macromolecules using heuristically selected MD derived parameters,[117] unfortunately, seamless coupling to the atomistic simulations remains a challenge for non-periodic systems.

Other methods based on computational or mathematical formalisms to systematically upscale or downscale computations, in time or in length, are stochastic based algorithms (e.g. Monte Carlo), averaging and homogenization.[118] Assuming that the evolution of the physical system can be described by probability density functions (pdf's), then the Monte Carlo simulation proceeds by randomly sampling these pdf's to arrive at a solution of the physical problem given a particular scoring metric. On the other hand, *homogenization* relies on replacing a highly heterogeneous material, characterized by the rapidly oscillating coefficients of a particular property from a finer scale, by an effective, homogeneous material which is characterized by constant (or slow) coefficients (in the same property) and a periodic cell corrector to account for scale changes while *averaging* takes a subset of the variables that evolve rapidly compared with the remainder in a system, and replaces these by their averaged effect. The major difficulty with a systematic approach to averaging and homogenization, when it is not directly associated with the physical structure of the system, becomes extracting physical insight from the resulting simplified equations that describe the systems behavior at different scales.

7.2.7 Multiparadigm Support in Multiscale Simulations

As multiscale simulation requirements extend into the billions of particles and microsecond's domain range it becomes imperative to use computational algorithms that are able to capture the necessary information at each scale with the least amount of algorithmic effort. Scale bridging methods that are efficient for finding chemical

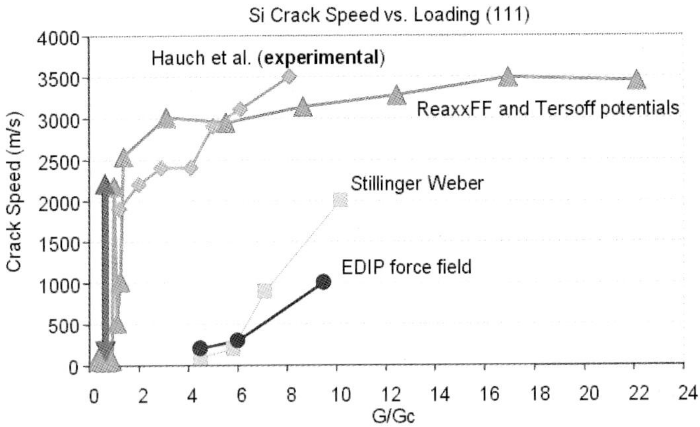

Figure 6. Silicon crack dynamics. At ~0.035% critical strain crack speed jumps from 0 to 3 km/sec).[119]

properties might not be so for finding conformational changes, or a single method might not be suitable for capturing the relative rates at which events of interest appear on a given time range. Instead of dealing with partitioned paradigms per scale (i.e. electrons of QM, atomic FF of MD, coarse grain FF of mesoscale dynamics, phenomenological parameters of macroscale) with fixed spatial or temporal scale ranges our approach implies a continuous description of the system, where, events are driven by the underlying physical phenomena and computed by appropriate computational regimes — each, capable of traversing several orders of magnitude in the time and length scales under a valid physical description, as a function of the available computational resource (as depicted by Fig. 1 — the shown box sizes do not imply a hard time or length limit but an indication of range and a function of the methods used and the computational resources available). Each regime can couple to others throw well-defined smooth transition-handshake regions.

This strategy is adopted by our Computational Materials Design Facility (CMDF) to precisely enable seamless mixing of paradigms for multiscale simulations. The Computational Materials Design Facility[a] (CMDF)[95] was developed to handle such a concept of multiscale, multiparadigm simulations. In CMDF computational domains can be automatically specified using several

[a] http://www.wag.caltech.edu/multiscale/multsicale_computations.htm.

physicochemical approaches:

- Atom types (e.g. method A $\rightarrow$ atom type Y and X–Y interactions)
- Strain (atomic) or stress (localization)
- Bond-orders (bond breaking/formation)
- Large strain/cohesive energy
- Frequency distribution (e.g. muli-rate sampling)
- Geometric criteria and information (e.g. interfaces, boundaries)

CMDF is a single image framework for mixing paradigms (ranging from electronic structure calculations at the QM level, through atomistic structure and Molecular Dynamics (MD) with reactive and non-reactive force-fields to Mesoscale and microstructure evolution using finite element analysis) with heterogeneous data and programming models and seamlessly coupling of length and time scales with minimum computational and error penalties. The ability to prepare a scales and paradigms bridging application leads to huge investments in time for every particular simulation setup. It uses the Python programming language to efficiently control the computational flow between disparate high-performance inline processing cores written in compiled languages (C/C++/Fortran) that carry out physicochemical calculations and computational science couplings.

We have successfully demonstrated the application of CMDF to numerous problems, for example, in Buehler *et al.*[71] the mixing of reactive (ReaxFF) and non-reactive potentials (Tersoff) lead to fundamental findings on the nature and dynamics of crack propagation in silicon under stress, a long-standing problem in the observed crack propagation in silicon (Si) crystals (see Fig. 7). This paradigm mixing is currently being demonstrated in protein active site reactivity and enzymatic catalysis.

7.3 MULTISCALE MODELING APPLICATIONS TO NANOMEDICINE

7.3.1 *Nanoscale Electronic Transport in Biosystems*

Progress in the conventional "top-down" lithographic approaches to downscaling will slow down once the fundamental limits of this technology have been reached.[121] This limit has become a combination of several factors including fabrication technologies, and characteristics such as interconnects density, power density and heat dissipation, ratio of doping to availability of majority carriers,

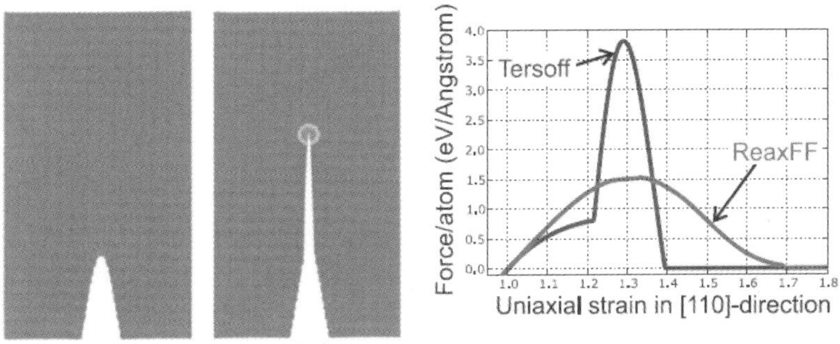

Figure 7. (Left) Pure Tersoff; (middle) Hybrid ReaxFF-Tersoff; (right) difference in large-strain behavior between Tersoff and ReaxFF. System size is 28K atoms. From Ref. 120.

gate oxide electron tunneling[122] in field effect transistors (when the oxide reaches 5–6 atomic layers of thickness) as well as quantum effects in short transistor channel regions with feature size below ~ 10 nanometer that lead to tunneling currents between source and drain. By 2020 device innovations are expected to bring about new performance increases or provide a means to keep the efficiency trend once predicted by Gordon Moore.[123] Alternatives technologies include molecular electronic devices and quantum devices.

Molecular electronic devices will be the chemistry based "bottom-up" counterpart to current top-down technology, where systems are built starting from molecules of proper electrical characteristics with self-assembly capabilities. Organic semiconductor materials have tremendous potential to transform and complement electronic devices and, in the context of nanomedical applications, become ubiquitous elements in the development of novel biometric, biocontrol and biomimetic solutions (potential applications include nano-transducers, sensors and actuators). An important property that must be realized is being able to control their electrical transport properties. This is becoming increasingly relevant through the advent of experimental techniques to self-assembly nanoelectronic systems over biological substrates from work pioneered by Rothemund.[125] Most organic materials are not crystals; rather, they are amorphous hence electrons cannot flow easily through an explicit conduction band due to the irregular orbital overlaps.

The possibility to synthesize organic molecules with desired structure and functionality drives the current design of molecular nano-devices. Wire, rectification, switch, and transistor functions have been demonstrated with molecules of various complexity.[126–131] In particular, the field of molecular electronics is dominated by the extraordinary ability to synthesize complex organic molecules (upwards of 300 atoms) as shown by Stoddart *et al.*[132–135] (see Fig. 8).

However, optimizing the operation of such devices, even the replacement of a few atoms often requires the complete chemical synthesis of the entire molecule to be carried out; molecular modeling techniques have provided a faster and cheaper way to predict and improve device characteristics[136–138] (e.g. drive capability [fan-out], switching characteristics, contact transitions, among others).

In studying electron transport properties for molecular electronics devices, a typical model involves determining the quantized conductance in the junction between the discrete molecule and continuous contacts which requires analyzing charge injection at the interfaces, charge transport through the molecule, charge modulation using external fields, and the nature of the electrostatic potential acting in finite voltages. Our groups' approach[124,139–142] uses the Landauer formalism[143] to associate conductance with electron tunneling transmission through the inter-electrode region. A general overview of the methodology involves: (1) determining the total energy (E) as a

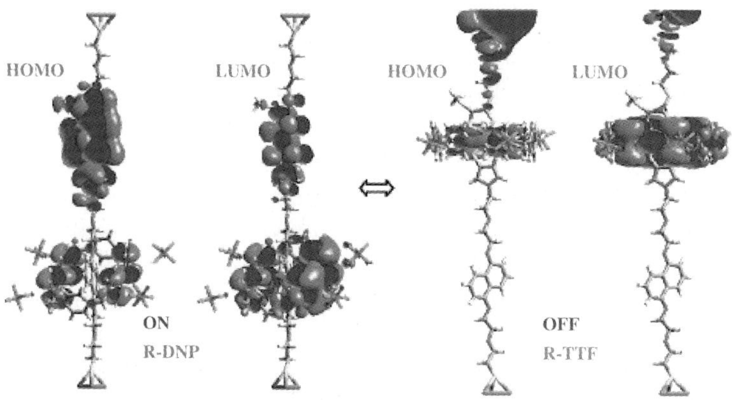

Figure 8. Molecular Orbitals of Au-rotaxane-Au switch. Green represents Au atoms, yellow is S, gray is C, red is oxygen, and white is hydrogen. From Ref. 124. Switching takes place via displacement of molecular ring between Au atoms. (See page 343 for color illustration.)

function of structure and composition incorporating electron contributions and chemical electrode potentials (for a two electrode device, $\mu_1 - \mu_2 = eV$) using DFT calculations with exchange correlations adjusted from QM calculations. The organic-metallic interactions are treated with a standard Dreiding FF[52] extended to include S-Au terms described via off-diagonal vdW energy terms. The selected configurations are then further optimized with the full quantum mechanical methods; (2) calculation of the transmission function using the matrix non-equilibrium Green's function method within the Landauer formalism. This allows calculating the probability of electrons tunneling through potential barriers, along continuous and discrete interfaces, in the presence of external potentials (given there is no conduction band). This involves estimating the probability of total electron transition (T) under a bias potential (V), from one electrode to the other using the corresponding chemical electrode potentials ($\mu_{1,2}$) per time unit (electron tunneling), and integrating over the energy landscape available for electron tunneling to determine the stationary current of the system,

$$I = \frac{2e}{h} \int_{-\mu_1}^{\mu_2} T(E, V) \left[f_1 (E - \mu_2) - f_2 (E - \mu_1) \right] dE. \qquad (7.10)$$

To study the linear and non-linear I–V regimes (see Fig. 9) differential conductance calculations are performed from $G = \partial I / \partial V$.

7.3.2 Nanoscale Thermal Transport in Biosystems

Predicting and controlling thermal transport at the nanoscale has numerous potential applications in nanomedicine, including thermoelectrical energy conversion, directional control of heat transfer, ablative therapy via selective heating of Drug Delivery Systems (DDS) or nanoparticles,[144–146] localized calorimetry of biomolecules to determine their binding properties (Proteins and DNA undergo structural changes and phase transitions at different temperatures), among others.

A significant part of current research efforts in thermal conduction is devoted to the development of nano superlattices with enhanced thermoelectric properties. With layer thicknesses on the order of the phonon mean free path ($\sim$1 nm), superlattices can have unique thermal transport (by phonons) properties, including an anisotropic thermal conductivity tensor, thermoelectric cooling

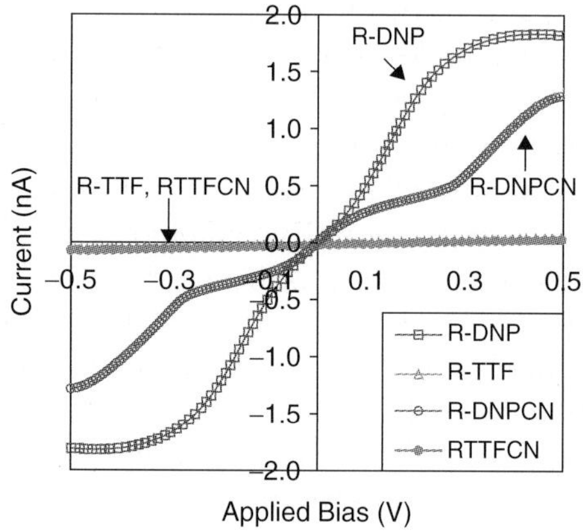

Figure 9. $I-V$ curve of Au-rotaxane-Au switch. (a) The four $I-V$ curves correspond to ring on DNP (R-DNP), on TTF (R-TTF), and for the modified version of the rotaxane bearing a CN group substituted on the DNP unit, R-DNPCN, and R-TTFCN states.[124]

(through the Peltier effect) and energy generation (through the Seebeck effect) — because thermal conductivity can be reduced while retaining electron transport properties, resulting in high values of the thermoelectric figure of merit — ZT.

Figure 10 depicts the multiscale strategy for calculating the thermoelectric power in nanometer scale materials. The fundamental

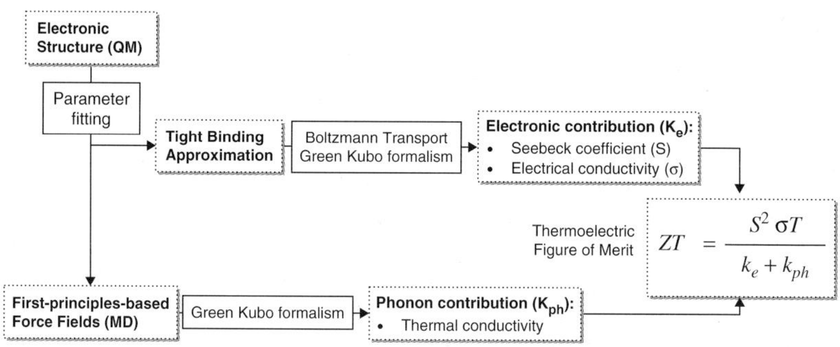

Figure 10. First-principles-based calculation of thermoelectric power in nanometer scale materials. S = thermoelectric power, T = absolute temperature, σ = electrical conductivity, and k_e to the electronic and k_{ph} phononic contributions to thermal conductivity, respectively.

goal is to maximize ZT, hence S must be large so that a small temperature difference can create a large voltage, σ must be large in order to minimize heating losses, and k must be small to reduce heat leakage and maintain a temperature difference.[16] Since nanometer scale porosity decreases permittivity and thermal conductivity as a result of the increased number of interfacial surfaces.[147] (e.g. when the system size becomes comparable to the phonon mean free path, the effective thermal conductivity can be as much as an order of magnitude smaller than the bulk value in insulators).

Coupling nanometer predictions of thermal properties to the mesoscale involves molecular dynamics (MD) to model the interaction of single phonons with grain boundaries (e.g. thermal conductivity tensor rates) and higher level phonon-grain boundary scattering models that incorporate homogenized rates from MD, along with system defects and geometry, and into Fourier's Law of heat conduction (e.g. using Kinetic or Dynamics Monte Carlo solutions).

7.3.3 Polymer-Based Nanosized Hydrogels (and Particles) and Dendrimers as Drug Delivery Systems (DDS)

Controlled nanosized ($<$ 200 nm) DDSs involving polymers[148] are used by tens of millions of people annually.[149] These are also being used as release systems for proteins, such as human growth hormone and interferon[148] showing how biomaterials can be used to positively affect the safety, pharmacokinetics and duration of release of important bioagents. These DDS have the ability to deliver a wide range of drugs to different areas of the body while maintaining the drug's bioavailability constant as long as desired (see Fig. 11). The main idea is to improve drug treatment efficiency via nanosized carriers capable of transporting and selectively releasing bioactive agents to treat specific illnesses while masking low water solubility, reducing rapid phagocytic and renal clearance, and eliminating systemic toxicity (e.g. conventional hydrophobic antitumor agent). Mayor issues addressed by DDSs are:

- *Decreased bioavailability* of drug payloads caused by circulatory and tissue-based macrophages or Reticuloendothelial uptake,
- *Selective targeting* through size dependent passive targeting or active targeting using receptor-mediated interactions, and
- *Controlled activation and drug release.*

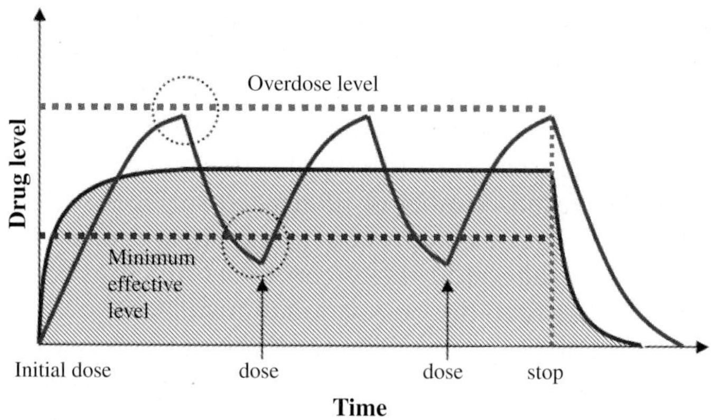

Figure 11. Drug levels in the blood with traditional drug dosing (saw curve) and controlled-delivery dosing (filled region).

Among the critical design factors for DDSs are:

- Size < 200 nanometer for capture of drug payload and renal clearance,
- Surface charge control for active targeting and macrophages avoidance,
- Surface hydrophobicity control for efficient protein adsorption and enhanced solubility *in vitro*,[b]
- Molecular selectivity to improve drug targeting,
- Biocompatibility and biodegradability,
- Degradation toxicity control of DDS and sub-species,
- Steric stabilization for polymer based DDSs to keep nanocarriers well dispersed and inhibit coagulation of suspensions.

Key to the successful use of a particular DDS is its drug release mechanism. The release rate of a drug quantity (M) per unit area of exposure (A), $d(M/A)/dt$ depends on the release mechanism. For example, the drug release rate from hydrogel pored matrices with dispersed drug can be modeled by an extended form of the steady-state Fick's Law diffusion equation:

$$\frac{d}{dt}\left(\frac{M}{A}\right) = \frac{1}{2}\left[(D\varepsilon/t)\, c_{sw}\,(2c_d - \varepsilon c_{sw})\right]^{\frac{1}{2}}\left(t^{-\frac{1}{2}}\right) \qquad (7.11)$$

[b] Some anti-cancer drugs are amphiphatic — e.g. Doxorubicin; others hydrophobic — e.g. Paclitaxel; and others hydrophilic — e.g. N-(phosphonoacetyl)-L-aspartate.

where D corresponds to the drug diffusion coefficient, c_s to the drug solubility in the water, c_d is the initial drug loading of the DDS and the drug concentration of the drug in the pore-filling liquid phase, respectively, and ε and τ correspond to the porosity and tortuosity of the matrix.

The primary release mechanisms currently under study include:

- **Photodynamic therapy** (PDT) which involves excited photosensitizer molecules that transfer energy to molecular oxygen to form highly reactive singlet oxygen which induces cell apoptosis. Different DDSs have been reported including Meta-tetra(hydroxyphenyl)chlorine (mTHPC) in sub-200 nanometer silica or poly(lactic-co-glycolic acid) PLGA hydrogel nanoparticles[150] and in dextran-methacrylate biodegradable hydrogels.
- **Thermosensitive therapy** which relies on temperature responsive properties such as phase transitions with temperature changes originated from a radiative source are used to control drug release from the DDS to cause cells apoptosis (through accumulation of drug inside cell nucleus via intercalation and interaction with topoisomerase II to cause DNA cleavage and cytotoxicity.[151] Examples of this technique have been reported in the literature for drug molecules like clonazepam and doxorubicin using nanocarriers based on (poly(N-isopropylacrylamide)-b-poly(e-caprolactone) — PNPCL with poly(N-isopropyacrylamide) — PNiPAAm shell surfaces to control temperature-dependent release mechanisms,[152] Poly(vinyl alcohol) — PVA nanosized hydrogel networked particles[153] and Gelatin-Polyacrylamide hydrogel networks.
- **Extracellular pH** relies on the metabolic environment of tumors where the vasculature is often insufficient to supply enough oxygen and nutrients for the fast proliferation of tumor cells. Under these hypoxic conditions, lactic acid is produced and there is hydrolysis of ATP (due to an energy-deficient environment), which lead to an acidic micro-environment — i.e. most tumors have lower extracellular pH (< 7.2) than the surrounding tissues and blood (7.5). Passive accumulation of colloidal conjugated pH responsive self-assembled nanosized polymer hydrogels or PEG-elated liposomes in solid tumor sites have been demonstrated by "enhanced permeation and retention (EPR)",[154–156] Sulfadimethoxine Pullulan acetate (PA) nanoparticles,[157] and

Polybutylcyanoacrylate (PBCA) nanoparticles have been used for brain cancer treatment[184] and more recently for liver cancer treatment[185].

Other release techniques include **Receptor-Mediated endocytosis**[c] and the combination of **hybrid mechanisms** based on the above. In Receptor-Mediated endocytosis the idea is to develop self-assembled nanometric DDS particles that are taken up by tumor cell receptors in vivo, and once inside the cell cytoplasm membrane the release mechanism deploys the bioactive agent (or appropriate antibodies) to destroy the tumor. These DDS may induce an immunological enhancement activity in the body, attach to the tumor cell by ligand-receptor mediated interaction, and release the anti-cancer drug in a controlled fashion. Hybrid systems would, in principle, incorporate different release and molecular recognition mechanisms on a single nanoparticle. Example DDSs capable of multifunction release mechanisms include Polymeric micelles[151] and dendrimers.[158,159]

Encapsulating the drug in the DDS and controlling its release to improve bioavailability, while keeping a balance between the effective and overdosing limits, involves parametrically controlling the DDS-drug complex mechanical and chemical properties. The multiparadigm, multiscale modeling and computation approach is contributing to such goal, allowing us to calculate a range of design parameters, including:

- Absorption and excitation spectra to determine drug/DDS molecule aggregation,
- Drug carrier weight ratios, drug loading efficiencies and drug contents as a function of hydrogel nanoparticle size,
- Drug particle diffusion coefficient,
- Singlet oxygen production efficiency and singlet oxygen diffusion out of the hydrogel,
- Thermodynamics stability from interfacial free energies,
- Hydrogel swelling as a function of pH (0–10) and T profile (25–40°C or starting at Lower Critical Solution Temperature — LCST),
- LCST as a function of molecular weight and co-monomers composition/block lengths,
- The drug release behavior (rates) as a function of varying T (above LCST) and block copolymer lengths,

[c] Cytoplasm membrane folds inward to form coated pits.

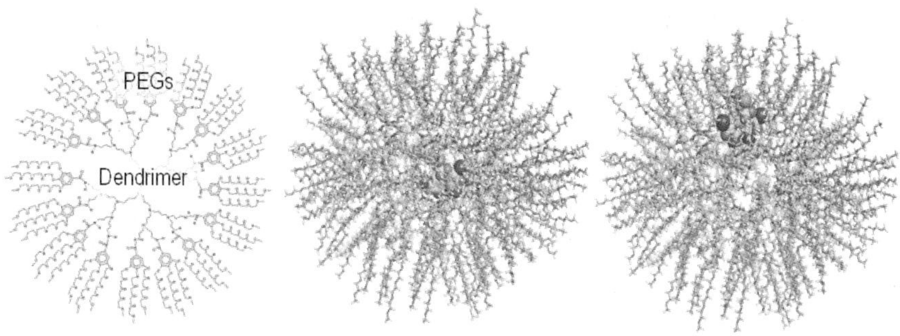

Figure 12. (Left) Schematic of PEGylated dendrimer, (center) Molecular models of dendrimer and drug molecule in core, (right) drug molecule diffusion from the core. (Image courtesy of Youyong Li, MSC-Caltech).

- DDS-drug self-assembly conditions (as a function of pH),
- Particle aggregation size and critical aggregation concentration as a function of pH,
- Drug loading efficiency and mean diameter of nanoparticles,
- Charge density and distribution of ionizable groups on the nanoparticle surface as a function of pH,
- Structural changes due to varying pH (nanoparticle morphology and interior structural changes),
- Release kinetics of drug molecules over a range of pH values.

7.3.3.1 Functionalized dendrimers

Dendrimers, that are branched polymers and have layered architecture, are one of the most promising synthetic polymers as they are finding applications in a variety of fields including biomedicine[160] for targeted drug-delivery. PEGylation of dendrimers overcomes most of their disadvantages like drug leakage, immunogenicity and toxicity.[161] We are currently using multiscale modeling to study PEGylated dendrimers of varying generations to determine their core's drug payload capabilities and alternative drug release mechanisms with respect to non-PEGylated systems. For this we consider the most energy favorable structures of the dendrimer in solution from MD-NVT simulations and calculate the diffusivity (stability) of drug molecules inside the hydrophobic dendrimer cavities as a function of temperature and time. Coarse-grain models are used to evaluate dendrimer–dendrimer interactions as well as dendrimer-tissue interactions and long-term drug release rates.

7.3.4 Polymer Hydrogel Networks in Tissue Engineering Scaffolding

Another area where biomaterials have had an impact is in Tissue Engineering (TE). By combining polymers with mammalian cells, it is now possible to make biomimetic skin for patients who have burns or skin ulcers, as well as polymer/cell combinations including corneas, cartilage, bone and liver.[162] Nanoparticle based biomaterials are also being used for the development of dental implants with optimized mechanical performance.

Cartilaginous tissue related disorders are the second leading cause of disability in the elderly population of the United States.[163] Unlike bone, liver, skin, and other tissues with high cell-turnover rates, cartilage is generally considered to have a limited capacity for self-repair.[164–167] Clearly, most of the properties found in cartilage depend on the dynamics of the three-dimensional nano electro-mechanical structure and the chemistry of the ECM.[168] Unfortunately, such properties remain largely unknown making the development of appropriate scaffold materials a scientific and technological challenge. Experiments have yet to probe the dynamic behavior of the biological ECM under *in-vivo* loads with accuracy[169] creating a gap between available experimental data and the theoretical models used to describe/predict its non-equilibrium and full equilibrium bio-mechanical properties. Furthermore, most theoretical contributions rely on empirically parameterized, from known (and limited) experimental data, finite element or continuum models of the ECM microstructure[169–179] (including biphasic and multiphasic theories developed to incorporate electrokinetic and transport behavior, e.g. Donnan osmotic pressures) while the scarce molecular simulations performed on known cartilage components[180] do not provide sufficient architectural details to evoke macroscopic function affinity. In general, previous studies have yet to point out the origins of viscoelasticity, surface tension, dynamics strain-stress relationships, and other important biomechanical properties of articular cartilage nor they provide accurate prediction of properties in the range of scales of interest to the material designer. In the same manner, hydrogel science and engineering has been dominated by studies on their experimental processing, and static characterization of their mechanical stability with very scarce molecular-level simulations[181–184] related mainly to the diffusion of water in the gel polymer. We have been applying the multiparadigm, multiscale *de novo* design strategy in the design and optimization of polymer hydrogels used in scaffold-supported cell therapies to promote the natural regeneration of cartilaginous tissue.

7.3.4.1 Preliminary findings

Our multiscale, multiparadigm approach provides an opportunity to develop the foundational knowledge needed to leapfrog the limitations of current polymer hydrogel scaffold materials design. Such advances are enabling the essential framework to:

- simulate the critical nano bio-mechanical properties of gel polymer networks, including the complex structural and electrokinetic phenomena responsible for mechanoregulation,
- develop an increased understanding of the fundamental mechanisms that regulate *in-vivo* performance,
- relate nano bio-mechanical properties of hydrogels (including non-linear strain-stress response, permeability, diffusion and electrokinetics as a function of temperature and pressure) to chemical polar/non-polar functionality and network architecture in contrast to those found in natural cartilage.

To assess the mechanical properties of synthetic hydrogels, our group has performed[185] preliminary studies using molecular nanoscale models of their structures to demonstrate how the stress-strain characteristic can be tailored for enhanced performance through double network (DN) polymerization. Using MD-NVT simulations, the stress-strain characteristics from uniaxial extension simulations (presented in Fig. 13a) are shown to be in good qualitative agreement with experimental observations.[186–188] We validated that by having independently cross-linked single hydrogel nets (SN) within an interpenetrating double net (DN), and:

- Controlling the molar ratio of the 1st w/r to the 2nd net to be highly cross-linked with a high Young's modulus (i.e. stiff + brittle), and
- the cross-linking density of 1st to the 2nd, where the 2nd net is loosely cross-linked (i.e. soft + ductile), then

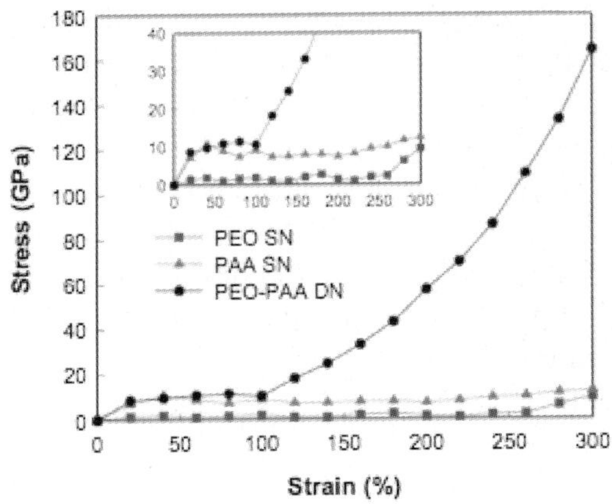

Figure 13. (a) Stress-strain curves of SN and DN hydrogels. We observed that the stress in the DN increases rapidly from ~ 100 % strain, which corresponds to ~ 250 % strain for the PEO SN. (b) The initial structure of the DN has ~ 80% strain in the PEO part and ~ 5% in the PAA. From Ref. 185.

the resulting DN results in a stiff but not brittle, ductile but not soft gel with strain-strain curves that can be tailored to the required cartilage scaffold properties. Figure 13a, shows the results for a PEO-PAA DN hydrogel; stress increases significantly after ~ 100% of strain (at ~ 111 Å of its physical dimension). From Fig. 13b, we see that ~ 250% in the PEO SN is exactly where the stress in the PEO SN starts increasing dramatically. We are currently investigating the molecular-level origin of fundamental thermodynamic properties (e.g. enthalpic and entropic contributions to the free energy) that give rise to pertinent bio-mechanical properties (e.g. viscoelastic properties) in synthetic hydrogel compounds (chondrocytes exhibit good cell viability within hydrophilic scaffolds like hydrogels,[189,190] using our bottom-up first-principles-based multiscale, multiparadigm modeling and simulation approach. We expect this will provide microscopic-level biomechanical predictive capabilities for the design of enhanced functional hydrogel scaffolds for cartilage tissue regeneration, by estimating quantities such as the equilibrium-swelling ratio (equilibrium water capacity), the compressive modulus and viscoelastic properties, electrokinetics rates for water and participating ions, failure modes, and their potential interdependencies as a function of controllable variables such as hydrogel crosslinking molecular weight, crosslinking density of

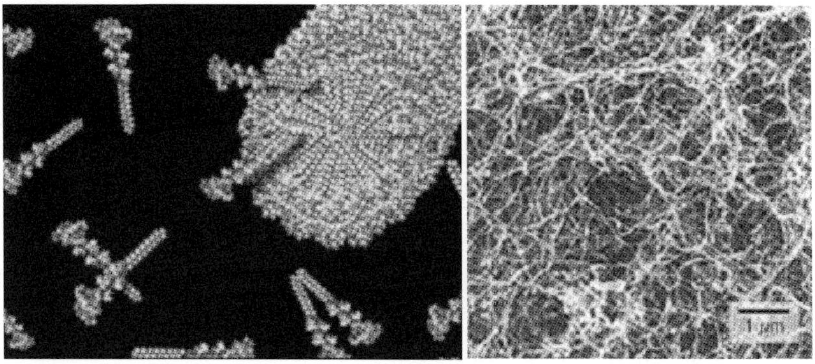

Figure 14. Self-assembled nanopeptide amphiphile molecules (left) for tissue regeneration. (Right) micrograph of nanofiber gel matrix formed from self-assembled nanopeptides.[191]

hydrogel components, degree of cyclization, intrinsic segmental flexibility of polymers, temperature, and other physical quantities.

7.3.4.2 Alternative nano-scaffolding systems

Bottom-up approaches to regeneration of tissue have been proposed and are currently being advanced using multiscale modeling aided computations. Most notably on the design of bioactive nanomaterials based on peptide amphiphiles[191] for regenerative medicine in a number of different targets (e.g. heart, spinal cord, bone, among others). The proposed family of peptide-amphiphile (PA) molecules, developed by the Stupp group at Northwestern University, self-assembles into high-aspect ratio nanofibers under physiological conditions that can be tailored to display bioactive peptide epitopes along each nanofiber's periphery. Self-assembly of proteins has been used by others[192–195] to develop novel biomaterials. The primary advantage of this method is that it starts from an aqueous solution of proteins that undergoes self-assembly-driven in situ gelation. Unfortunately, little has been studied on the effects of these nanosystems on cell structure, their bio structural mechanics, and degradation kinetics as scaffolding material for cartilage regeneration.

7.3.5 Enzyme Catalysis

Enzymes are crucial biomolecules that catalyze biochemical reactions critical for the growth and survival of living cells. Catalyzed reactions are critical for many biological pathways like signal transduction, metabolism and cell regulation. An understanding of how enzymes work can help in the design of biomaterials with unique catalytic properties. During the catalysis process, the enzymes also undergo small or large conformational changes and some amino acids play critical roles by strongly interacting with the substrate molecules. Modeling such a system accurately requires a combined multiscale-multiparadigm approach because the reactive substrates/products and critical enzyme residues need to be treated quantum mechanically due to the involvement of bond formation and bond breakage.

7.3.5.1 Computational approach to enzymatic catalysis

This enzymatic catalysis is an ideal case for a multiscale and multiparadigm framework, because electrons need to be considered for the substrate undergoing reaction, the final product and all key enzyme residues interacting with the substrate and product (Fig. 15).

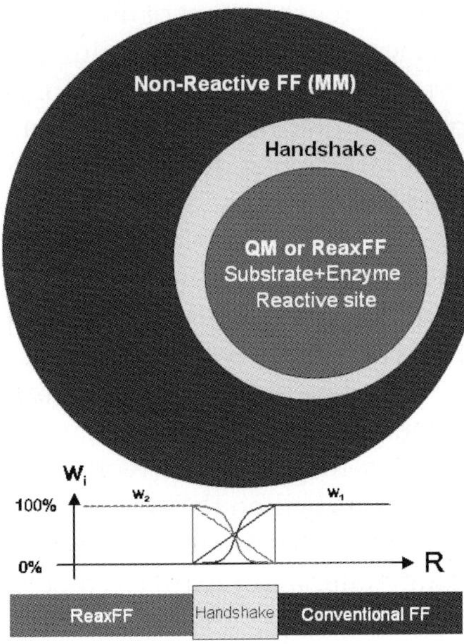

Figure 15. Enzyme region organizations in the QM/MM or ReaxFF/MM framework. Handshake region between reactive and non-reactive regimes can be treated using various methods as described in Sec. 7.2.7.

This is achieved using QM level calculations or, for improved efficiency, the ReaxFF discussed earlier. The remainder of the enzyme can be treated by using non-reactive force fields described earlier. Such a scheme is commonly referred to as the QM/MM approach,[196] where the enzymatic center is treated quantum mechanically and the remaining system is treated using molecular mechanics. The QM/MM framework has been implemented using a range of methods reviewed in detail recently,[197] all of which share the multiscale-multiparadigm philosophy. Below, a couple of example studies highlight the results that can be obtained using such an approach and also open challenges.

7.3.5.2 Selected examples and major challenges

The synthesis of aromatic amino acids in bacteria, fungi and plants involves the catalytic Claisen rearrangement of chorismate to prephenate by the enzyme Chorismatemutase. This reaction is an important case system for testing theories on the origin of enzymatic catalysis, enzyme–substrate reaction pathways and the structural role played by many residues in the stabilization of the transition-state (TS). It has been studied in the absence and presence of the enzyme using a coupled QM/MM method by Mulholland and collaborators.[198] They suggest electrostatic TS stabilization to be the major contributor to catalysis in this reaction. An estimation of the free energy required to generate 'near attack conformations' (NACs)[199] and contributions of variations in the environment on the reaction barrier[200] and contributions of variations in the environment on the reaction barrier.[201] The NAC configurations are similar to the TS, which makes them bind equally well to the enzyme. Such enzymes can be used to test and design biomaterials with customized catalytic properties.

Based on the system at hand, the interface between the QM region and the MM region plays a critical role in the quality of the calculation. This entails that covalent bonds across the QM/MM boundary be treated properly and forces especially the long-range electrostatic be accounted for appropriately in both QM and MM regions. The QM/MM methods are currently being extended to describe excited-state dynamics and reactivity of photoactivated biological systems but challenges remain to not only properly account for nonadiabatic effects but also on interfacing excited states from QM regions with inherently ground-state description of MM regions.

Alternatively, we have recently expanded the ReaxFF parameterization to aqueous phase chemistry; this allows us to dynamically simulate enzyme and DNA-reactions over much larger regions of the macromolecule. To ensure that ReaxFF provides a reliable description for the reactive events in enzymes, we fitted the ReaxFF parameters to a wide range of QM-data relevant to reactive and non-reactive events in protein systems (e.g. hydrogen transfer barriers, dissociation energies, etc.).

Using ReaxFF, we have successfully modeled peptide hydrolysis catalyzed by a catalytic triad as in serine proteases (Figs. 16 and 17). As such catalytic cycles usually occur in the milliseconds to seconds time scale and can thus not be observed in picoseconds to nanoseconds simulations, we have accelerated the catalytic reactions by imposing restraints on the bonds that are broken or formed. This study has demonstrated that the reaction including breaking and

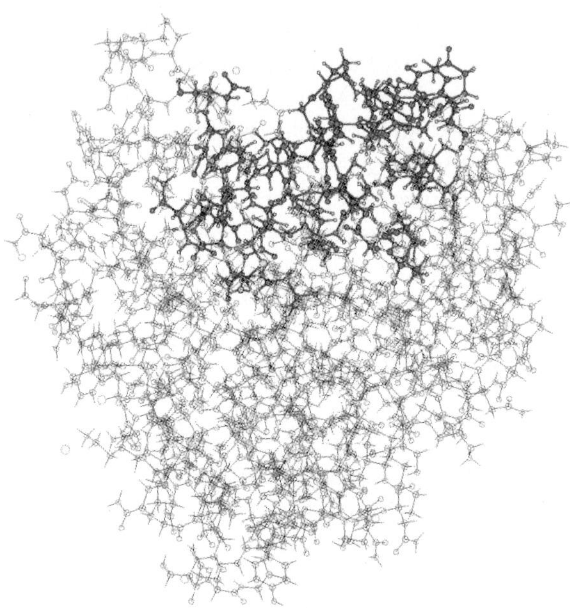

Figure 16. Region of the protease enzyme treated with ReaxFF (non-transparent atoms) and region kept fixed (transparent atoms) during the protease MD.

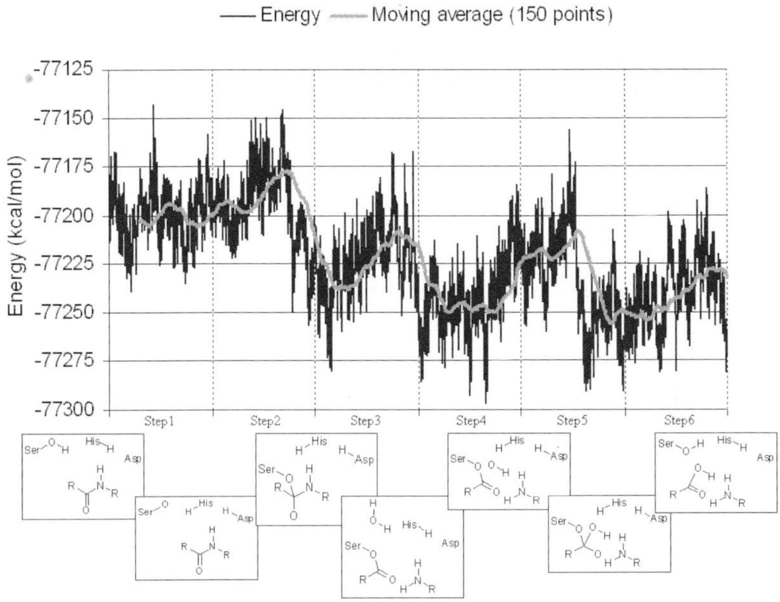

Figure 17. Energy monitored over the course of peptide hydrolysis by a catalytic triad. On the bottom, educts, intermediates along the reaction pathway, and products are shown.

formation of several bonds could be simulated and that the energy barriers of ~25 kcal/mol per step are overcome under physiological conditions. In this simulation the entire binding site (all atoms within 5.0 Å of the substrate; about 600 atoms) was treated with ReaxFF, while all atoms outside this region were fixed (see Fig. 16). Using our multiparadigm approach, ReaxFF can be efficiently coupled with non-reactive force fields (using CMDF), allowing full MD simulations of enzymatic activity, both, at the reaction sites and away from these. Instead of directly including QM in the minimization or dynamics, which is cumbersome due to the substantial computational expense mismatch between QM and FF-methods, we use the QM to derive ReaxFF parameters and then use ReaxFF to simulate the dynamics of the enzyme reactions.

7.3.6 Protein Activation

The activation of proteins, which leads to their biological function, usually involves small or large conformational changes that mechanically drive their function. Commonly found protein activation mechanisms can provide a robust way of artificially inducing or altering biological function. Activation of an important class of membrane proteins (*G* protein coupled receptors or GPCRs), which are used by cells to convert an extracellular signal into an intracellular signal through a conformational change, plays a critical role in many disease processes. These receptors mediate responses to hormones, neurotransmitters, senses of sight, smell, taste, and even mechanical stress. All of these receptors share a 7-helical topology where the helices span the transmembrane region of the cells seven times. This also makes them the richest set of proteins that are pharmaceutical targets. Their activation mechanism is beginning to unravel[202] and will provide numerous avenues for nanomedical interventions through selective activation or deactivation of cellular responses using novel techniques. Due to lack of structural information about these receptors (only two GPCRs have experimental structures: Bovine Rhodopsin[203] and Human Beta-2 Adrenergic Receptor,[204] their structures need to be predicted using first-principles based methods.

Our group has developed such structure prediction methods and applied successfully to many GPCRs.[205]

7.3.6.1 Multiscale computational approach to protein activation

Many applications in computational science rely on the ability to efficiently search the conformational space of macromolecules, including: free energy perturbation calculations,[206] structure prediction and determination by X-ray crystallography or NMR spectroscopy. Proteins sample this conformational space very differently in their active and inactive forms. Approximate searches for the global minimum of a structure conformation space can be reached using simulated annealing,[207] from a set of different conformations generated from molecular dynamics and Monte Carlo procedures. On the other hand, high temperature TAD algorithms for structure determination offer a clearly defined objective function for defining the merit of the sampling strategy, i.e. direct agreement with experimental diffraction data. Accelerated dynamics provides another promising route to sampling longer time scales and conformational changes.[208]

The time scale of activation of most proteins including GPCRs lies in the millisecond or higher time scales. This has been the major bottleneck in computationally exploring activation mechanisms. Any all-atom molecular dynamics aimed at probing conformational changes that lead to activation is intractable due to the short time step (~ 2 fs) that is necessary for proper dynamics. This necessitates the use of coarse grain representations which fuses custom selected heavy atoms into beads which still retain torsional degrees of freedom that drive conformational changes in proteins. This will allow one to take longer time steps (~ 30–40 fs) during the dynamics and coupled to the fact that number of particles in the system is now reduced, it allows for dynamics sampling in the microseconds range. TAD algorithms coupled to coarse grained FFs are expected to provide the most gain in sampling necessary and as codes become more efficient and computer power gets cheaper, the time scale of sampling will approach the millisecond range which will then begin to sample conformations critical for activation mechanisms.

7.3.6.2 Example problem and major challenges

Many human cancers have implicated members of the Epidermal Growth Factor Receptor (EGFR). The extracellular regions of these proteins undergo major conformational changes upon activation that are believed to be ligand induced.[209] Figure 18 shows the EGFR inactive monomer in panel A, ligand (EGF) induced active EGFR homodimer in panel B, and an activation model in panel C. Such large conformational changes pose a major computational challenge which when surmounted (using approaches mentioned in the preceding section) are expected to yield valuable information about activation mechanisms that can help in the design of better drugs.

The biggest bottleneck in studying such protein dynamics has been adequate conformational sampling. No algorithm exists that can guarantee completeness in conformational search. This makes it imperative to design experiments that can guide simulation to pick out relevant conformational spaces to sample activation related protein structures. In order to reach the millisecond time-scales and beyond using atomically resolved simulations several key points need to be made. The formulation stiffness of long-term MD simulations rely on low-order symplectic integrators and fast solvers for the EOM. Symplectic integrators are often implicit, and require more function evaluations and smaller time steps to produce the same computational accuracy as compared to standard non-symplectic

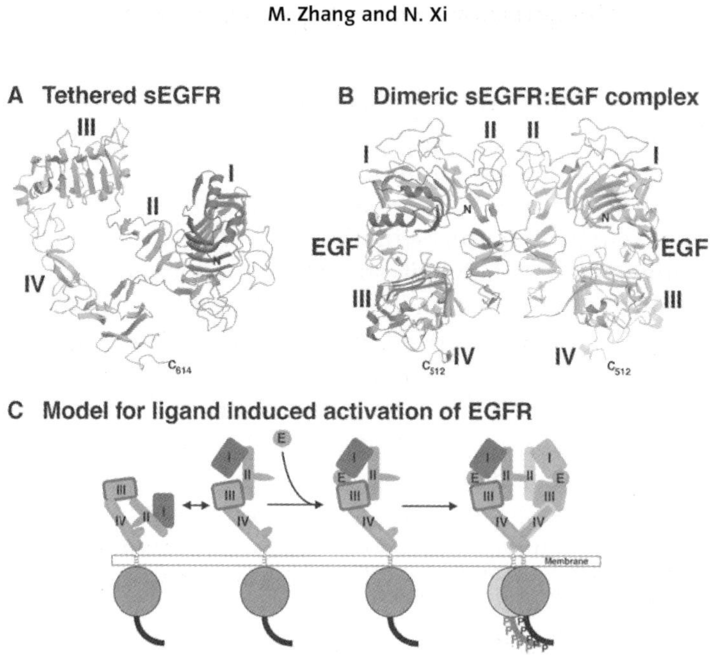

Figure 18. EGF ligand induced activation of EGFR that converts the inactive monomer (A) into the active homodimer (B). A possible activation model is shown in (C). From Ref. 209.

integrators,[210] leading to a dichotomy that can be resolved using the later for computational efficiency. We have also used variable time-step integrations schemes based on energy-variance but it should be noted that symplectic integrators do not necessarily remain symplectic under such conditions.

7.4 CONCLUDING REMARKS

In recent years there has been a significant advance in the design of time-lower bound algorithms, corresponding iso-efficiently scalable implementations, and a steady increase in computational capabilities. This has led to an exponential increase in the number of particles that can be simulated using first-principles methods as well as force field based methods (refer to Fig. 19). The advent of Petascale computing will bring about an unprecedented possibility for tackling open problems with inherently multiscale phenomena including structure-function relationship in sufficiently large biomolecular complexes (billions of atoms), understanding macromolecular interactions (milliseconds and beyond), and in

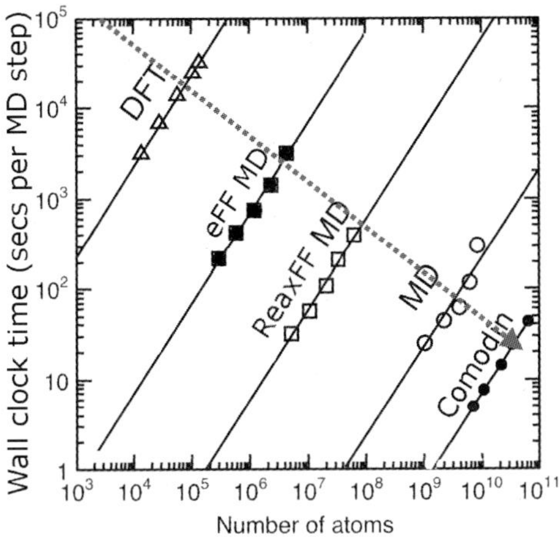

Figure 19. Relative qualitative computation times of QM (DFT-QMC), eFF, reaxFF, atomistic MD and coarse-grain Comodin MD algorithms. Lines show estimated ideal scaling, and dots, triangles, and circles extrapolated results per method (DFT, MD adapted from Ref. 211). Red line depicts predicted time versus length scaling for 1 K processor simulation. (See page 343 for color illustration.)

developing accurate *in silico* characterization and predictive design tools of bioinspired and biomimetic devices and systems. Theory and computation will also aid in the development of appropriate nanoscale control strategies of potential devices and systems were an increasingly important requirement is to enable access to high resolution information with minimum space and time computational complexity (e.g. systems with real-time deadlines).

ACKNOWLEDGEMENTS

The research described herein covers multiple areas for which support has been granted in the past, or for which continuing funding exists, from different agencies including the National Science Foundation (NSF) under Grant No. 0727870, and the National Institutes of Health (NIH) under grant No. R21-MH073910-01-A1.

REFERENCES

1. NIH (National Institutes of Health, 2007).
2. Wikipedia (2007).

3. R. A. Freitas, *J. Comput. Theor. Nanos.* **2**, 1 (2005).
4. S. D. Caruthers, S. A. Wickline, and G. M. Lanza, *Current Opinion in Biotechnology* **18**, 26 (2007).
5. S. M. Moghimi, A. C. Hunter, and J. C. Murray, *Faseb Journal* **19**, 311 (2005).
6. M. C. Roco, ed., *National Nanotechnology Initiative — Past, Present, Future*, 2nd Ed. (2007).
7. D. F. Emerich, and C. G. Thanos, *Biomolecular Engineering* **23**, 171 (2006).
8. Y. Y. Liu, H. Miyoshi, and M. Nakamura, *International Journal of Cancer* **120**, 2527 (2007).
9. V. N. Uversky, A. V. Kabanov, and Y. L. Lyubchenko, *Journal of Proteome Research* **5**, 2505 (2006).
10. J. O'Neill, M. Manion, P. Schwartz, and D. M. Hockenbery, *Biochimica Et Biophysica Acta-Reviews on Cancer* **1705**, 43 (2004).
11. J. Tersoff, and D. R. Hamann, *Physical Review B* **31**, 805 (1985).
12. J. Tersoff, and D. R. Hamann, *Physical Review Letters* **50**, 1998 (1983).
13. K. Vonklitzing, G. Dorda, and M. Pepper, *Physical Review Letters* **45**, 494 (1980).
14. M. Ben-Nun *et al.*, *Faraday Discussions*, 447 (1998).
15. A. Jaramillo-Botero, *Dekker Encyclopedia of Nanoscience and Nanotechnology*, M. D. Publishers, Ed. (Marcel Dekker Publishers, 2004).
16. A. Boukai, K. Xu, and J. R. Heath, *Advanced Materials* **18**, 864 (2006).
17. T. Cagin, A. Jaramillo-Botero, G. Gao, and W. A. Goddard, *Nanotechnology* **9**, 143 (1998).
18. R. A. Freitas, *Nanomedicine* (Landes Bioscience, 2007), Vol. I: Basic Capabilities.
19. A. Warshel, *Computer Modeling of Chemical Reactions in Enzymes and Solutions*, pp. XIV, 236 p, Wiley, New York (1991).
20. M. Karplus, and J. Kuriyan, *Proceedings of the National Academy of Sciences of the United States of America* **102**, 6679 (2005).
21. P. Derreumaux, and N. Mousseau, *Journal of Chemical Physics* **126**, (2007).
22. L. de Broglie (1924).
23. E. Schrödinger, *Annalen Der Physik* **79**, 489 (1926).
24. A. Messiah, Ed., *Quantum Mechanics*, Vol. 1 (Reprinted by Dover Publications, 1999), Vol. 1.
25. M. Born, and R. Oppenheimer, *Annalen Der Physik* **84**, 0457 (1927).
26. M. J. Frisch *et al.* (Gaussian, Inc., Wallingford CT, 2004).
27. M. W. Schmidt *et al.*, *Journal of Computational Chemistry* **14**, 1347 (1993).

28. Jaguar (Schrodinger, Inc, Portland, OR, 1991–2000).
29. G. Kresse, and J. Hafner, *Physical Review B* **47**, 558 (1993).
30. R. Dovesi*et al.*, U. O. Torino, Ed. (Torino, 2006).
31. M. D. Segall*et al.*, *Journal of Physics-Condensed Matter* **14**, 2717 (2002).
32. P. A. Schultz (Sandia National Laboratories, Albuquerque, 2007).
33. J. Harvey (Bristol, 2001).
34. C. Moller, and M. S. Plesset, *Physical Review* **46**, 0618 (1934).
35. E. A. Carter, and W. A. Goddard, *Journal of Chemical Physics* **88**, 3132 (1988).
36. R. A. Friesner, *Proceedings of the National Academy of Sciences of the United States of America* **102**, 6648 (May 10, 2005).
37. W. A. Goddard, T. H. Dunning, W. J. Hunt, and P. J. Hay, *Accounts of Chemical Research* **6**, 368 (1973).
38. B. H. Greeley *et al.*, *Journal of Chemical Physics* **101**, 4028 (1994).
39. D. J. Tannor *et al.*, *Journal of the American Chemical Society* **116**, 11875 (1994).
40. T. Bredow, and K. Jug, *Theoretical Chemistry Accounts* **113**, 1 (2005).
41. P. Hohenberg, and W. Kohn, *Physical Review B* **136**, 864 (1964).
42. W. Kohn, and L. J. Sham, *Physical Review A* **140**, 1133 (1965).
43. Z. Y. Li, W. He, and J. L. Yang, *Progress in Chemistry* **17**, 192 (2005).
44. W. M. C. Foulkes, L. Mitás, R. J. Needs, and G. Rajagopal, *Rev. Mod. Phys.* **73**, 33 (2001).
45. X. J. Chen, J. M. Langlois, and W. A. Goddard, *Physical Review B* **52**, 2348 (1995).
46. R. Car, and M. Parrinello, *Physical Review Letters* **55**, 2471 (1985).
47. H. Q. Ding, N. Karasawa, and W. A. Goddard, *Journal of Chemical Physics* **97**, 4309 (1992).
48. H. Q. Ding, N. Karasawa, and W. A. Goddard, *Chemical Physics Letters* **196**, 6 (1992).
49. J. W. Ponder, and D. A. Case, *Protein Simulations* **66**, 27 (2003).
50. S. J. Weiner, P. A. Kollman, D. T. Nguyen, and D. A. Case, *Journal of Computational Chemistry* **7**, 230 (1986).
51. A. D. MacKerell *et al.*, *Journal of Physical Chemistry B* **102**, 3586 (1998).
52. S. L. Mayo, B. D. Olafson, and W. A. G. III, *J. Phys. Chem.* **94**, 8897 (1990).
53. W. L. Jorgensen, D. S. Maxwell, and J. TiradoRives, *Journal of the American Chemical Society* **118**, 11225 (1996).
54. M. Karplus, and J. A. McCammon, *Nature Structural Biology* **9**, 646 (2002).

55. C. M. Cortis, and R. A. Friesner, *Journal of Computational Chemistry* **18**, 1570 (1997).
56. W. C. Still, A. Tempczyk, R. C. Hawley, and T. Hendrickson, *Journal of the American Chemical Society* **112**, 6127 (1990).
57. V. S. Bryantsev, M. S. Diallo, and W. A. Goddard, *Journal of Physical Chemistry A* **111**, 4422 (2007).
58. Y. H. Jang *et al.*, *Journal of Physical Chemistry B* **107**, 344 (2003).
59. A. C. T. van Duin, S. Dasgupta, F. Lorant, and W. A. Goddard, *Journal of Physical Chemistry A* **105**, 9396 (2001).
60. A. Strachan, E. M. Kober, A. C. T. van Duin, J. Oxgaard, and W. A. Goddard, *Journal of Chemical Physics* **122** (2005).
61. A. C. T. van Duin *et al.*, *Journal of Physical Chemistry A* **107**, 3803 (2003).
62. S. S. Han, A. C. T. van Duin, W. A. Goddard, and H. M. Lee, *Journal of Physical Chemistry A* **109**, 4575 (2005).
63. Q. Zhang *et al.*, *Physical Review B* **69** (2004).
64. K. D. Nielson, A. C. T. van Duin, J. Oxgaard, W. Q. Deng, and W. A. Goddard, *Journal of Physical Chemistry A* **109**, 493 (2005).
65. S. Cheung, W. Q. Deng, A. C. T. van Duin, and W. A. Goddard, *Journal of Physical Chemistry A* **109**, 851 (2005).
66. N. Chen, M. T. Lusk, A. C. T. van Duin, and W. A. Goddard, *Physical Review B* **72** (2005).
67. H. B. Su, R. J. Nielsen, A. C. T. van Duin, and W. A. Goddard, *Physical Review B* **75** (2007).
68. K. Chenoweth, S. Cheung, A. C. T. van Duin, W. A. Goddard, and E. M. Kober, *Journal of the American Chemical Society* **127**, 7192 (2005).
69. A. Strachan, A. C. T. van Duin, D. Chakraborty, S. Dasgupta, and W. A. Goddard, *Physical Review Letters* **91** (2003).
70. A. C. T. van Duin, Y. Zeiri, F. Dubnikova, R. Kosloff, and W. A. Goddard, *Journal of the American Chemical Society* **127**, 11053 (2005).
71. M. J. Buehler, A. C. T. van Duin, and W. A. Goddard, *Physical Review Letters* **96**, 4 (2006).
72. W. A. Goddard *et al.*, *Topics in Catalysis* **38**, 93 (2006).
73. J. Ludwig, D. G. Vlachos, A. C. T. van Duin, and W. A. Goddard, *Journal of Physical Chemistry B* **110**, 4274 (2006).
74. J. T. Su, and W. A. Goddard, *Physical Review Letters* **99** (2007).
75. W. Goddard III, paper presented at the WTEC Workshop, Arlington, VA, 8–9 May 1997–1998.
76. J. P. Ryckaert, G. Ciccotti, and H. J. C. Berendsen, *J. Comput. Phys.* **23**, 327 (1977).

77. W. F. Van Gunsteren, and H. J. C. Berendsen, *Molecular Physics* **34**, 1311 (1977).
78. H. C. Andersen, *J. Comput. Phys.* **52**, 24 (1983).
79. V. Krautler, W. F. Van Gunsteren, and P. H. Hunenberger, *J. Comput. Chem.* **22**, 501 (2001).
80. B. Hess, H. Bekker, H. J. C. Berendsen, and J. G. E. M. Fraaije, *J. Comput. Chem.* **18**, 1463 (1997).
81. S. Miyamoto, and P. A. Kollman, *J. Comput. Chem.* **13**, 952 (1992).
82. D. S. Bae, and E. J. Haug, *Mech. Struct. Mach.* **15**, 359 (1987).
83. A. Voter, Sorensen, paper presented at the Materials Research Society Symposium 1999.
84. A. Voter, Sorensen, *Journal of Chemical Physics* **112**, 9599 (2000).
85. A. Fijany, A. Jaramillo-Botero, T. Cagin, and W. Goddard III, paper presented at the Parallel Computing: Fundamentals, Applications and New Directions, 1998.
86. A. Fijany, T. Cagin *et al.*, *Advances in Engineering Software* **29**, 441 (1998).
87. R. A. Abagyan, and A. K. Mazur, *Journal of Biomolecular Structure & Dynamics* **6**, 833 (1989).
88. A. K. Mazur, and R. A. Abagyan, *Journal of Biomolecular Structure & Dynamics* **6**, 815 (1989).
89. D. S. Bae, and E. J. Haug, *Mech. Struct. Mach.* **15**, 481 (1988).
90. D. S. Bae, J. G. Kuhl, and E. J. Haug, *Mech Struct Mach* **16**, 249 (1988).
91. A. Jaramillo-Botero, and A. C. I. Lorente, *J. Parallel Distr. Com.* **62**, 1001 (2002).
92. N. Vaidehi, A. Jain, and W. A. Goddard, *Journal of Physical Chemistry* **100**, 10508 (1996).
93. A. Jain, N. Vaidehi, and G. Rodriguez, *J. Comput. Phys.* **106**, 258 (1993).
94. A. M. Mathiowetz, A. Jain, N. Karasawa, and W. A. Goddard, *Proteins* **20**, 227 (1994).
95. A. Jaramillo-Botero, Y. Liu, and W. A. Goddard (2006).
96. R. Featherstone, *Int J Robot Res* **2**, 13 (1983).
97. A. Fijany, T. Cagin, A. Jaramillo-Botero, and W. Goddard, *Adv. Eng. Softw.* **29**, 441 (1998).
98. H. Nyquist, *P. Ieee.* **90**, 280 (2002).
99. C. E. Shannon, *P. Ieee.* **72**, 1192 (1984).
100. J. Shelley, M. Shelley, R. Reeder, S. Bandyopadhyay, and M. I. Klein, *Journal of Physical Chemistry B* **105**, 4464 (2001).
101. R. D. Groot, *Langmuir* **16**, 7493 (2000).
102. R. D. Groot, and K. L. Rabone, *Biophysical Journal* **81**, 725 (2001).
103. P. Jedlovszky, *Molecular Physics* **93**, 939 (1998).

104. P. Jedlovszky, I. Bako, G. Palinkas, T. Radnai, and A. K. Soper, *Journal of Chemical Physics* **105**, 245 (1996).
105. C. F. Lopez, P. B. Moore, J. C. Shelley, M. Y. Shelley, and M. L. Klein, *Computer Physics Communications* **147**, 1 (2002).
106. S. J. Marrink, and A. E. Mark, *Biophysical Journal* **87**, 3894 (2004).
107. S. J. Marrink, and A. E. Mark, *Journal of the American Chemical Society* **125**, 15233 (2003).
108. S. J. Marrink, A. H. de Vries, and A. E. Mark, *Journal of Physical Chemistry B* **108**, 750 (2004).
109. S. J. Marrink, H. J. Risselada, S. Yefimov, D. P. Tieleman, and A. H. de Vries, *Journal of Physical Chemistry B* **111**, 7812 (2007).
110. V. Molinero, and W. A. Goddard, *Journal of Physical Chemistry B* **108**, 1414 (2004).
111. N. Vaidehi, and W. A. Goddard, *Journal of Physical Chemistry A* **104**, 2375 (2000).
112. T. Cagin *et al.*, *Computational and Theoretical Polymer Science* **11**, 345 (2001).
113. J. Elezgaray, and M. Laguerre, *Computer Physics Communications* **175**, 264 (2006).
114. J. Hunger, and G. Huttner, *Journal of Computational Chemistry* **20**, 455 (1999).
115. J. Hunger *et al.*, *European Journal of Inorganic Chemistry*, 693 (1998).
116. E. B. Tadmor, M. Ortiz, and R. Phillips, *Phil. Mag. A* **73**, 1529 (1996).
117. Y. Y. Tang *et al.*, *Biophysical Journal* **91**, 1248 (2006).
118. G. Pavliotis, and A. Stuart, Eds., *Multiscale Methods: Averaging and Homogenization* (2008).
119. M. J. Buehler, H. Tang, A. C. T. van Duin, and W. A. Goddard, *Physical Review Letters* **99** (2007).
120. M. J. Buehler, A. C. T. van Duin, and W. A. Goddard, *Physical Review Letters* **96** (2006).
121. I. T. R. F. Semiconductors, International Technology Roadmap for Semiconductors, 2006 update (ITRS, 2006).
122. Y. Taur *et al.*, *Proceedings of the Ieee* **85**, 486 (1997).
123. G. E. Moore, *Proceedings of the Ieee* **86**, 82 (1998).
124. W. Q. Deng, R. P. Muller, and W. A. Goddard, *Journal of the American Chemical Society* **126**, 13562 (2004).
125. P. W. K. Rothemund, *Abstracts of Papers of the American Chemical Society* **231** (2006).
126. C. P. Collier *et al.*, *Science* **285**, 391 (1999).
127. E. W. Wong *et al.*, *J. Am. Chem. Soc.* **122**, 5831 (2000).

128. C. P. Collier *et al.*, *Science* **289**, 1172 (2000).
129. C. P. Collier *et al.*, *J. Am. Chem. Soc.* **123**, 12632 (2001).
130. J. L. Atwood, and A. Szumna, *Chem. Comm.* **8**, 940 (2003).
131. M. A. Reed, C. Zhou, C. J. Muller, T. P. Burgin, and J. M. Tour, *Science* **278**, 252 (1997).
132. I. Aprahamian, W. R. Dichtel, T. Ikeda, J. R. Heath, and J. F. Stoddart, *Organic Letters* **9**, 1287 (2007).
133. C. P. Collier *et al.*, *Journal of the American Chemical Society* **123**, 12632 (2001).
134. W. R. Dichtel, J. R. Heath, and J. F. Stoddart, *Philosophical Transactions of the Royal Society a-Mathematical Physical and Engineering Sciences* **365**, 1607 (2007).
135. E. W. Wong *et al.*, *Journal of the American Chemical Society* **122**, 5831 (2000).
136. J. W. Choi *et al.*, *Chemistry-a European Journal* **12**, 261 (2005).
137. S. S. Jang *et al.*, *Journal of the American Chemical Society* **127**, 14804 (2005).
138. S. S. Jang *et al.*, *Journal of the American Chemical Society* **127**, 1563 (2005).
139. W. Q. Deng, A. H. Flood, J. F. Stoddart, and W. A. Goddard, *Journal of the American Chemical Society* **127**, 15994 (2005).
140. W. Q. Deng, X. Xu, R. Muller, M. Blanco, and W. A. Goddard, *Abstracts of Papers of the American Chemical Society* **225**, U708 (2003).
141. T. Ikeda *et al.*, *Chemistry-an Asian Journal* **2**, 76 (2007).
142. Y. H. Kim, S. S. Jang, and W. A. Goddard, *Applied Physics Letters* **88** (2006).
143. S. Datta, *Electronic Transport in Mesoscopic Systems*, H. Ahmad, M. Pepper, eds., Cambridge University Press (1995).
144. Z. B. Ge, Y. J. Kang, T. A. Taton, P. V. Braun, and D. G. Cahill, *Nano Letters* **5**, 531 (2005).
145. L. R. Hirsch *et al.*, *Proceedings of the National Academy of Sciences of the United States of America* **100**, 13549 (2003).
146. D. P. O'Neal, L. R. Hirsch, N. J. Halas, J. D. Payne, and J. L. West, *Cancer Letters* **209**, 171 (2004).
147. D. G. Cahill *et al.*, *Journal of Applied Physics* **93**, 793 (2003).
148. R. Langer, *Nature* **392**, 5 (1998).
149. R. Langer, *Scientific American* **288**, 50 (2003).
150. D. Gao, H. Xu, M. A. Philbert, and R. Kopelman, *Angewandte Chemie International Edition* **46**, 2224 (2007).
151. N. Nasongkla *et al.*, *Nano Lett.* **6**, 2427 (2006).
152. C. Choi, S. Y. Chae, and J. W. Nah, *Polymer* **47**, 4571 (2006).

153. J. K. Li, N. Wang, and X. S. Wu, *J Control Release* **56**, 117 (1998).
154. H. Maeda, J. Wu, T. Sawa, Y. Matsumura, and K. Hori, *J Control Release* **65**, 271 (2000).
155. S. Modi, J. P. Jain, A. J. Domb, and N. Kumar, *Curr Pharm Design* **12**, 4785 (2006).
156. M. Yokoyama, and T. Okano, *Adv. Drug. Deliver. Rev.* **21**, 77 (1996).
157. K. Na, and Y. H. Bae, *Pharmaceut Res.* **19**, 681 (2002).
158. Y. Choi, and J. R. Baker, *Cell Cycle* **4**, 669 (2005).
159. O. L. P. De Jesus, H. R. Ihre, L. Gagne, J. M. J. Frechet, and F. C. Szoka, *Bioconjugate Chemistry* **13**, 453 (2002).
160. C. C. Lee, J. A. MacKay, J. M. J. Frechet, and F. C. Szoka, *Nature Biotechnology* **23**, 1517 (2005).
161. V. Gajbhiye, P. V. Kumar, R. K. Tekade, and N. K. Jain, *Current Pharmaceutical Design* **13**, 415 (2007).
162. J. P. Vacanti, and R. Langer, *Lancet* **354**, Si32 (1999).
163. R. S. Langer, and J. P. Vacanti, *Scientific American* 86 (1999).
164. A. I. Caplan, M. Elyaderani, Y. Mochizuki, S. Wakitani, and V. M. Goldberg, *Clinical Orthopaedics and Related Research*, 254 (1997).
165. W. T. Green, *Clinical Orthopaedics and Related Research*, 237 (1977).
166. F. H. Silver, and A. I. Glasgold, *Otolaryngologic Clinics of North America*, 847 (1995).
167. C. J. Wirth, and M. Rudert, *Arthroscopy*, 300 (1996).
168. V. C. Mow, A. Ratcliffe, and A. R. Poole, *Biomaterials*, 67 (1992).
169. J. L. Ronsky *et al.*, *Journal of Biomechanics* **28**, 977 (1995).
170. P. Bursac, C. V. McGrath, S. R. Eisenberg, and D. Stamenovic, *Journal of Biomechanical Engineering-Transactions of the Asme* **122**, 347 (2000).
171. G. A. Ateshian *et al.*, *Transport in Porous Media*, 5 (2003).
172. K. A. Athanasiou, M. P. Rosenwasser, J. A. Buckwalter, M. Olmstead, and V. C. Mow, *Tissue Engineering*, 185 (1998).
173. G. S. Beaupre, S. S. Stevens, and D. R. Carter, *Journal of Rehabilitation Research and Development* **37**, 145 (2000).
174. P. S. Donzelli, R. L. Spilker, G. A. Ateshian, and V. C. Mow, *Journal of Biomechanics* **32**, 1037 (1999).
175. E. H. Frank, and A. J. Grodzinsky, *Journal of Biomechanics*, 629 (1987).
176. D. P. Fyhrie, and J. R. Barone, *Journal of Biomechanical Engineering-Transactions of the ASME*, 578 (2003).
177. F. Guilak, W. R. Jones, H. P. Ting-Beall, and G. M. Lee, *Osteoarthritis and Cartilage* **7**, 59 (1999).
178. V. Mow, and X. E. Guo, *Annual Review of Biomedical Engineering* **4**, 175 (2002).

179. J. M. Huyghe, and J. D. Janssen, *Int. J. Eng. Sci.* **35**, 793 (1997).
180. M. D. Buschmann, and A. J. Grodzinsky, *Journal of Biomechanical Engineering-Transactions of the Asme* **117**, 179 (1995).
181. Y. Tamai, H. Tanaka, and K. Nakanishi, *Mol. Simulat.* **16**, 359 (1996).
182. Y. Tamai, H. Tanaka, and K. Nakanishi, *Macromolecules* **29**, 6761 (1996).
183. Y. Tamai, H. Tanaka, and K. Nakanishi, *Macromolecules* **29**, 6750 (1996).
184. J. Mijovic, and H. Zhang, *J. Phys. Chem. B* **108**, 2557 (2004).
185. S. S. Jang, W. A. Goddard, and M. Y. S. Kalani, *Journal of Physical Chemistry B* **111**, 1729 (2007).
186. Y. Tanaka, J. P. Gong, and Y. Osada, *Prog. Polym. Sci.* **30**, 1 (2005).
187. A. Nakayama *et al.*, *Adv. Funct. Mater.* **14**, 1124 (2004).
188. J. P. Gong, Y. Katsuyama, T. Kurokawa, and Y. Osada, *Adv. Mater.* **15**, 1155 (2003).
189. S. J. Bryant, T. T. Chowdhury, D. A. Lee, D. L. Bader, and K. S. Anseth, *Ann. Biomed. Eng.* **32**, 407 (2004).
190. S. J. Bryant, K. S. Anseth, D. A. Lee, and D. L. Bader, *Journal of Orthopaedic Research* **22**, 1143 (2004).
191. D. A. Harrington *et al.*, *J. Biomed. Mater. Res. A* **78A**, 157 (2006).
192. J. Cappello *et al.*, *Journal of Controlled Release* **53**, 105 (1998).
193. W. A. Petka, J. L. Harden, K. P. McGrath, D. Wirtz, and D. A. Tirrell, *Science* **281**, 389 (1998).
194. J. Kisiday *et al.*, *P. Natl. Acad. Sci. USA* **99**, 9996 (2002).
195. S. G. Zhang, *Nature Biotechnology* **21**, 1171 (2003).
196. A. Warshel, and M. Levitt, *Journal of Molecular Biology* **103**, 227 (1976).
197. H. M. Senn, and W. Thiel, *Atomistic Approaches in Modern Biology: From Quantum Chemistry to Molecular Simulations* **268**, 173 (2007).
198. K. E. Ranaghan *et al.*, *Organic & Biomolecular Chemistry* **2**, 968 (2004).
199. K. E. Ranaghan, and A. J. Mulholland, *Chemical Communications*, 1238 (2004).
200. B. Szefczyk, A. J. Mulholland, K. E. Ranaghan, and W. A. Sokalski, *Journal of the American Chemical Society* **126**, 16148 (2004).
201. F. Claeyssens, K. E. Ranaghan, F. R. Manby, J. N. Harvey, and A. J. Mulholland, *Chemical Communications*, 5068 (2005).
202. B. K. Kobilka, *Biochimica Et Biophysica Acta-Biomembranes* **1768**, 794 (2007).
203. K. Palczewski *et al.*, *Science* **289**, 739 (2000).
204. V. Cherezov *et al.*, *Science* **318** (2007).
205. W. A. Goddard and R. Abrol, *Journal of Nutrition* **137**, 1528S (2007).
206. D. L. Beveridge, and F. M. Dicapua, *Annual Review of Biophysics and Biophysical Chemistry* **18**, 431 (1989).

207. S. Kirkpatrick, C. D. Gelatt, and M. P. Vecchi, *Science* **220**, 671 (1983).
208. D. Hamelberg, J. Mongan, and J. A. McCammon, *Journal of Chemical Physics* **120**, 11919 (2004).
209. K. M. Ferguson, *Biochemical Society Transactions* **32**, 742 (2004).
210. J. M. Sanz-Serna, *Acta Numerica*, Vol. 1992, pp. 243–286, Cambridge University Press (1992).
211. A. Nakano *et al.*, *Computing in Science & Engineering* **3**, 56 (2001).

Chapter Eight

Game Theoretical Formulation of the Immune System and Inspiration for Controlled Drug Delivery Application

Sharon Bewick, Mingjun Zhang and William R. Hamel

8.1 INTRODUCTION

The media is rife with examples of conflict at all levels, from gang-related street crime through government activities against terrorism to military operations in foreign countries. While we are all familiar with warfare at this level, we often overlook the remarkable similarities between social battles and biological conflict. In fact, the natural world is, in many ways, controlled by the same principles of competition and cooperation that inspire combat between individual people, social and religious groups, and entire governments. Unfortunately, despite efforts of scientists, sociologists and mathematicians, quantitatively predictive models of competition at any level are scarce.

In the fields of economics and ecology, Game Theory has emerged as one of the dominant frameworks from which to view conflict and cooperation. Game Theory has not, however, been as widely applied to cellular biology. Microbiology has been a predominantly experimental field for many years, and thus Game Theory's

Nanomedicine: *A Systems Engineering Approach* edited by Mingjun Zhang & Ning Xi

www.panstanford.com
978-981-4241-36-6

limited application is likely due to skepticism towards quantitative models in general. That said, Game Theory is, at the moment, a less defined mathematical framework than standard techniques like differential equations. In particular, the general application of Game Theory has been frustrated by controversy over both implementation and interpretation. These controversies have likely contributed to the scarcity of Game Theoretic models in cell biology as well.

Given the underlying principles of biological adaptation that have shaped cellular processes, though, we believe that with sufficiently careful definitions of the conflicts being modeled, and sufficiently detailed experimental measurements, there is no reason why microbiology should be any less amenable to a Game Theoretical approach than ecology, especially when applied to specific instances of cellular conflict like those which occur in the immune system. Applying game theory to cellular systems, and in particular, to the immune system, is appealing for several different reasons. First, by studying conflict at the cellular level we may be able to affect immunological competitions to our advantage, which will have significant applications in areas such as disease control, tumor suppression and controlled drug delivery. In addition, analysis of microbiological systems and comparison between experiment and theory will likely shed light on many of the interpretive difficulties associated with Game Theory itself. As a result, applying Game Theory to the immune system may help in terms of the interpretation and implementation of Game Theory itself, at least as far as its application to cellular biology and microbiological systems.

8.2 THE IMMUNE SYSTEM

One of the most striking examples of conflict on the cellular level occurs between the vertebrate immune system and the multitude of pathogens, from viruses to multicellular parasites, which battle for control of tissues within the vertebrate host. If we can develop an understanding of this conflict, and in particular of the complex interplay between the defensive strategies of the host immune system and the offensive strategies of the invading pathogen, it will likely lead to significant advances in therapeutic drug delivery and treatment of disease, both on an individual and a population-wide scale. We begin, therefore, by outlining the basic functions and components of the vertebrate immune system, paying particular attention to the different combative roles played by the various cell types and their associated signaling molecules.

The size of the vertebrate immune system is constrained both by spatial considerations and by resource costs associated with its maintenance. As a result, it would be impossible for the immune system to permanently harbor enough separate defense elements to effectively and specifically combat every strain of every pathogen that the host might encounter in its lifetime. That said, once a pathogen *has* been identified within the host system, the host can fine-tune and expand its response to that pathogen without encountering the same limitations on space and resource availability. Unfortunately, the process of fine-tuning and expansion incurs its own cost in terms of time. Biological systems cannot change instantaneously, thus while the host system adapts its response to specifically counter the invading pathogen, the pathogen has a chance to replicate and spread, gaining a foothold in the host system, and thus making it more difficult to eliminate.

Clearly, then, there is a trade-off such that what the immune system loses in speed it gains in specificity and vice versa. For vertebrates, this trade-off has resulted in the existence of two separate, and yet interconnected, branches of the immune system. *Innate immunity* provides a front-line response to invading pathogens. Unfortunately, while components of the innate immune system help to suppress pathogen numbers, they frequently lack the specificity required to fully clear the pathogen from the host. As a result, vertebrates have evolved what is termed *adaptive immunity* as well. Adaptive immunity develops more slowly than innate immunity, but is designed to recognize specific antigens (surface proteins, products of the pathogen lifecycle, etc.) that are characteristic to individual pathogens, or even pathogen strains.

8.2.1 The Innate Immune System

As suggested above, innate immunity provides a general defense that develops within seconds of pathogen entry into the host.[1] The most important components of innate immunity are summarized in Table 1, along with their functions.

The innate immune system responds to signals that are common to a wide variety of different infections. These signals can be structures that are similar or conserved in a broad class of pathogens, or

Table 1. Components of the innate immune system.

Component	Function
Monocytes	Mature to macrophages
Macrophages	Phagocytosis, antigen presentation, inflammation (reside in tissue)
Dendritic cells	Immature: phagocytosis Mature: antigen presentation (most efficient APC cell type)
Neutrophils	Phagocytosis, antigen presentation, inflammation (reside in blood, but may migrate to tissue)
Basophils	Inflammation
Eosinophils	Defense against larger parasites (e.g. helminthes)
Natural killer (NK) cells	Kill host cells with reduced expression of class I MHC proteins on their surface by secreting perforin/granzymes
Mast cells	Inflammation
Platelets	Clotting (restores physical barriers) and inflammation
Complement proteins	Regulation of immune function (e.g. opsonization, triggering formation of the *membrane attack complex*)

else they can be symptoms of distress produced by host tissue itself. A commonly cited example of the former is the lipopolysaccharide (LPS) coat found on the outer surface of all gram-negative bacteria. LPS is recognized by a protein known as "Toll-like receptor 4" which can be found on the surface of a variety of different immune cell types. In contrast, distress signals can include anything from the presence of molecules, like histamines, that are secreted by damaged cells, to the absence of surface proteins, like major histocompatibility complexes (MHC), that are found on all healthy cells.

The initial defensive measure taken by the innate immune system is the construction and maintenance of both physical and chemical barriers, which make it difficult for pathogens to gain entry into the host system. When viruses and bacteria do manage to cross these barriers, the innate immune system employs other combative measures including:

(1) *opsonization* (or protein coating) of the pathogen, which prevents further spread of the pathogen, and marks the pathogen for clearance from the system.

(2) *phagocytosis* (or ingesting) and lysis (or digesting) of the pathogen, which again removes the immediate threat from the system.
(3) *cytotoxic killing* of infected host cells, whereby NK cells that recognize aberrant MHC production secrete toxic chemicals which destroy the afflicted host cell to prevent subsequent infection of neighboring tissue.

Many of the chemical signals used and secreted by components of the innate immune system are common to pathways in the adaptive immune system as well. Moreover, the phagocytotic cells of the innate immune system often display, on their surfaces, small protein sequences, or antigens, of the pathogens that they have consumed. These phagocytes, known as antigen presenting cells (APC), actively seek out and present their antigens to cells of the adaptive immune system in order to stimulate an adaptive immune response. Therefore it is clear that the role of the innate immune system is not limited to pathogen clearance, or suppression, but includes a signaling aspect as well, and thus is also integral to the development of an adaptive immune response.

8.2.2 The Adaptive Immune System

Unlike the innate immune response, which begins within seconds of pathogen invasion, the adaptive immune response requires time to develop. Once in place, however, an adaptive immune response can effectively target characteristic properties of a particular pathogen or pathogen strain. This specificity allows for a more aggressive attack on the pathogen, and is usually sufficient to clear the pathogen from the system. The most important components of adaptive immunity are summarized in Table 2, along with their functions. The two predominant cell types responsible for adaptive immunity are known as B-lymphocytes and T-lymphocytes. B-lymphocytes are specifically designed to combat extra-cellular pathogens, while T-lymphocytes defend against intra-cellular invaders. While there is only one type of B-lymphocyte, T- lymphocytes are divided into further sub-classes known as $CD8^+$ T-lymphocytes and $CD4^+$ T-lymphocytes. The hierarchical organization of the T-lymphocytes, and their selective patterns of activation in response to various types of pathogen invaders will be discussed in more detail in Sec. 8.2.2.2.

Table 2. Components of the adaptive immune system.

Component	Function
B-lymphocytes	Produce antibodies, antigen presentation
$CD8^+$ T-lymphocytes	Recognize class I MHC-antigen complexes on the surface of infected host cells and differentiate into CTL cells
Cytolytic T-lymphocytes (CTL)	Kill host cells with class I MHC-antigen complexes on their surface by secreting perforin/granzymes
$CD4^+$ T-lymphocytes	Recognize class II MHC-antigen complexes on the surface of APC cells and differentiate into helper T-cells
helper T-lymphocyte (TH1)	Secrete IL-2, IL-3, IFN-γ, TNF-β, TNF-α and GM-CSF, stimulate T-cell immunity and phagocytotic killing, are proinflammatory
helper T-lymphocyte (TH2)	Secrete IL-3, IL-4, IL-5, IL-6, IL-9, IL-10, IL-13, TNF-α and GM-CSF, stimulate antibody production by promoting B-cell growth and differentiation
Antibodies	IgM: initial antigen recognition, activates complement system IgD: bound to membrane of naïve B cells with no previous antigen exposure IgG: high affinity antigen binding, stimulates phagocytosis, neutralization through binding to antigen IgA: dominant in secretions IgE: immune response to parasites, stimulates mast cells and basophils to secrete inflammatory agents

8.2.2.1 B-Lymphocytes

B-lymphocytes are produced and mature in the bone marrow. Once mature, they migrate to the lymphatic system where, by circulating through the network of lymph and blood vessels, they can effectively patrol the entire vertebrate body. Each B-lymphocyte has unique cell surface proteins known as B-cell receptors (BCRs), which differ from B-lymphocyte to B-lymphocyte as a result of somatic (non-germline) recombination of host genes. BCRs are capable of recognizing and binding to bacterial and viral antigens; however, since all B-lymphocytes differ, only a handful of B-lymphocytes present in the host immune system at any given time will bind to an invading pathogen with sufficient strength to stimulate a response.

When a B-lymphocyte is stimulated by recognition of a matching antigen, it begins to proliferate through a process known as clonal expansion. As a result of this proliferation, the host immune system produces a multitude of B-lymphocytes with BCRs specific to the stimulating antigens on the invading pathogen. In addition, some of these antigen-specific B-lymphocytes differentiate into specialized cells that secrete proteins known as antibodies. Antibodies have receptor regions identical to the BCRs on the B-lymphocytes that produced them (BCRs are actually a class of cell-surface bound antibodies themselves). As a result these free-floating antibodies also bind strongly to their target antigens. This has the effect of neutralizing the pathogen so that it cannot effectively replicate or spread. Moreover, by coating pathogens with antibodies, the immune system essentially marks the pathogens for destruction and elimination by other components of the immune system.

8.2.2.2 T-Lymphocytes

Similar to B-lymphocytes, T-lymphocytes are produced in the bone marrow; however, rather than maturing there, they migrate to the thymus, where they undergo a process of negative selection that eliminates any T-lymphocytes which could potentially cross-react with host tissue. Like B-lymphocytes, each T-lymphocyte has its own set of unique surface proteins, called T-cell receptors (TCRs). Unlike BCRs, which recognize antigens on the surfaces of free, or extra-cellular, pathogens, TCRs can only bind to antigens that have already been subsumed and processed by host cells. As a result, T-lymphocytes effectively combat intracellular pathogens. In general, there are two different forms of intracellular pathogens — pathogens that are inside host cells as a result of host cell infection, and pathogens that are inside antigen presenting cells (APCs) as a result of phagocytosis. Both of these intracellular pathogens are recognized by TCRs, however the T-lymphocytes supporting the TCRs are different in each case.

When a pathogen infects, or is phagocytosed by a host cell, the host cell splices some of the pathogen's antigenic proteins into smaller segments, which are then bound to host proteins known as major histocompatibility complexes (MHCs). Once formed, entire antigen-MHC complexes migrate to the surface of the infected cell or APC, where they are displayed for subsequent T-lymphocyte recognition. The TCRs on the surfaces of the T-lymphocytes bind to the entire antigen-MHC complex, thus recognition by T-lymphocytes is determined not only by the shape and physicochemical properties of the antigen, but also by the nature of the MHC to which the antigen is bound.

In general, there are two types of MHC proteins — those known as class I MHCs and those known as class II MHCs. Class I MHCs are produced by *all* host cells. As suggested in Sec. 8.2.1, even healthy host cells are coated with class I MHCs. (In the case of healthy cells, however, these MHCs complex with proteins produced by the host itself. Self-MHC complexes are not recognized by T-lymphocytes as a result of the negative selection process that occurs during T-lymphocyte maturation in the thymus.) Class I MHCs can only be recognized by a subset of chemically distinct T-lymphocytes known as $CD8^+$ T-cells. When a specific antigen-class I MHC complex is recognized by a specific $CD8^+$ TCR, it triggers proliferation of that particular $CD8^+$ T-lymphocyte. Therefore, as with B-lymphocytes, stimulation of $CD8^+$ T-lymphocytes results in a rapid expansion of $CD8^+$ T-cells with specificity for the invading pathogen. During this proliferation process, some of the $CD8^+$ T-lymphocytes differentiate into specialized killer cells known as cytotoxic T-lymphocytes (CTLs). CTLs have recognition proteins on their surfaces that are identical to the TCRs on their parent $CD8^+$ T-cells, thus CTLs maintain a high specificity for antigen-MHC complexes characteristic of the invading pathogen. When CTLs encounter antigen-MHC complexes that they recognize on the surfaces of infected host cells, the CTLs are stimulated to release toxic chemicals which destroy both the host cell and the pathogens that it harbors.

Class II MHCs are only produced by specialized APC cells from the host immune system. Antigens bound to class II MHCs are recognized by a different subset of chemically distinct T-lymphocytes known as $CD4^+$ T-cells. As with B-lymphocytes and $CD8^+$ T-lymphocytes, when a specific antigen-class II MHC complex is recognized by a specific $CD4^+$ T-lymphocyte, the $CD4^+$ T-lymphocyte is stimulated to proliferate and differentiate. In the case of $CD4^+$ T-lymphocytes, differentiation results in the formation of specialized cells known as helper T-lymphocytes. There are several different types of helper T-lymphocytes. These different types of helper T-cells perform different immunological functions. Helper T-lymphocytes of the type 1 (TH1) variety help to stimulate $CD8^+$ T-lymphocyte proliferation and phagocytotic killing of the pathogen by macrophages. In contrast, helper T-lymphocytes of the type 2 (TH2) variety, promote B-lymphocyte proliferation and antibody production.

8.2.3 A Complex Network of Interactions: Regulating Immune System Behavior

One aspect which makes studying the vertebrate immune system both challenging and appealing is the impressive level of complexity that it exhibits. The different components of innate immunity and the different components of adaptive immunity interact not only within their respective systems, but also across system boundaries, thus cells of the innate immune system can affect actions taken by cells of the adaptive immune system and vice versa. While the components of the immune system interact in many different ways, most interactions can be classified as either direct, requiring cell-cell contact, or indirect, via the production and detection of signaling molecules. The natures of these direct and indirect interactions depend on both circumstance and the components involved. In general, both stimulatory and inhibitory interactions are possible. Stimulatory interactions are beneficial in that they allow for a rapid build-up of an immune response towards an invading pathogen. Nevertheless, an excessive immune response can be costly in terms of both resources and damage to host tissue. As a result, inhibitory mechanisms must also be in place in order to suppress an immune response when it is out of control.

A good example of immune system cells that exchange information via direct contact is the interaction between APC cells and T-lymphocytes. As discussed in Sec. 8.2.2.2, when a pathogen is phagocytosed by an APC, the APC cell membrane displays antigenic peptides from the pathogen as part of an antigen-class II

MHC complex. This complex is recognized by the TCRs on specific T-lymphocytes, stimulating those T-lymphocytes to undergo clonal expansion. Frequently, clonal expansion is further stimulated when additional proteins on the APC cell membrane bind to additional protein on the T-lymphocyte (for example, when CD80 on the APC binds to CD28 on the T-lymphocyte). These *costimulatory* signals up-regulate the immune response and are often necessary in order to mount a successful attack on an invading pathogen.

Other proteins expressed on the APC and T-lymphocyte cell membranes, however, provide inhibitory responses. For example, the interaction between CD86 on the surface of APC cells and CD152 on the surface T-lymphocytes results in the suppression of T-lymphocyte clonal expansion. Mice that are genetically engineered to lack CD152 die as a result of extreme T-lymphocyte activity. Clearly, the action taken by a T-lymphocyte upon contact with an APC cell is controlled by a tightly regulated set of stimulatory and inhibitory signals which are, themselves, fine-tuned in response to the invading pathogen and the current state of the host system. Similar mechanisms exist between other cell types known to exchange information upon direct contact and recognition.

In addition to interacting via direct cell-cell contact, many cells of the immune system synthesize and secrete small molecules known as *cytokines* which serve to coordinate their activities and the activities of the other immune system components. Many cytokines are known as *interleukins* (IL) since they represent communication between the leukocytes, or white blood cells of the immune system. For historical reasons, however, certain cytokines have special names like "tumor necrosis factor" (TNF) and colony stimulating factor (CSF). As with information exchange via direct contact between cells, information exchange via signaling molecules can have both stimulatory and inhibitory effects. For example, interleukin-14 (IL-14) stimulates B-cell proliferation while at the same time inhibiting antibody secretion.

From the above discussion, it should be clear that the variety and interconnectedness of immune cell interactions provide the immune system with a vast network of pathways through which to send and receive information, exert control and regulate response. Figure 1 shows a partial summary of the interactions between common immune system components. For a more complete list, however, the reader is referred to a standard text book in immunology.[2]

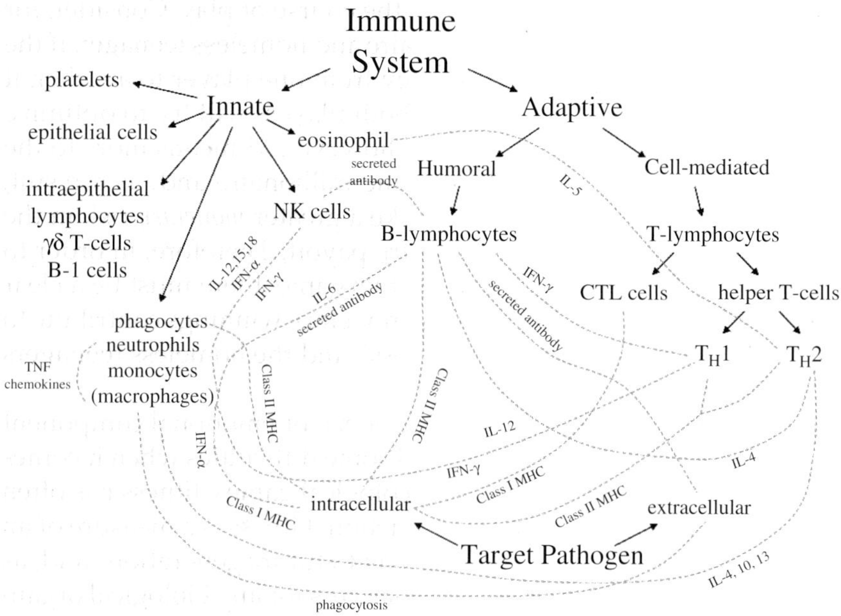

Figure 1. Interactions between common immune system components.

8.3 GAME THEORY

Game Theory, formalized in the 1940's by von Neumann and Morgenstern, provides a mathematical framework for the logical analysis of conflict (and cooperation). In a game, two or more players choose courses of action, or *strategies*, which jointly determine the outcome of the conflict. In order to analyze any such game, the game must be clearly defined. This requires a complete list of all possible strategies for each player and an associated *payoff* to each player that depends on both his choice of strategy and the strategies chosen by every other player in the game.

One of the fundamental requirements of Game Theory, and one of the reasons why it can be problematic when applied to certain "games", is the assumption that all players behave rationally at all times. Said differently, a game theoretical analysis relies on players *always* trying to maximize their respective payoffs, regardless of the situation. Particularly in social games, this requires careful definition of the payoff function itself, since the emotional and circumstantial significance of an outcome adds a dimension beyond the "obvious" payoff values won and lost during the course of play. Consider, for instance, a game between a millionaire and homeless teenager. If the game involves the transfer of money from one player to another, it might be tempting to assume that both players will try to optimize their total *monetary* profits. Clearly, however, \$1 means more to the homeless teenager than it does to the millionaire and, as a result, the millionaire may be willing to take a greater *monetary* risk in the hopes of achieving a higher *monetary* payoff. Therefore, in order to fully define the payoff function for this game, there must be a clear understanding of how *both* risk *and* monetary winnings contribute to the millionaires definition of "success" and the homeless teenagers definition of "success".

For biological games, there is no social or emotional component to consider. There are, however, still some difficulties when it comes to defining a payoff function. For ecological games, fitness has often been used as the ultimate payoff function. Fitness is a measure of an organism's ability to pass his genes on to future generations and, as such, is an obvious goal, or mark of success for any biological organism engaged in conflict. Unfortunately, defining an organism fitness can be challenging, because we currently lack an understanding of exactly how and why certain factors contribute to biological fitness, and biological success. Despite these difficulties, on the time scale of an infection, it is reasonable to assume that fitness of the pathogen is closely correlated with its ability to reproduce and create copies of itself, while the fitness of the host is closely correlated with its ability to clear the pathogen quickly with limited damage to host tissue. Therefore, in games between a host immune system and an invading pathogen, the payoff function should be defined in terms of the number of pathogen replicates and the total loss of host tissue and host resources.

Let us return to Game Theory for a moment, and consider further the assumption of rational behavior. Obviously, Game Theory fails to be predictive if a player behaves irrationally, throwing caution to the wind and purposely trying to lose the game, or at least making no effort to win it. Within the assumption of rational behavior, however, is the additional postulate that all players *know*, or can figure out, what the rational course of action should be. Game Theory fails when a player cannot think through the steps of the game clearly enough to deduce his own optimal strategy. Players can be fooled into suboptimal behavior and, in fact, frequently are in social games.

In biological games, however, we assume that through the process of adaptation, both players in the game have been fine-tuned in terms of defending against one another. As a result, most biological strategies, and particularly those that appear to have stabilized over time, are almost certainly near-optimal.

There are many different formulations which could potentially be used to model games between a vertebrate immune system and an invading pathogen. In general, games themselves can be zero or non-zero sum, while their solutions can be either pure strategy solutions or mixed strategy solutions. For non-zero sum games, solutions can be further subdivided into those which are Pareto optimal, and those which are not. We will consider the various different ways of formulating game theory, and the various different solutions associated with each of the formulations in Sec. 8.3.2. First, however, it is worthwhile to discuss why Game Theory is such an appealing mathematical technique to employ when modeling the immune system.

8.3.1 *Why Choose Game Theory to Model the Immune System?*

From the previous discussion, it should be clear that biological games, even more than economic, military, or social games, are amenable to the implicit assumptions required for a game theoretical analysis. We have, therefore, answered the question "Why *can* the immune system be modeled using Game Theory?" Still, that is not the same as answering the question "Why *should* the immune system be modeled using Game Theory?"

Certainly trying to unravel the intricate workings of the immune system through mathematical analysis is nothing new. In fact, in recent years there has been an explosion in the number of quantitative models that have been proposed to explain different aspects of the immune response. For example, T- and B-lymphocyte receptors have been analyzed using string models,[3] geometric models[4] and random energy models,[5] while ordinary differential equations (ODEs) have been used to explain everything from T-cell response during antiviral therapy,[6] to immunological memory[7,8] (the ability of the immune system to better defend against reinvasion by a previously encountered pathogen).

players in the game have diametrically opposed agendas, or payoff functions. In other words, whatever one player wins, the other player loses and vice versa. As a result, the sum of the payoff to player one and the payoff to player two will equal zero for every possible outcome of the game. In what follows we show, through example, how zero sum games can be formalized and analyzed for a particular host-pathogen interaction.

Example 8.1. During the course of an infection, certain viruses mutate towards increasing glycosylation of their protein envelopes, apparently allowing them to escape recognition by host antibodies. Despite the apparent advantage of glycosylation in terms of immune system evasion, early infections involving these pathogens are still often dominated by viruses with bare protein coats. Therefore it has been suggested that while glycosylation facilitates escape from an antibody attack, it also disrupts an initial step in the infection process.[9] One possible explanation is that glycosylated viruses, although they can effectively evade the adaptive immune system, actually experience increased pressure from the innate immune system.

Formulation. The first step in formulating a game theoretical model for the system described above is the definition of all possible strategies that can be played by each of the parties involved. Since adaptive immunity takes time to develop as a result of biological and physical constraints (see Sec. 8.2), we rule out any strategies that rely on an early adaptive immune response. Moreover, if the host system is limited by resources, or the need to minimize excessive tissue damage, it is conceivable that the immune system cannot mount both a strong innate immune response and a strong adaptive immune response simultaneously. This leaves only two strategies for the host immune system. The immune system can (a) mount a strong innate immune response throughout the entire course of the infection, or (b) mount a strong innate immune response, but then switch to a strong adaptive immune response when adaptive immunity becomes available. The virus, on the other hand, has four strategies: (a) infect in a glycosylated state and remain glycosylated, (b) infect in a bare state and remain bare, (c) infect in a glycoylated state and mutate to a bare state, and (d) infect in a bare state and mutate to a glycosylated state.

In addition to defining the strategies for each player, a game theoretical analysis also requires construction of a payoff function, or payoff matrix. In our example, a good candidate for the payoff function *to the virus* is the average number of successful transmission events during the early period of the infection. If we assume that transmission probability is a linear function of viral load, then it is very reasonable to further assume that the total host suffering is negatively correlated with the total number of transmission events as well. Therefore, the game can be interpreted as zero-sum, with the payoff *to the host* being the negative value of the payoff to the virus.

The next challenge, of course, is assigning numerical values to the total number of transmission events that occur during the early stages of infection. This must be done for each of the strategy combinations defined above, and thus requires either very detailed empirical evidence, or a reasonable mathematical model. To generate the payoff matrix in Fig. 2a, we have used a model that assumes simple logistic growth of the virus modified by a killing term that is dependent on both the state of the host and the state of the virus. In addition, we have allowed for both mutations in the viral population after a fixed number of viral replications, and a time-dependent switch between innate and adaptive immune responses in the host.

	innate/innate	innate/adaptive
glycosylated/ glycoslyated	3.63	3.27
bare/bare	5.13	2.04
glycosylated/bare	3.68	1.71
bare/glycoslyated	5.10	3.81

(a)

	innate/innate	innate/adaptive
bare/bare	5.13	2.04
bare/glycoslyated	5.10	3.81

(b)

	innate/adaptive
bare/bare	2.04
bare/glycoslyated	3.81

(c)

Figure 2. Game matrices for the basic glycosylated vs. bare game.

Solution. Game Theoretical methods can be used to analyze any game with a payoff matrix like the one defined in Fig. 2a. In general it is easiest to begin the analysis by searching for *pure strategy* or *saddle-point* solutions. Pure strategy solutions, when they exist, can be found using the Dominance Principle. According to the Dominance Principle, a player will never choose a strategy if it is outperformed by another strategy for every possible choice of action that may be taken their opponent.

Consider the viral strategy "glycosylated/bare" in the game described above. Regardless of how the immune system chooses to defend, the virus can gain more by playing "bare/glycosylated" than it can by playing "glycosylated/bare". As a result, there is no reason why the virus should *ever* play the strategy "glycosylated/bare". Similarly, "bare/bare" outperforms "glycosylated/glycosylated" independent of whether the immune system chooses an "innate/innate" response or an "innate/adaptive" response, thus the virus never profits by choosing to enter glycosylated and remain glycosylated. Both the "glycosylated/bare" strategy and the "glycosylated/glycosylated" strategy are said to be *strictly dominated*. Any strategy that is strictly dominated can be removed from the game since it should never be played. This leaves the reduced game matrix shown in Fig. 2b.

Clearly the strategy "bare/bare" outperforms the strategy "bare/glycosylated" if the host immune system chooses an "innate/innate" response, while the opposite is true if the host immune system chooses an "innate/adaptive" response. Because neither viral strategy in the reduced game matrix above is strictly dominated, neither can be safely removed. It is, however, possible to further reduce the game matrix by considering host strategies. Keeping in mind that the host wants to maximize its own payoff (which is the same as minimizing the payoff *to the virus*) it is clear from the game matrix in Fig. 2b that the host should choose the strategy "innate/adaptive" over the strategy "innate/innate" regardless of the action taken by the virus. "Innate/innate" is thus *strictly dominated* and can be dismissed. The reduced payoff matrix that results is shown in Fig. 2c, from which it is obvious that the virus will choose the strategy "bare/glycosylated" and win a total of 3.81 transmission events. Notice that 3.81 is not the maximum number in the original payoff matrix. Rather, the maximum number was 5.13. The virus, however, cannot achieve the absolute maximum of 5.13

transmission events unless the host makes a strategic mistake. Similarly, 3.81 is not the minimum number in the original payoff matrix either. Rather, the minimum number was 1.71. The host, however, cannot limit the virus to the absolute minimum of 1.71 transmission events unless the virus makes a strategic mistake. In other words, if both host and virus act rationally, and also assume that their opponent will act rationally, the end result will lead to a payoff of 3.81 to the virus and −3.81 to the host. Furthermore, the virus will choose the "bare/glycosylated" strategy which is observed experimentally, while the host will choose the "innate/adaptive" response.

The analysis of the above game leads to what is known as a *pure strategy* solution. This terminology stems from the fact that both players can optimize their success in the game by playing single strategy with certainty. Another name for a pure strategy solution is a *saddle-point* solution, since it is the solution which maximizes player one's payoff along player one's strategies and minimizes player one's payoff along player two's strategies. Not all two-person zero-sum games, however, have pure strategy solutions, as illustrated in the next example.

Example 8.2. Consider a simplified version of the previous problem where mutation on the time-scale of the infection has fitness associated costs that make it unfavorable for the virus. This could be the case, for instance, with larger DNA viruses whose genomes appear to be less tolerant to mutation.[10]

Formulation. If the virus can't mutate, it only has two viable strategies. It can (a) assume a glycosylated state, or (b) assume a bare state. The payoff matrix in Fig. 3 was generated for this modified game using a model and a payoff function similar to the one described for the original game, though with slightly different parameters.

Unlike the original game, the simplified game cannot be solved using the Dominance Principle. In terms of viral strategies, neither

	innate/innate	innate/adaptive
glycosylated	2.64	2.80
bare	5.13	2.05

Figure 3. Game matrix for the simplified glycosylated vs. bare game.

"bare", nor "glycosylated" are strictly dominated. Similarly, in terms of immune system strategies, neither "innate/innate" nor "innate/adaptive" are strictly dominated either. As a result, this new game has no pure strategy solution, meaning that the players in the game cannot play a single strategy with certainty and still maximize their payoff against a rational opponent.

Consider, for instance, what would happen if the virus always appeared in its "bare" state. In that case, the host immune system could limit the virus to a payoff of 2.05 by defending with an "innate/adaptive" response. But if the immune system always defended with an "innate/adaptive" response, the virus could improve its payoff by assuming a "glycosylated" attack. In contrast to a constant "bare" attack, a constant "glycosylated" attack, is best defeated if the immune system adopts a permanent "innate/innate" strategy. However, if the immune system consistently plays an "innate/innate" strategy, then the virus can maximize its payoff by always attacking in a "bare" state. Notice how this circular argument suggests a continual cycling through the strategies. Again, this is because there is no saddle-point, or equilibrium and, as a result, players limited to pure strategies are caught in a endless loop continually trying to capitalize on knowledge of what the other players will do.

Cycling through such loops, however, is not an optimal strategy. In games that lack pure strategy solutions, a player's best option is to "mix up", or randomize, their attacks and defenses. By randomizing, a player prevents their opponents from ever knowing exactly what strategy will be played and, accordingly, it becomes impossible to stage a preemptive strike.

While players engaged in a game without a pure strategy solution can improve their payoffs by randomizing their actions, choosing strategies *completely* at random does not lead to an optimal solution either. Rather, a game without a saddle-point solution is best played by randomizing *only* in the short term, while maintaining a long term statistical order to the strategy choices that are made. This can be done by assuming *fixed probabilities* for the chance of choosing any possible pure strategy action. Players that assign fixed probabilities wisely can optimize their average payoff in the face of a rational opponent who is doing the same. This leads to a new type of equilibrium, known as a *mixed strategy solution*. Naturally, mixed strategy solutions only have meaning if the game is played repeatedly according to the same rules. Although the interpretation of mixed strategies in "one-off" games is problematic, most biological situations naturally assume repeated play. In a game between the immune system and a pathogen invader, for example, there are always multiple encounters between identical pathogen replicates and identical immune cells that have been generated through the process of clonal expansion.

Solution. Returning to the game described by the payoff matrix in Fig. 3, we can search for mixed strategies as follows. First we consider the virus. The virus can play the strategy "bare" with any probability, P_{bare}, between 0 and 1. $P_{\text{bare}} = 0$ means that the virus always chooses a "glycosylated" attack, while $P_{\text{bare}} = 1$ means that the virus always chooses a "bare" attack. For any $0 < P_{\text{bare}} < 1$, the virus will sometimes choose a "glycosylated" attack and other times choose a "bare" attack. Next we consider the defense of the host. If the host plays an "innate/innate" response, then they payoff to the virus will be 2.64 if $P_{\text{bare}} = 0, 5.13$ if $P_{\text{bare}} = 1$, and some number *between* 2.64 and 5.13 if $0 < P_{\text{bare}} < 1$. Similarly, if the host plays an "innate/adaptive" response, then the payoff to the virus will be 2.80 if $P_{\text{bare}} = 0$, 2.05 if $P_{\text{bare}} = 1$, and some number *between* 2.80

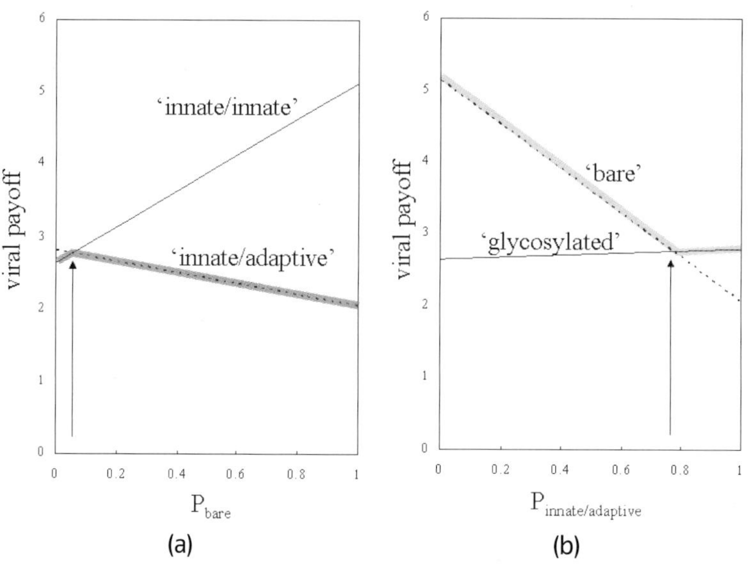

Figure 4. Viral payoffs for mixed strategy solutions.

and 2.05 if $0 < P_{\text{bare}} < 1$. Figure 4a shows the payoff to the virus as a function of P_{bare} for both an "innate/innate" response and an "innate/adaptive" response on the part of the immune system.

The host immune system is, of course, trying to limit the viral payoff as much as possible. Notice, however, that even if the host immune system could predict the value of P_{bare} chosen by the virus and then play according to this knowledge, the immune system could not hold the virus to any less than the payoff marked by the red line on Fig. 4a. The red line, then, represents the worst case scenario for the virus, or the virus's minimum payoff. The virus can maximize its minimum payoff by choosing $P_{\text{bare}} = 0.0464$ as shown by the arrow in the figure.

A similar argument can be made for the immune system. The immune system can choose the "innate/adaptive" response with any probability, $P_{\text{innate/adaptive}}$, between 0 and 1. $P_{\text{innate/adaptive}} = 0$ means that the immune system never switches to an adaptive response, $P_{\text{innate/adaptive}} = 1$ means that the immune system always switches to an adaptive response and $0 < P_{\text{innate/adaptive}} < 1$ means that the immune system sometimes switches to an adaptive response and other times maintains an innate response. If the virus attacks in its "bare" form, then the payoff to the virus will be 5.13 if $P_{\text{innate/adaptive}} = 0$, 2.05 if $P_{\text{innate/adaptive}} = 1$ and some number *between* 5.13 and 2.05 if $0 < P_{\text{innate/adaptive}} < 1$. Similarly, if the virus attacks in its "glycosylated" form, then the payoff to the virus will be 2.64 if $P_{\text{innate/adaptive}} = 0$, 2.80 if $P_{\text{innate/adaptive}} = 1$ and some number *between* 2.64 and 2.80 if $0 < P_{\text{innate/adaptive}} < 1$. Figure 4b shows the payoff to the virus as a function of $P_{\text{innate/adaptive}}$ for both a "bare" attack and a "glycosylated" attack on the part of the virus.

The virus, of course, would like to maximize its own payoff. Notice from Fig. 4b, however, that even if the virus could predict the value of $P_{\text{innate/adaptive}}$ and then choose its strategy according to this knowledge, the virus could not maximize its payoff beyond the blue line shown in the figure. This blue line, then, represents the worst case scenario for the host immune system, or the virus's maximum payoff. The host immune system can try to minimize the viruses maximum payoff (and thus maximize its own minimum payoff) by choosing $P_{\text{innate/adaptive}} = 0.7685$ as shown by the arrow in the figure.

Therefore, for the game described by the payoff matrix in Fig. 3, the virus's optimal mixed strategy is to play "bare" 4.64 % of the time (and thus "glycosylated" 95.36% of the time), while the host immune system's optimal mixed strategy is to play "innate/adaptive" 76.85% of the time (and thus "innate/innate" 23.15% of the time). If both players stick to their optimal mixed strategies, the virus will win a total payoff of 2.76, while the payoff to the host immune system will be −2.76.

8.3.2.2 Two-Player non-zero-sum games

The assumption that the host and the virus have diametrically opposed agendas, and thus that the game is zero-sum is, of course, a simplifying approximation. Certainly, while the host suffers as a result of infection by the pathogen, there is not necessarily a direct relationship between host suffering and the size of the pathogen population. Therefore, it is possible that while the pathogen tries to play the game in order to optimize its total population, the host tries to play the game in order to limit some combination of the total viral population, the total damage to host tissue, and the total loss of host resources. When the two players in a game do not have strictly opposite goals, then the sum of the payoff to player one and the payoff to player two will *not* equal zero for every possible outcome of the game. This leads to what's known as a two-player non-zero sum game.

Example 8.3. As a simple example, consider what happens when the virus's payoff is dependent on its transmission rate, while the host's payoff is dependent on both the virus's transmission rate and a cost associated with mounting the particular defense strategy that is chosen.

Formulation. Specifically, notice the change in the payoff matrix from Fig. 2a if we assume that the energy and resources required to stage an adaptive response are twice as costly as the energy and resources required to simply maintain an innate response. Defining C as the cost of maintaining an innate response, a new payoff matrix can be generated as shown in Fig. 5a.

Solution. Any game with a payoff matrix like the one shown in Fig. 5a can be analyzed using Game Theoretical techniques. As with zero-sum games, each players tries to maximize his respective payoff. Unlike zero-sum games, however, in non-zero sum games, maximizing the payoff to one player is not necessarily the same as minimizing the payoff to the other. Consider what happens to the

	innate/innate	innate/adaptive
glycosylated/ glycoslyated	3.63, -3.63 - C	3.27, -3.27 - 2C
bare/bare	5.13, -5.13 - C	2.04, -2.04 - 2C
glycosylated/bare	3.68, -3.68 - C	1.71, -1.71 - 2C
bare/glycosylated	5.10, -5.10 - C	3.81, -3.81 - 2C

(a)

	innate/innate	innate/adaptive
glycosylated/ glycoslyated	3.63, -6.73	3.27, -9.47
bare/bare	5.13, -8.23	2.04, -8.24
glycosylated/bare	3.68, -6.78	1.71, -7.91
bare/glycosylated	5.10, -8.20	3.81, -10.01

(b)

	innate/innate
glycosylated/ glycoslyated	3.63, -6.73
bare/bare	5.13, -8.23
glycosylated/bare	3.68, -6.78
bare/glycosylated	5.10, -8.20

(c)

Figure 5. Game matrices for the glycosylated vs. bare game with an additional host cost incurred for adaptive immunity.

payoff matrix in Fig. 5a, for instance, if $C = 3.1$. The new payoff matrix for $C = 3.1$ is shown in Fig. 5b.

In the game defined by the payoff matrix in Fig. 5b, the virus chooses between the four possible glycosylation states in an attempt to maximize the viral payoff, which is the first number of each ordered pair in the table. In contrast, the host chooses whether or not to switch to an adaptive response in an attempt to maximize the host payoff, which is the second number of each ordered pair in the table. Notice that the higher additional cost associated with mounting an adaptive immune response results in a greater *host payoff* for the "innate/innate" strategy, regardless of what the virus does. In other words, the "innate/adaptive" strategy is now strictly dominated, and thus the host never does better by choosing to switch to an adaptive response. As with zero-sum games, when a strategy is strictly dominated, it can be dismissed, thus the payoff matrix in Fig. 5b is reduced to the payoff matrix in Fig. 5c.

Given that the host will play an "innate/innate" strategy, the virus can now choose between its four possible glycosylation states in an attempt to maximize its own payoff. Clearly, this leads to a "bare/bare" attack. As a result, the solution to the game described in Fig. 5b suggests that the virus should adopt a "bare, bare" strategy, the host immune system should mount an "innate/innate" response, the payoff to the virus will be 5.13, and the payoff to the host will be -8.23. This solution is comparable to the saddle-point solution or the pure strategy solution in zero-sum games, however in non-zero sum games it is termed a Nash equilibrium point.

One important aspect of non-zero sum games is that they can incorporate elements of cooperation. That is not to say that the two players actively agree to help one another. Rather, one player, in doing what is best for himself, can actually choose a strategy that helps the other player as well. Such cooperation is impossible in zero-sum games, since any gain for one player is automatically a loss for the other and vice versa. While cooperation does not occur in all non-zero sum games, it is certainly evident in the game described by Fig. 5b. Notice how, for any viral strategy, the immune system's choice to play "innate/innate" leads to greater payoffs for *both* the host immune system *and* the virus. In other words, the strategy that is most beneficial to the host is also most beneficial to the virus.

Unfortunately, with implicit cooperation comes a conceptual difficulty that does not occur with zero-sum games, and that has led to much debate in terms of the interpretation of game theoretic analysis.

Example 8.4. As an example, consider a payoff matrix similar to the one shown in Fig. 2a, however this time assume that the host tissue is damaged by the glycosylated virus significantly more than it is damaged by the bare virus.

Formulation. This can be represented as a cost, X, to the host for every viral strategy that involves glycosylation. The new payoff matrix is shown in Fig. 6a, and then again in Fig. 6b for the value $X = 4$.

Solution. By considering the host payoffs, it is clear that the strategy "innate/innate" is strictly dominated, and thus can be removed to give the reduced payoff matrix shown in Fig. 6c. Assuming that the host plays an "innate/adaptive" response, the virus would do best by choosing a "bare/glycosylated" attack. This would give the virus a payoff of 3.81 and the host a payoff of -7.81, and is the Nash

	innate/innate	innate/adaptive
glycosylated/ glycoslyated	3.63, -3.63 - X	3.27, -3.27 - X
bare/bare	5.13, -5.13	2.04, -2.04
glycosylated/bare	3.68, -3.68 - X	1.71, -1.71 - X
bare/glycosylated	5.10, -5.10 - X	3.81, -3.81 - X

(a)

	innate/innate	innate/adaptive
glycosylated/ glycoslyated	3.63, -7.63	3.27, -7.27
bare/bare	5.13, -5.13	2.04, -2.04
glycosylated/bare	3.68, -7.68	1.71, -6.71
bare/glycosylated	5.10, -9.10	3.81, -7.81

(b)

	innate/adaptive
glycosylated/ glycoslyated	3.27, -7.27
bare/bare	2.04, -2.04
glycosylated/bare	1.71, -6.71
bare/glycosylated	3.81, -7.81

(c)

Figure 6. Game matrices for the glycosylated vs. bare game with an additional host cost incurred for glycosylated attack.

equilibrium point for the game described by the payoff matrix in Fig. 6b.

While there was no difficulty finding the Nash equilibrium in this game, its interpretation is slightly troublesome. A closer look at the entire payoff matrix in Fig. 6b reveals that while the Nash equilibrium dictates that the virus play "bare/glycosylated" and the host immune system play "innate/adaptive", both parties would do better if the virus chose a "bare/bare" strategy, and the host immune system chose an "innate/innate" response. In that case, the virus would improve its payoff from 3.81 to 5.13, while the host immune system would improve its payoff from −7.81 to −5.13. The solution defined by the payoffs (5.13, −5.13) is known as a Pareto optimal solution. A solution is Pareto optimal if there is no other outcome that makes every player at least as well of, and at least one player better off. Notice that the Nash equilibrium point in the game from Fig. 6b is *not* a Pareto optimal solution, because both parties can do better if the virus switches to a "bare/bare" attack and the immune system switches to an "innate/innate" response. In general, not all Nash equilibrium points are Pareto optimal, nor are all Pareto optimal points Nash equilbria.

When Nash equilibrium points are not Pareto optimal, the obvious question is why the two players wouldn't choose the Pareto optimal solution over the Nash equilibrium. That question has plagued Game Theory from its earliest beginnings and, at present, there is really no satisfactory, universal answer. In some games, the Pareto optimal solution requires that the two players trust each other, because if either cheats, the other loses more than he would have by adhering to the Nash equilibrium strategy. In other games, like the one in Fig. 6b, however, this is not the case. It is possible that by applying Game Theory to the immune system some insight will be gained into how natural systems resolve this seeming contradiction between different types of equilibrium points.

8.3.2.3 Time dependence in game theory

In all of the previous sections, the implicit assumption has been that conflict between the host immune system and the pathogen occurred as a single event in time, albeit with the possibility of repetitive play in the case of mixed strategy solutions. In other words, the basic two-player games that have been discussed so far assume that each player chooses his strategy at the beginning of the game and then follows through with it regardless of what the other player is doing. In reality, however, players can often alter their strategies in response to the actions of the other players.

The easiest way to represent choices that depend on the time course of the game, and the previous decisions that have been made by the parties during the course of play is to use what's known as a game tree.

Example 8.5. As an example, consider the game described by the payoff matrix in Fig. 5b, however this time, suppose that the host immune system can choose whether or not to switch to an adaptive response based on knowledge of the state of the invading virus, while the virus can choose whether or not to mutate based on knowledge of the attack that has been mounted by the immune system.

Formulation. The game tree for this scenario is shown below in Fig. 7a. In a game tree, time progresses to right, with each decision point represented by a node, and each decision option represented

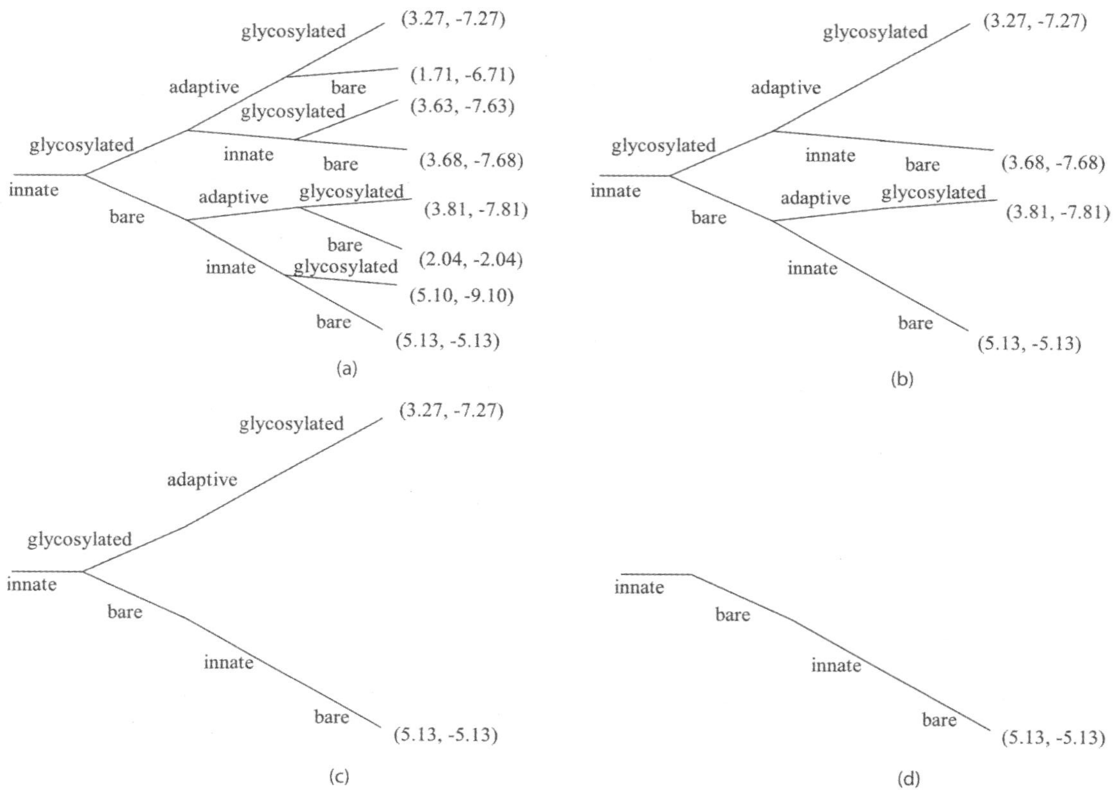

Figure 7. Game tree for the glycosylated vs. bare game with an additional host cost incurred for glycosylated attack.

by a branch stemming from that node. In the game tree above, there are three different times at which decisions occur. The first decision is made by the virus at point A, the second decision is made by the immune system at point B, and the third decision is made by the virus again, at point C.

Solution. Games that are represented by game trees can be solved using a method known as *backwards induction*. Essentially, backwards induction involves working backwards from the possible payoffs assuming that at each step the player making the decision will choose his strategy according to whatever optimizes his payoff. For example, in the game tree below, the virus makes the last decision. Dependent on how the game progressed before the final viral decision, the virus could find itself at any of the four possible C nodes. Assuming that the virus knows which node it is on, though, it will choose its strategy so as to maximize its payoff which, similar to the payoff matrix in Fig. 6b, is the first number in each ordered pair. Strategies that the virus does not choose can then be removed leaving the game tree in Fig. 7b.

Working further backwards in time (towards the left hand side of the tree), the host immune system makes the next decision at B. Again, dependent on previous events, the host immune system could find itself at either B node. Assuming that the host has knowledge of which B node it is on, though, the host will choose its strategy so as to optimize its own payoff, which is the second number in each ordered pair. Again, strategies that are not chosen by the host can be removed, leaving the reduced game tree in Fig. 7c.

There is now only one decision left in the game tree, which is the virus's choice at A to invade with either a glycolsylated protein coat, or a bare protein coat. Since the virus makes the decision at A, the virus will choose the strategy which maximizes its own payoff, or the first number in each ordered pair. Removing the alternate choice gives the final game tree in Fig. 7d, which is then the solution to the game.

In other words, assuming that the virus and the host immune system make time dependent strategy choices and can monitor each other's actions, the optimal solution to the game occurs when the virus invades in a "bare" state and remains bare, while the host immune system begins with an innate response, and maintains that innate response throughout the entire course of the infection. It is interesting to note that by formulating the game as a game tree with time-dependent choices, the virus and the host arrive at the Pareto optimal solution. In other words, when the players have access to more information and, in particular, to time-dependent information reflecting the choices of their opponents, the Pareto optimal solution arises naturally.

While game trees are an effective way to formulate games with time-dependent strategy decisions that occur at discrete time steps, in reality many conflict scenarios involve players that constantly alter their behavior in response to both the current state of the game, and the past actions of their opponents. When strategy choices can be made continuously through the entire course of the game, the game is typically analyzed using what is known as *Differential Game Theory*. Differential Game Theory is an extension of the game tree method that takes the limit of infinitely small time steps. Despite the mathematical complications introduced, the fundamental approach used to solve differential games is identical to the method of backwards induction used to analyze game trees.

8.3.2.4 Three-Player games and controlled drug delivery

While many Game Theoretical applications consider games with only two players, it is possible to incorporate three or more players into the same basic framework. When studying the host immune system and its interactions with invading pathogens, these additional players could be either multiple pathogens that co-infect the host at the same time, or else a player administering drugs in the hopes of curing the infection.

The latter of these two possibilities is particularly interesting with respect to controlled drug delivery strategies. Much of the recent research emphasis on controlled drug delivery strategies has focused on the development of novel clinical techniques that will allow for drug release over specific timed intervals, in precise locations of the body. In addition to having the physical ability to control and manipulate the kinetics and diffusion of different drug molecules, however, successful controlled drug delivery strategies must also rely on knowledge of exactly where, when and at what concentrations, certain drug molecules will be most effective.

In this respect, the drugs, and the physicochemical properties of the drugs, are not independent of their host environment, including the presence and actions of the virus. In other words, a model for controlled drug delivery which only considers the drug itself, will have limited predictive capabilities. Specifically, it will fail to fully describe any situation in which the behavior of the drug is significantly influenced by either the interplay between the drug and the host immune system, or the drug and the invading pathogen.

Game Theory, of course, eliminates this limitation, since it is a formulation which simultaneously considers the optimization problems of all players involved. Therefore, with game theory, the response of the pathogen to the drug can be incorporated when designing optimal strategies for administering a particular drug.

Although three-player games may prove useful for analyzing drug delivery strategies and drug regimens, three-player games are much more difficult to interpret with Game Theoretic tools, since they require finding equilibrium points in a three-dimensional matrix. As a result, it may prove to be easier to incorporate drug delivery into the basic two player game through the introduction or scaling of parameters in the equations describing viral dynamics and immune cell response.

Provided that the drug parameters are under control of the host, one can formulate drug-delivery as a three-player game that has been reduced to a two-player game as a result of an alliance between the host immune system and the player administering the drug. In that case, the drug and the host immune system will necessarily strive towards a common goal, with the drug parameters tuned in a time-dependent manner so as to optimally fight against the invading pathogen. The effects of drugs on viral dynamics and immune cell response have been successfully included into several different ODE models; however, to our knowledge, these models have not considered the optimal tuning of drug delivery parameters in the context of a game or conflict situation. The game theoretical formulation will help to resolve this aspect of controlled drug delivery purpose.

8.4 SUMMARY

In this chapter, we have attempted to provide an introduction to both immune system biology and Game Theory. Our goal has been to show how Game Theory offers an alternative mathematical approach with which to study and interpret host-pathogen interactions and immune system defense. We have given several generic examples illustrating the ways in which different game theoretical formulations could be applied to conflict between the immune system and an invading pathogen. While these examples are not specific to any particular disease, we believe that by applying the same fundamental methodology to a certain well studied pathogens, a Game Theoretical analysis will provide insight into both the behavior of that pathogen and the immune system response that the pathogen elicits.

Future directions for this work involve its application to specific pathogen varieties, as well as the inclusion of more detailed and more complex immune system mechanisms. We also hope to incorporate drugs and controlled drug delivery into the basic Game Theoretical models in order to better understand how to optimize treatment for specific diseases like AIDS and malaria. In summary, then, we believe that Game Theory, because it accounts for the essentially competitive nature of the host-pathogen interaction, is an ideal framework from which to consider pathogen invasion and immune system defense. Moreover, since pathogens and the immune system are necessarily interdependent, we strongly believe that any analysis which only considers one or the other is likely to misinterpret observed behavior because it does not account for the complex and crucial interplay between the two parties. In contrast, since the focus of Game Theory is the constant battle between the immune system and an invading pathogen, we fully expect Game Theory to offer insight into aspects of pathogen and immune cell behavior that have thus far been difficult to explain using other theoretical methods.

REFERENCES

1. P. Zipfel, R. Würzner, and C. Skerka, Complement evasion of pathogens: Common strategies are shared by diverse organisms, *Molecular Immunology*, **44**, 3850–3875 (2007).
2. G. B. Pier, J. B. Lyczak, and L. M. Wetzler, *Immunology, Infection and Immunity*, ASM Press (2004).
3. J. D. Farmer, S. A. Kauffmann, N. H. Packard, and A. S. Perelson, The immune system, adaptation, and machine learning, *Physica D*, **22**, 187–204 (1986).

4. L. A. Segal, and A. S. Perelson, Computations in shape-space: A new approach to immune network theory, in *Theoretical Immunology, Part two* (1988).
5. S. A. Kauffmann, E. D. Weinberger, and A. S. Perelson, Theoretical Immunology, Part one, in *Maturation of the Immune Response via Adaptive Walks on Affinity Landscapes* (1988).
6. D. Wodarz, Killer cell dynamics: Boosting Immunity against immunosupressive infections, in *Mathematical and Computational Approaches to Immunology*, Springer (2007).
7. D. Wodarz and M. A. Nowak, Immune responses and viral phenotype: Do replication rate and cytopathogenicity influence virus load? *Journal of Theoretical Medicine*, **2**, 113–127 (2000).
8. D. Wodarz, K. M. Page, R. A. Arnaout, A. R. Thomsen, J. D. Lifson and M. A. Nowak, A new theory of cytotoxic T-lymphocyte memory: Implications for HIV treatment, *Philos. Trans. R. Soc. Lond. B Biol. Sci.* **355**, 329–343 (2000).
9. S. A. Frank, *Immunology and Evolution of Infection Disease*, Princeton University Press (2002).
10. J. W. Drake, B. Charlesworth, D. Charlesworth, and J. F. Crow, Rates of spontaneous mutation, *Genetics*, **148**, 1667–1686 (1998).

Color Index

CHAPTER 1

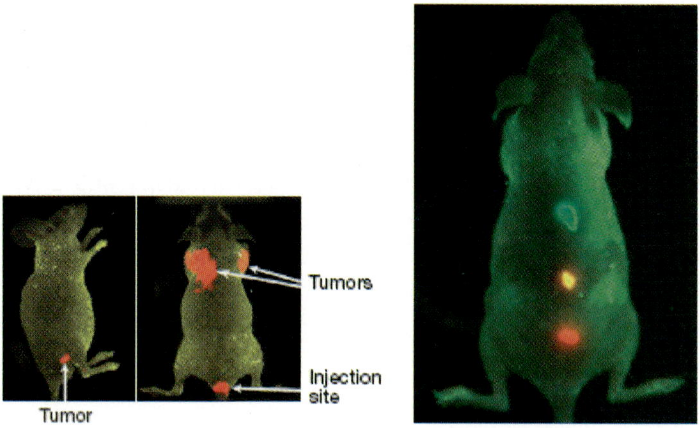

(a) (b)

Figure 3. Imaging in live animals using quantum dots (QDs). (a) Molecular targeting and *in vivo* imaging using antibody-(QD) conjugate. (b) *In vivo* imaging of multicolored QD-encoded microbeads. Reprinted from Ref. 62 with permission from Nature Publishing Group.

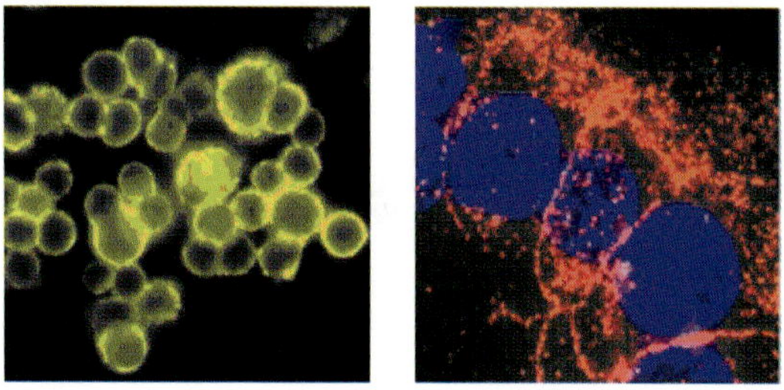

(a) (b)

Figure 4. Detection of cancer marker Her2 *in vitro* with QD-streptavidin. (a) Her2 detected on surface of free cells using QD 560-streptavidin (yellow). (b) Her2 detected on a section of mouse mammary tumor tissue using QD 630-streptavidin (red). Reprinted from Ref. 21 with permission from Nature Publishing Group.

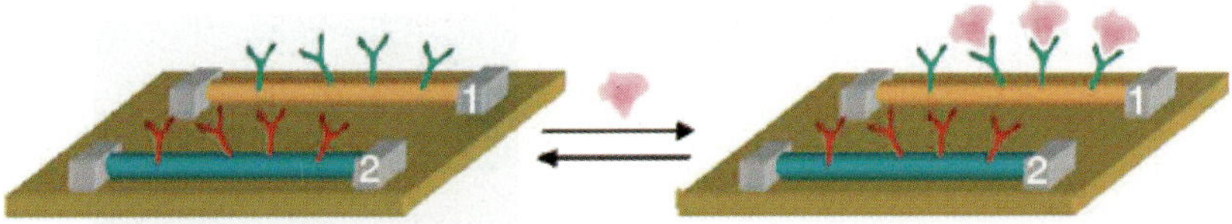

Figure 5. Schematic showing two nanowire devices, 1 and 2, within an array, where nanowires were modified with different (1, green; 2, red) antibody receptors. Reprinted from Ref. 86 with permission from Nature Publishing Group.

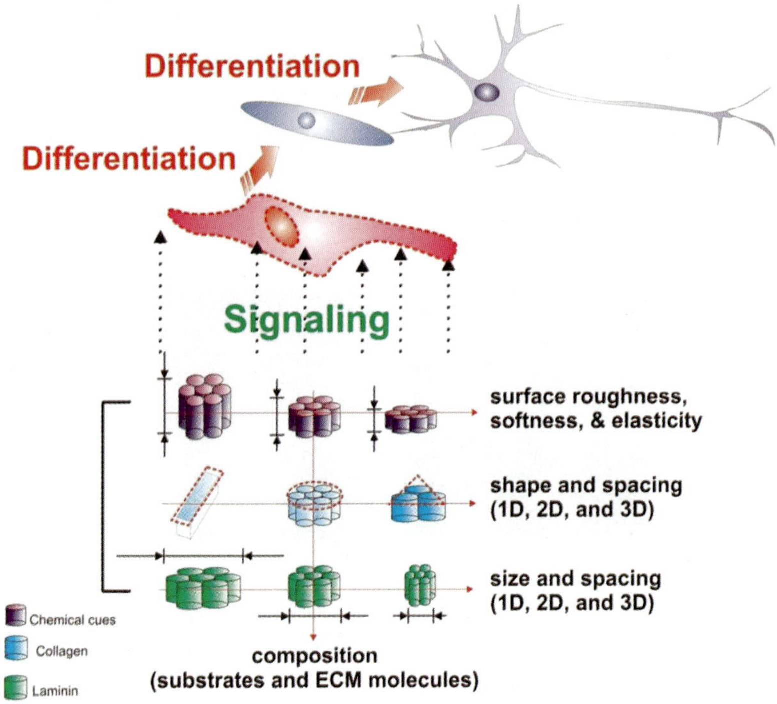

Figure 9. Application of micro-/nanoscale surface engineering in stem cells. Micro- and nanostructures that interact with stem cells at the molecular level can be utilized to control stem cell fate.

CHAPTER 5

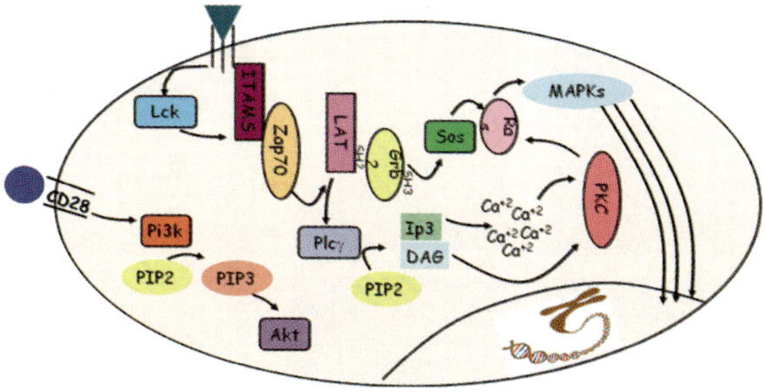

Figure 2. *Diagram of a signaling pathway: T cell signaling.* Extracellular ligands (represented by the green triangle and blue circle) bind to membrane spanning receptor proteins, initiating a cascade of protein modification and phosphorylation events (represented by curved black arrows). While most of the signaling molecules are proteins with enzymatic activity, some are proteins whose main function is to bind other proteins (ITAMS is one example), and others are nonproteins, such as the calcium ions. The end response depicted here is the effect of MAPK members on activity in the nucleus, inducing changes in genetic regulatory pathways.

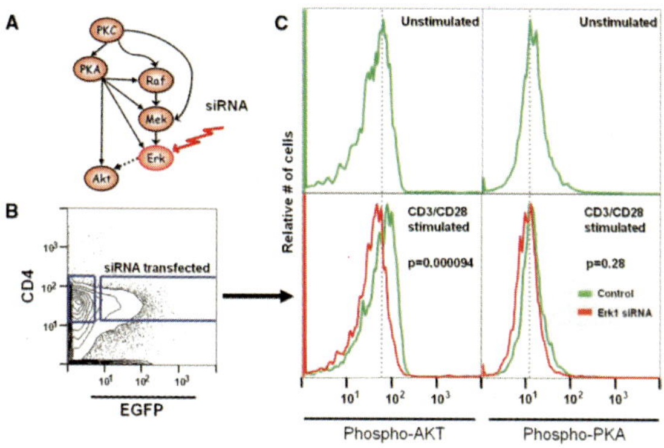

Figure 9. Validation of model prediction. (A) The model predicts that an intervention on Erk will affect Akt, but not PKA. (B) To test the predicted relationships, Erk1 and Erk2 were inhibited using siRNA in cells stimulated with antibody to CD3 (anti-CD3) and anti-CD28. (C) Amounts of Akt phosphorylation in transfected CD4 cells [enhanced green fluorescent protein (EGFP) cells] were assessed, and amounts of phosphorylated PKA are included as a negative control. When Erk1expression is inhibited, phosphorylated Akt is reduced to amounts similar to those in unstimulated cells, confirming our prediction (P 0 0.000094). PKA is unaffected (P 0 0.28).

CHAPTER 6

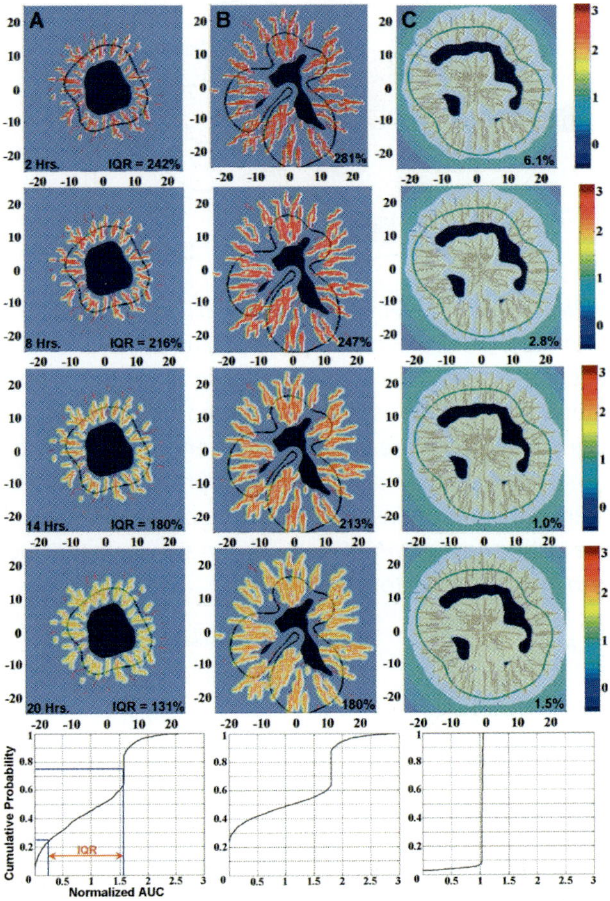

Figure 6. DNA-bound AUC at four times (rows: 2, 8, 14, and 20 h) post bolus initiation for three two dimensional simulated baseline tumor lesions (columns). I and II are doxorubicin, while III is cisplatin. Results are normalized to average lesion AUC at the time taken to enable comparison of distribution heterogeneities. Thick black contours are tumor boundaries. Thin red curves are vasculature. Dark regions are necrotic areas. Each unit represents 200 μm. Bottom probability distributions show final AUC distribution at 20 h. A concise measure of heterogeneity is given by the inter-quartile range (IQR), depicted in the lower left graph and explained in the text. Although AUC in host tissue is also shown in plots, the analysis considers only DNA-bound drug in viable lesion. Reprinted with permission from Sinek *et al.*, *J. Theor. Biol.* (in press). Copyright © Elsevier.

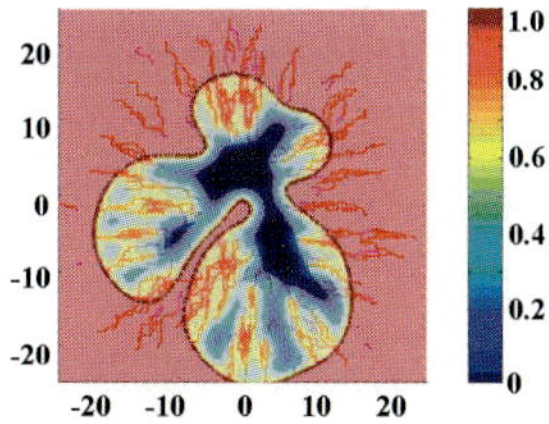

Figure 7. Contour plot shows cell substrate distribution in Lesion B demonstrating significant heterogeneity. Other lesions are similar. Reprinted with permission from Sinek *et al.*, *J. Theor. Biol.* (in press). Copyright © Elsevier.

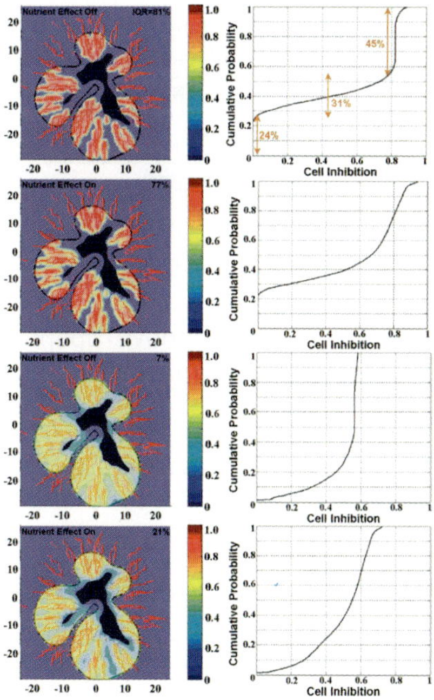

Figure 8. (Upper block) Cell growth inhibition profile of Lesion II at baseline settings with and without the substrate effect after bolus administration depicted in Fig. 2. Probability plot and IQR are now of inhibition distribution and are not normalized with respect to any average. Although the IQR indicates decreased heterogeneity with the substrate effect, both the color distribution plot and the probability plot indicate increased heterogeneity as is evidenced by the broadening of the curve. (Lower block) The same experiment, except with doxorubicin penetration increased. Now both the plots and IQR show increased heterogeneity. The appropriate IC50 is used in each experiment. Reprinted with permission from Sinek *et al.*, *J. Theor. Biol.* (in press). Copyright © Elsevier.

CHAPTER 7

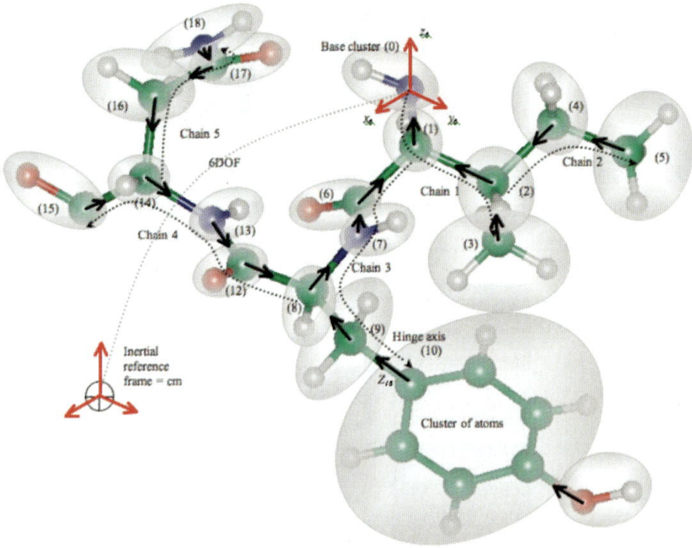

Figure 3. Constrained molecular multibody (Tripeptide with 54 atoms and 162DOF is reduced to 19 clusters and [19+6]DOF); implicit constraints result in molecular rigid bodies shown in grey.

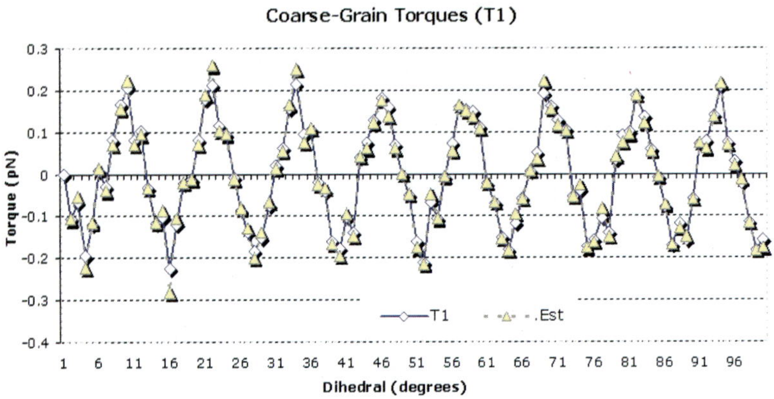

Figure 5. NVT @300 constrained MD using predicted (energies) torques in internal coordinates from the hybrid GA-gradient module in CMDF for a polymer chain (single polymer base torsion shown: atomistic Dreiding force field, blue/yellow: GA-gradient estimated). Fitness function based on RMSD measure calculated from structure derived from Dreiding force field calculations.

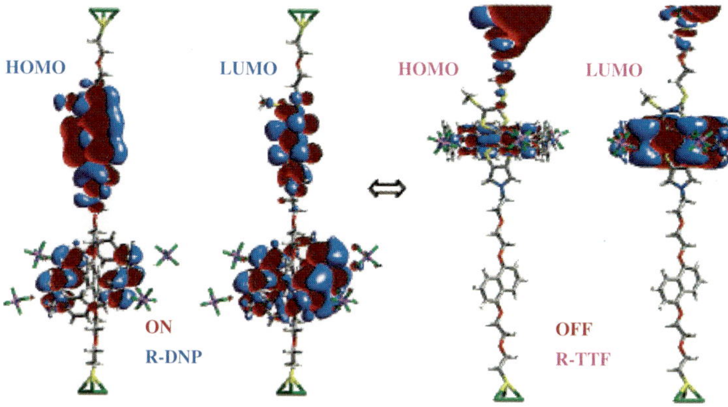

Figure 8. Molecular Orbitals of Au-rotaxane-Au switch. Green represents Au atoms, yellow is S, gray is C, red is oxygen, and white is hydrogen. From Ref. 124. Switching takes place via displacement of molecular ring between Au atoms.

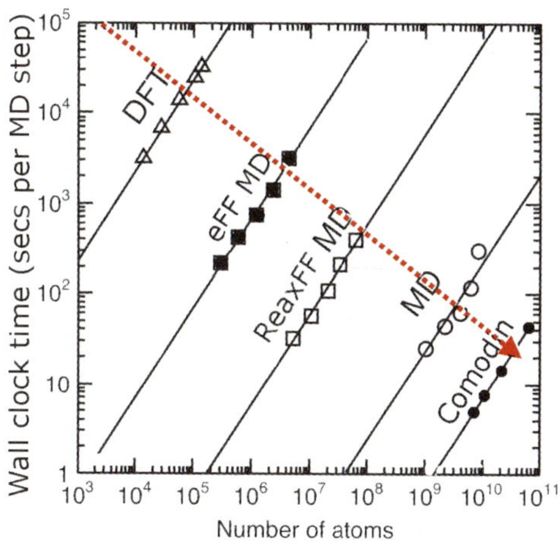

Figure 19. Relative qualitative computation times of QM (DFT-QMC), eFF, reaxFF, atomistic MD and coarse-grain Comodin MD algorithms. Lines show estimated ideal scaling, and dots, triangles, and circles extrapolated results per method (DFT, MD adapted from Ref. 211. Red line depicts predicted time versus length scaling for 1 K processor simulation.

Subject Index

Nanomedicine: *A Systems Engineering Approach* edited by Mingjun Zhang & Ning Xi

www.panstanford.com
978-981-4241-36-6